More praise for *Cats' Paws and Catapults*

"[Steven Vogel] writes with unusual recognition of the needs of the inexpert reader."

—Philip Morrison, *Scientific American*

"Unceasingly, [Vogel] advocates the fun of science."

—Peter Gorner, *Chicago Tribune*

"If [D'Arcy] Thompson's work defined the classical period in the science of form, the field has just entered its renaissance. One of the leaders of the resurgence is the zoologist Steven Vogel."

—Tyler Volk, *The Sciences*

"[F]ew scientific books show; most scientific books tell, recite strange names, and wallow in complexity. But over the years I have come to expect that Steven Vogel will always show me science. . . . Any of [Vogel's] books could entertain an expert or an amateur."

—Mike May, *American Scientist*

"Who is the better technologist, Mother Nature—source of seashells, spider webs, and birds' wings—or the human engineer—creator of skyscrapers, nylon, and airplanes? This engrossing question lies at the heart of a fine new book by Steven Vogel, an expert in biomechanics with a flair for genial philosophizing."

—Samuel Florman, author of *The Civilized Engineer* and *The Existential Pleasures of Engineering*

CATS' PAWS and CATAPULTS

Mechanical Worlds of Nature and People

STEVEN VOGEL

Illustrated by Kathryn K. Davis
with the author

W. W. NORTON & COMPANY
New York • London

Printed in the United States of America
First published as a Norton paperback 2000

For information about permission to reproduce selections from this book, write to Permissions, W. W. Norton & Company, Inc., 500 Fifth Avenue, New York, NY 10110.

The text and display of this book were composed in Adobe Garamond
Desktop composition by Tom Ernst
Manufacturing by the Haddon Craftsmen, Inc.
Book design by BTD / Mary A. Wirth

LIBRARY OF CONGRESS CATALOGING-IN-PUBLICATION DATA
Vogel, Steven, 1940–
Cats' paws and catapults : mechanical worlds of nature and people / by Steven Vogel ; illustrated by Kathryn K. Davis.
p. cm.
Includes bibliographical references and index.
ISBN 0-393-04641-9
1. Biomechanics. 2. Mechanics. I. Title.
QH513.V64 1998
571.4'3—dc21 97-44807
CIP

ISBN 0-393-31990-3 pbk.

W. W. Norton & Company, Inc., 500 Fifth Avenue, New York, N.Y. 10110
www.wwnorton.com

W. W. Norton & Company Ltd., 10 Coptic Street, London WC1A 1PU

1 2 3 4 5 6 7 8 9 0

For Jane

Contents

Preface

Life is what biology's about. Technology is something else altogether. Or so I believed before I got into a kind of biology that's about technology as well as life. More to the point, it—biomechanics—looks at the technology of life, at the mechanical world of nature. Sometimes that world resembles the mechanical world that we humans have created. But sometimes the two differ strikingly. This book compares those technologies. It's about the ordinary things and creatures around us; it intends, immodestly, to change the way you look at your surroundings—at least a little. It has some other missions as well.

I've come to realize that engineers are as curious about our world as we are about theirs. Some suspect that a look at organisms might help them create designs and fabricate devices. How could a biologist disagree? But shifting from one world to the other isn't a trivial matter, and the traveler needs a road map and a guidebook. This book tries to provide them by introducing biomechanics in a point-by-point comparison with the more familiar world of our own technology.

At the same time I want to inject an element of sobriety into our romantic view of living things. The elegance of natural design seduced a lot of us into becoming biologists. Nature does what she does very well indeed. But—and here's the rub—why should she do so in the best possible way? And why should she provide a model for what we want to do? I want to ruffle our tendency to view nature as the gold standard for design and as a great source of technological breakthroughs.

Beyond that, I want to argue that natural design provides no honest foil for skewering human technology. In getting together the material for this book, I've repeatedly bumped into an antitechnological literature—nature-worshiping, engineer-bashing tracts. Its authors take a thoroughly unrealistic view of our contemporary situation and prospects and lay blame inappropriately; hanging the social consequences of technology on engineers amounts to hanging airport congestion on the Wright brothers.

About this last mission I cheerfully admit personal bias. I have no formal background in engineering and only a primitive knowledge of the underlying physics and mathematics. I couldn't have done what science and writing I've done without the unfailing generosity and support of engineers. For nearly forty years, and at several institutions, they've explained things, steered me out of cul-de-sacs, pulled my foot out of my mouth before I published nonsense, and suggested accessible source material; in short, they've done everything imaginable to welcome me to their domain.

Doing this book has treated me to an intellectual feast. I indulged my fondness for building things by making a version of an ancient Egyptian drill, and I had a pretext to lay hands on (and read) the first book printed on paper derived from wood pulp. I had an excuse to use nine libraries on the Duke campus (and several elsewhere), not to mention interlibrary loans, CD-ROMs, government documents, various on-line databases, old newspapers on microfiche, and network news groups.

I remain daunted by the clear relevance and separate sophistication of anthropology, archaeology, paleontology, economics, architecture, geometry, geography, law, and the histories of science, technology, exploration, domestication, and culture. All the complex interconnections bring to mind the biomechanical problems of keeping your finger in the air while your ear's to the ground and of keeping your feet on the ground while your head's in the clouds.

I've drawn quite shamelessly on professional colleagues and other friends. In no other project have I received such a treasure of useful suggestions, ideas, and examples. I'm especially indebted to Matthew Healy, Michael LaBarbera, Catherine Loudon, Jane Vogel, and Stephen Wainwright, each of whom read the entire manuscript in first draft and made copious but always kind and tactful suggestions. In addition, useful ideas emerged from conversations with David Alexander, Michael Blum, Richard Burian, Steven Churchill, Ruth Day, Martha Dunham, Betsey Dyer, Shelley Etnier, Robert Full, Margaret Hivnor, Diane Kelly, Peter Klopfer, Daniel Lieberman, Dan Livingstone, Anne Moore, M. Patricia Morse, Bruce Nicklas, Francis Newton, Fred Nijhout, George Pearsall, Charles Pell, Henry Petroski, Jeffrey Podos, Michael Reedy, Knut Schmidt-Nielsen, Kalman Schulgasser, John Sharpe, Robert Teer, Edward Tenner, Lloyd Trefethen, John Wourms, and many other people of whom I lack a proper list. I'm grateful to a host of helpful librarians, especially Richard Hines and David Talbert.

Edwin Barber, my editor at W. W. Norton, has done more for my writing than anyone since my student days. In particular, he has given detailed guidance as well as general admonishment in the struggle for spontaneity and the battle against academic pretentiousness and obfuscation.

Finally, I'd like to pronounce words of passionate appreciation and advocacy for libraries where what you need—or don't know you need until you see it—sits on end user–accessible shelves. Or their electronic equivalents.

Several of the figures derive from previously published material. For permitting redrawing, I gratefully acknowledge the following copyright holders. Figure 2.4: Columbia University Press (Fig 8.5 of B. D. Dyer and R. A. Obar, *Tracing the History of Eucaryotic Cells*); Figure 4.15: Academic Press, Inc. (Fig. 2 of J. T. Finch and A. Klug, *J. Mol. Biol.* 15: 344 and Fig. 5.4 of R. E. F. Matthews, *Plant Virology,* 3rd ed.); Figure 5.12: Dr. Mimi Koehl (Figure I-9 of Duke University Ph.D. dissertation); Figure 8.4, American farm and Darrieus rotor windmills: Van Nostrand-Reinhold, Inc. (Figs. 22, 42 of F. R. Eldridge, *Wind Machines,* 2nd ed.); Figure 10.10: Dr. Vance A. Tucker (Cover, *Science,* 14 Nov. 1969); and Figure 13.3: John Wiley and Sons, Inc. (Fig. 5.19 of M. E. Rosheim, *Robot Evolution*). For permitting direct copying of Figures 11.1 and 13.2, I acknowledge the copyright holders, Gordon and Breach Scientific Publishers, Inc. (Fig. 27 of J. Kastelic et al., *Connective Tissue Research* 6: 11) and Dr. William M. Kier (Fig. 11 of Duke University Ph.D. dissertation), respectively. For allowing redrawing of the mollusk shell of Figure

6.8 (from S. W. Wise and W. W. Hay, p. 427 in *Trans. Amer. Micr. Soc.* 87), I thank Dr. Vicki Pearse, editor, *Invertebrate Biology.* For loaning the *Arctium* for Figure 12.12, I thank Dr. Robert Wilbur, director, Duke University Herbarium. Sam (Soft and Mellow) Cat suggested the title's feline allusion; he never confuses the mouse of one technology with that of the other and remains ever hopeful that the printer will emit something better than paper.

STEVEN VOGEL
Durham, North Carolina

CATS' PAWS
and
CATAPULTS

Chapter 1

NONCOINCIDENT WORLDS

When some of us were much younger—for me the late 1940s—we read *Flash Gordon* every Sunday in the comics. With the casual confidence of kids, we assumed that space travel to extraterrestrial civilizations was just around the corner. Mr. Gordon seemed as close to our world as George Washington and a lot closer than Julius Caesar. While *Star Trek*, with Mr. Spock, is leagues ahead of *Flash Gordon* in sophistication, reaching other beings has become a much dimmer prospect. At fault is neither the loss of our youthful certitude nor any inferiority of Spock to Gordon. What's happened is that the arrival of the space age, like any reality, has brought its potion of sobriety, its diminution of innocence. Space travel has proved a lot trickier—and more expensive—and extraterrestrial civilizations a lot more remote than we ever imagined.

Nonetheless, the success of *Star Wars* and *Star Trek* and the continued popularity of science fiction testify that the allure remains. Much of the appeal obviously turns on their views of alternative cultures, things in short supply here on earth. Ours is a single world. The civilizations of the

Orient and those of Europe and Africa have interacted for more than a thousand years, and extensive intermingling with those of the Americas has now gone on for five centuries. Human technology may have become vastly more complex, but it has lost diversity and frozen into a stereotype. The global convergence leaves no Atlantis in the offing. Meanwhile we're ever more doubtful that we'll bump into any other technology.

A shame, perhaps, but not all that bad. We do have an alternative technology as a mirror in which to view our own—the technology of organisms, the result of evolution by natural selection over the past few billion years. Life forms a technology in every proper sense, with a diversity of designs, materials, engines, and mechanical contrivances of every degree of complexity.

As systems to compare we could ask for nothing better than nature's designs and human inventions. Nature's technology occurs on the surface of the same planet as that of human culture, so it endures the same physical and chemical limitations and must use the same materials. But nature copes and invents in a way fundamentally different from what we do. At the very least, the rate at which she alters herself is glacial by our cultural standard.

The very shapes of the two technologies differ dramatically. Just look around you. Right angles are everywhere: the edges of this page, desk corners, street corners, floor corners, shelves, doors, boxes, bricks, and on and on. Then look at field, park, or forest. Where are the right angles? Absent? No, but rare, which raises questions. Why so few right angles in nature? Why do civilizations find them so serviceable?

Natural and human technologies differ extensively and pervasively. We build dry and stiff structures; nature mostly makes hers wet and flexible. We build of metals; nature never does. Our hinges mainly slide; hers mostly bend. We do wonders with wheels and rotary motion; nature makes fully competent boats, aircraft, and terrestrial vehicles that lack them entirely. Our engines expand or spin; hers contract or slide. We fabricate large devices directly; nature's large things are cunning proliferations of tiny components. One can easily continue; indeed, much of this book is simply an exploration of such contrasts—of their mechanical aspects in particular.

At some very basic level all of us recognize how different are the products of humans and of nature. Artists take advantage of that subconscious perception to jar us by depicting one culture using the forms of the other. The cubists draw human faces with flat sides and harsh, straight edges, often at right angles to each other. Salvador Dalí, in painting, and

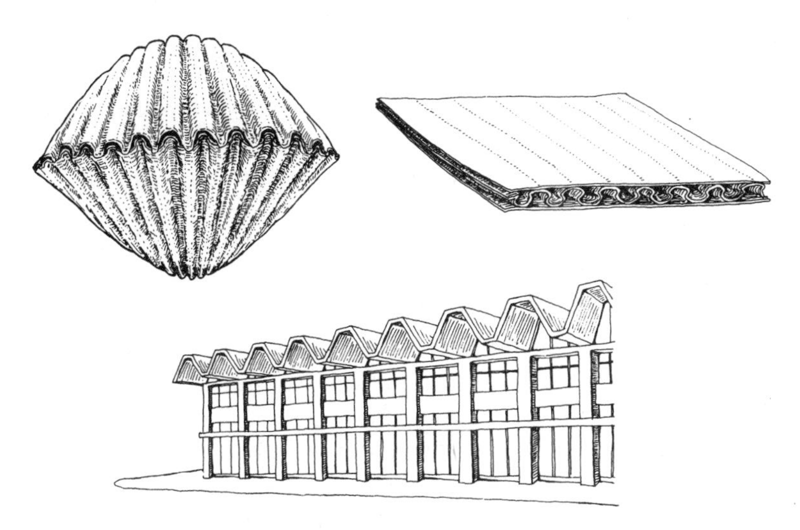

FIGURE 1.1. *Corrugated or fanfold surfaces as cheap routes to stiffness: scallop shell, corrugated paperboard, and ridge and valley roof.*

Claes Oldenburg, in sculpture, re-create the hard objects of manufactured technology—watches, engine blocks, and such—with the natural world's lack of rigidity. The incongruity startles, intentionally and dependably.

But one can easily make too much of these differences. Both bicycle frames and bamboo stems take advantage of the way a tube gives better resistance to bending than a solid rod. A spider extends its legs by increasing the pressure of the fluid inside in much the same way that a mechanical cherry picker extends to prune trees or deice planes. Both technologies construct things using curved shells (skulls, eggs, domed roofs), columns (tree trunks, long bones, posts), and stones embedded in matrices (worm tubes, concrete). Both use corrugated structures (as in Figure 1.1) to get stiffness without excessive mass—whether the shell of the scallop, one of the rare swimmers among bivalve mollusks, or the stiffening structures of doors, packing boxes, and aircraft floors, or fan-folded paper and occasional roofs. Both catch swimming or flying prey with filters through which fluid flows—whether spiders or whales, gill-netting fishers or mist-netting birders.

I care very much about the ways of engineering, but by profession and predilection I'm a biologist, not an engineer. The implied equivalence between the two fields is a little misleading, though; the two are not merely opposite sides of the same fence. The biologist studies something that exists: nature, in all its splendor. The engineer, by contrast, creates. Further, the engineer's successes have more immediate impact than those of the biologist, and failure exacts penalties far beyond the approbation of a few peers.[1]

In practice, calling my calling biology isn't specific enough since the mechanical interests of most biologists don't go beyond keeping their scientific equipment cooperative. So we who look at nature's mechanical aspects call our field biomechanics[2] (or, as on a reimbursement check I once received from the American society of the same, biomechantics—whether in high spirits, high dudgeon, or pure accident). Awkwardly, something called biotechnology looms large on the contemporary scene. As it happens, biotechnology in its current sense is quite a different endeavor and will play almost no role here. Biotechnology is largely a synthetic rather than our analytic activity; in addition, it mainly focuses on the molecular and microscopic while we care more about the mechanical and macroscopic. By contrast, what goes by the name biophysics is analytic enough, but it's similarly rooted in a molecular domain.

The biomechanic ought to make a candid, if unflattering, admission at the outset. Simple logic suggests examining nature's mechanical technology as a first step toward both creating and understanding human technology—at least to establish the range of possibilities. Who can deny that nature got here first? In fact, the shoe is almost always on the other foot. The biomechanic usually recognizes nature's use of some neat device only when the engineer has already provided us with a model. Put another way, biomechanics mainly still studies how, where, and why nature does what engineers do.

One might reasonably expect any proper biologist to find natural, human-free systems impressive, even aesthetically standard-setting. And often we do; without our affection for nature, we might have chosen other ways to spend our time, so almost without exception biologists are biophiliacs. But loving nature is not at all the same as finding her perfect, a gold standard for design. A lot of famous and otherwise estimable people have viewed nature as some Edenic perfection of process and product. Using short quotations out of context may be a little unfair, but as a biologist I cringe at statements such as those that follow. I can forgive the ancients more easily than my contemporary post-Darwinians.

If one way be better than another, that you may be sure is Nature's way. (Aristotle, fourth century B.C.E.)[3]

Human ingenuity may make various inventions, but it will never devise any inventions more beautiful, nor more simple, nor more to the purpose than Nature does; because in her inventions nothing is wanting and nothing is superfluous. (Leonardo da Vinci, fifteenth century)[4]

Sources of hydraulic contrivances and of mechanical movements are endless in nature; and if machinists would but study in her school, she would lead them to the adoption of the best principles, and the most suitable modifications of them in every possible contingency. (Thomas Ewbank, mid-nineteenth century)[5]

One handbook that has not yet gone out of style, and predictably never will, is the handbook of nature. Here, in the totality of biological and biochemical systems, the problems mankind faces have already been met and solved, and through analogues, met and solved optimally. (Victor Papanek, contemporary)[6]

This casual attitude toward nature's automatic excellence can't be casually dismissed. For one thing, it implies that the engineer or entrepreneur who copies nature will leap ahead of those plodders who rely on mere human ingenuity. For another, it appeals all too persuasively to those who blame engineers for the ills of the modern world.[7] I find neither attitude attractive. (But engineer bashing has receded somewhat, even as an antiscientific community has flourished; maybe some of its acolytes of the sixties and seventies got hoisted in the eighties and nineties by their undeniable affection for personal computers.)

So is this a book about copying nature? Emphatically not. As we'll see, on surprisingly few occasions has copying proved useful. Indeed, felicitous transfer of bits and pieces should not be expected. We're dealing with separate contexts of mechanical design, each system uniquely integrated by its own elements of internal harmony and consistency. Moreover, one of these systems, even though it includes ourselves (we are, after all, creatures of nature), is a far stranger mechanical technology than commonly realized. So nature's version needs special attention, and thus a biologist-biomechanic feels compelled to write a book about technology—or, better, about technologies.

Chapter 2

TWO SCHOOLS OF DESIGN

Almost anything starts from a plan, whether a blueprint, template, some macromolecular chemical code, or just a scheme held in mind. But no plan is without antecedent, whether in an individual's mysterious alchemy of experience or as the result of innumerable ancestral adjustments. And neither human nor natural technology represents a single act of creation. But nowhere do they diverge more than in how their plans originate, in the processes we might call design.

Nature's process is that mechanism Darwin uncovered, evolution by natural selection. Human technology springs from what is variously called invention, discovery, development, or planning. A little confusingly, the word "evolution" has recently been associated with human technological progress.[1] Sometimes that implies a kind of selective process, but most often it just alludes to incremental change, with things building one upon another.

THE NATURE OF NATURAL SELECTION

Oddly enough, the familiar act of human creation is harder to explain in acceptably scientific terms than the way nature creates her devices. The very everyday character of human creativity, though, to anyone who has drawn a picture, written a poem, or baked a cake allows us to evade any precise formulation. By contrast, the intuition recoils when we are faced with evolution by natural selection. It can have direction and can perhaps even make progress, yet do so with no semblance of planning. Its information is carried and dealt with by molecules, and molecules lack proper perceptual reality. And its time course ordinarily exceeds anything in our direct personal experience. Even if its reality is now beyond serious doubt, the whole thing just seems overwhelmingly unlikely.

What concerns us here is its mechanism, for in that specific mechanism called natural selection lie both the power and the disabilities of the evolutionary process. Putting that mechanism as a series of observations and interconnected statements (see the following box) should emphasize the persuasive underlying logic as well as offset a lot of misconceptions and mystical notions of nature's perfection.[2]

1. **Observations:**
   - **a. Every organism can produce more than one offspring, so populations, if unrestrained, will increase continuously.**
   - **b. Every organism needs some minimum amount of material from the environment to survive and reproduce.**
   - **c. The material available to a population of organisms is finite in extent, restraining its increase.**

2. **Consequence of a, b, and c:**
   - **d. A population in a given area will rise to some maximum size.**

3. **Consequences of a and d:**
   - **e. For a population at this maximum size, more individuals will be produced than the environment can support.**
   - **f. Some individuals will not be able to survive and reproduce.**

4. Further observations:

g. Individuals within populations vary in ways that affect their success in reproduction.

h. At least some of this variability is inherited; individuals resemble their parents more than they do more distantly related individuals.

5. Consequences of e through h:

i. Characteristics that increase the number of an individual's surviving offspring will be more prevalent in the population in the next generation.

The final statement, of course, encapsulates evolution by natural selection. Notice that nowhere in the scheme does "design" appear. Using the word as a noun raises no hackles, but making it an honest verb is nearly impossible. To design ordinarily requires a designer. In evolution, though, change happens as a blind result of selection for whatever improves reproductive success. In this sense, nothing in nature is "designed" or has "purpose." Nonetheless, most conspicuous bits of biological structure do serve specific functions. How else could they increase reproductive success? The ear of an animal that's preyed upon enables it to hear a predator's approach, to take evasive action, and perhaps to live and breed. So in another sense, the one relevant here, the ear's "design" certainly does have a "purpose." A seminal work in biomechanics has provoked neither controversy nor offense with the name *Mechanical Design in Organisms.*[3]

In short, design in nature is a process of generating variability (item g above—mutation and recombination, to get technical about it) and then selecting those variants that are beneficial. Naturally, most variations are either neutral or detrimental. The late George Beadle, who received a Nobel prize for his work in genetics, used the analogy of a typist repeatedly copying a page of manuscript. Each copy is proofread for errors; if an error appears, the copy is discarded—except for the rare error that improves the prose. If one of these appears, then the new version becomes the model to be copied. So nature's process is inefficient but inexorable, given a supply of errors and some kind of selection. Random errors yield nonrandom change, and anticipation or planning needn't occur.

This thoroughly stupid process of natural selection can generate results we'd hardly characterize as progress. Consider the consequences of a mammalian mating system in which males have some innate preference for large-breasted females. With remorseless logic, selection will supply increasingly well-endowed females, quite beyond any utility for nursing neonates and well into the range of mechanical awkwardness. Perhaps the example might not be hypothetical; even with our prothoracic cultural bias, breast reduction surgery is far from uncommon.

THE LIMITATIONS ON EVOLUTIONARY CHANGE

The dazzling diversity of the living world too easily disguises the fact that the evolutionary process faces constraints far more severe than anything impeding human designers. We biologists recognize these constraints, but we don't often rise above our natural chauvinism and make enough public noise about them.

Every organism must grow from an initially smaller to an ultimately larger size. Nature, in effect, must transmute a motorcycle into an automobile while providing continuous transportation. The need for growth without loss of function can impose severe geometrical limitations. Consider the possible shapes of mollusks, a widespread and diverse group that includes scallops, slugs, snails, and squid. Mollusk shell doesn't grow, so a shell cannot enlarge except by adding incrementally to edges and inner surfaces. For most shapes, such incremental additions would quickly lead to awkward changes in proportions, as in Figure 2.1. For example,

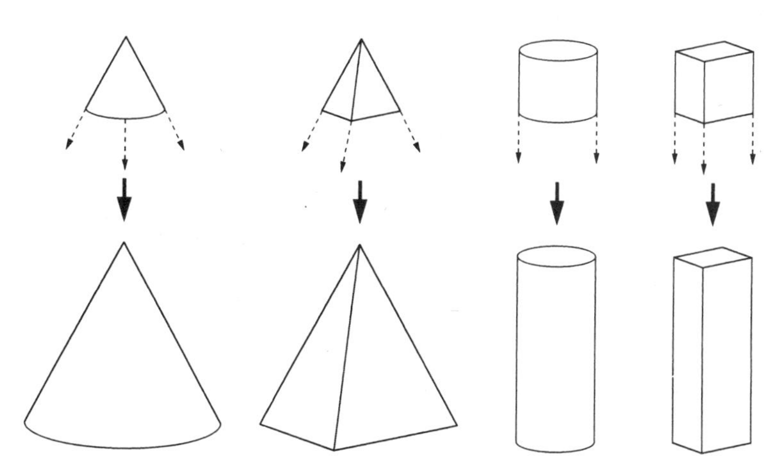

FIGURE 2.1. *Increasing size by adding to an edge. Cone and pyramid retain their shapes, while cylinder and rectangular solid are disproportionately elongated.*

enlarging a cylindrical shell by lengthening would make it relatively skinnier and more vulnerable to breakage; a long rod is more easily broken than a short one, as you can show with a strand of dry spaghetti. Compensatory internal thickening would reduce the relative volume available for the critical guts and gonads. Hollow cones are much better than cylinders or most other shapes if what matters is growth in size without simultaneous change in shape. And mollusk shells are basically cones, whether single cones, as in the snail- and nautiluslike forms, or doubly conical, as in the clamlike forms. Hypothetical mollusk shells can be generated as conic derivatives with a computer, something first done by the paleontologist David Raup thirty years ago.[4] The constraint on their shapes is so severe that almost all the forms of real shells can be produced by a fairly simple program.

Life can't easily pause periodically for renovation, although such things as the pupal stage of insects come close to a growth hiatus. Arthropods in general—insects, spiders, and crustaceans, mainly—grow fitfully, periodically shedding their outside casings as well as some of the inside equipment. Molting thus provides an alternative to the edgewise growth of molluscan skeletons. But it restricts structural possibilities also, surely precluding many useful internal pillars, trusses, and the like. In addition, while insects are consummate fliers, no insect molts functioning wings, so the flying stage is always a final, nongrowing one; little flies are never baby flies. Finally, molting costs material and imposes periods of mechanical vulnerability.[5]

One major group of animals has reconciled support and growth. We vertebrates have a skeletal system that can grow and remodel itself continuously. By contrast with mollusk shell and arthropod cuticle, bone is a living tissue, a complex and unusual accomplishment of fishes, frogs, birds, and people. A growing skeleton may be the greatest vertebrate innovation, the central item in our success as moderate-size to large creatures.

Organisms must also reproduce, so they (the females, in species with separate sexes) need the full equipment to make more of their kind. That's adding a requirement that an automobile tow behind it a factory for making automobiles or at least making motorcycles. Again, some partial evasions are known among complexly colonial creatures. Among bees and ants, for instance, workers are nonreproductive. Among colonial coelenterates (the group that includes jellyfish and sea anemones) only a few specialized polyps of the colony—what looks to us like an individual is really a colony—can have offspring. Even so, the reproductive units are still recognizable organisms—queen bees and reproductive polyps. No organism has invented a proper out-of-body progeny production plant.

After growth and reproduction comes dispersal. Sometimes the three may be done by a single form of an organism, as in creatures like us that have simple life histories. In other cases elaborate metamorphoses may separate dramatically different forms. We're the unusual ones; metamorphoses and the resulting complex life histories characterize most plants, many insects, and almost all of the great diversity of marine invertebrates. We humans tend to take mobility for granted, but it's tough for a tree, an oyster, a parasitic worm, or a sponge to get from here to there. Each of these discharges special samples of itself to spread its kind. In many barnacles and clams, tiny, swimming, predatory larvae convert into sessile, reproductive adults that feed by filtering microorganisms from water. In butterflies, leaf-munching growing crawlers become nectar-sipping reproductive fliers. Such conversions must seriously constrain the range of possible designs.

Other limitations are imposed by what we might call informational constraints. The plan for making any organism has a problem that deserves more attention. In this era of bytes and computers, we recognize information as something quantitative and measurable. The basic unit, the bit, resolves the uncertainty in a choice between two equally probable alternatives, the information you gain when you look at which way a flipped coin lands.[6] In essence, plans are stores of information. To construct a human or similar animal, a fertilized egg has available around 10^{10} (1 followed by 10 zeros) bits of information in its DNA. That may sound like a lot—until one realizes that each of us has about 10^{14} cells, a number no less than 10,000 *times* greater. So 10,000,000,000 bits isn't such a big deal. As we've learned from our computers, two-dimensional representations—graphics—absorb far more memory than mere text. Whether a picture is worth a thousand words, it surely uses as much disk space. Organisms, though, are *three*-dimensional, and details as fine as a millionth of a millimeter are important. Specifying that much detail should require a truly vast store of information, many millions of times greater than the 10^{10} bits in egg or sperm. Thus the shape of an organism has to be set by, relatively speaking, a very sketchy set of plans. That the chickens in a flock look identical cannot result from identical specification of every detail of each.

This shortage of information clearly underlies a lot of biological design. Back in 1950 a prescient physicist, Horace R. Crane,[7] predicted

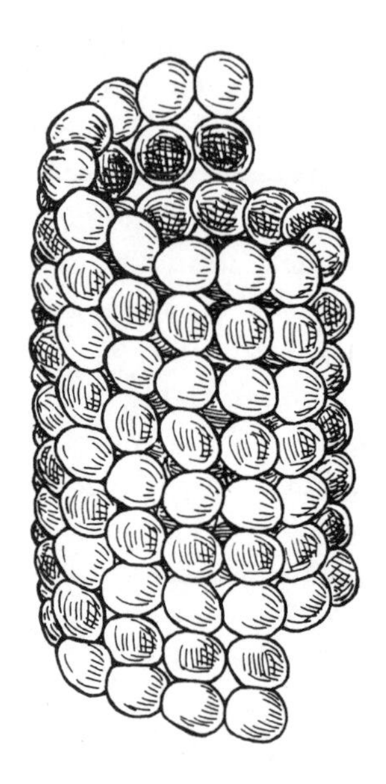

FIGURE 2.2. *A helix of identical wooden blocks and a model of a microtubule. The latter has thirteen elements per turn, each consisting of a pair of protein molecules. In either case, each element is in a position equivalent to that of every other.*

that a lot of subcellular structures (he didn't know which) would turn out to be helical in form, not because helices necessarily worked best but because they could be assembled with especially simple instructions. A helix can be built from identical subunits (as a wall is built from identical bricks); also, every subunit is inserted in exactly the same way as every other, as in Figure 2.2. If you know how to install one subunit, you know how to do the rest. Crane anticipated not only the double helix of DNA but its supercoiling (a helix of helices), the so-called alpha helix of parts of many proteins, and, on a larger scale, helical microtubules and microfilaments important in maintaining the shape and motility of cells. Microtubules and microfilaments have a remarkable capacity for self-assembly; if all the components are put together (with perhaps a bit of the formed structure as a starter), they ordinarily fall into place without any need for mold or scaffolding or, more important, for any additional information.

Building large organisms out of lots of cells is probably made necessary by that shortage of information. Cells may look diverse, but they all have a lot in common; if you can build one kind, you need only a little more information (relatively, of course) to build all the others. Furthermore, in the development of each individual, one group of instructions can set more than one structure. In humans, hand size is an excellent predictor of foot size; before stretch fabrics were common, the salesperson wrapped a sock around your fist to tell whether it would fit your foot. A single alteration of the genetic material—a mutation—ordinarily affects both sides of the body of an animal. A mutant fruit fly doesn't have one white eye; it has two. Beyond these informational economies are others. The hearts and lungs of all of us are in the same places, but at some level of detail the locations of our parts are unpredictable. Anatomy students learn the names of the large blood vessels, but the small ones stay blessedly anonymous—simply because their arrangement varies from one person to the next.

Nor does evolution easily manage fundamental change. For one thing, the variation that provides its raw material—genetic mutations and the juggling of characters accomplished by sexual recombination—consists mainly of small changes. Natural selection merely tests changes, such as a little thicker fur or a little longer ear for possible reproductive advantage. For another, natural selection operates on every individual. Thus any change must yield a fairly immediate advantage if it's to increase its representation in a population. Much less useful is a structure whose reproductive advantage is realized only when some other change appears.[8] As the evolutionary biologist Richard Dawkins eloquently argues (without controversy), the evolutionary process is a tinkerer rather than a proper designer, a perpetual modifier rather than a creative innovator.[9]

Quite obviously major innovations have occurred, and biologists have exercised lots of ingenuity devising possible scenarios in which such innovations don't require long periods of elaboration of yet functionless structures. Birds, bats, and insects fly with flapping wings. Since no small protowing makes a proper aircraft, we worry about how wings got started in each of these lineages.[10] Did a running birdlike creature make longer hops with outstretched feathered arms? Or did such appendages permit longer jumps from branch to branch? Or did extended appendages—longer "arms"—help shed heat produced by extended runs? Both flying squirrels and gliding lizards have membranous skin between fore and hind limbs. Are these creatures reasonable models for birds or bats at a stage before active, powered flight?

The evolutionary process has its hands tied in yet another way. Every organism is a product of its particular evolutionary history. Such a history limits design far more than ensuring that today's disk will work in yesterday's computer. It's tempting to assume that every organism is optimally attuned to its personal circumstances as a result of its lengthy evolution, but it's profoundly wrong. Ancestry traps an organism. Consider a few fine features that appear only in one lineage.

Earlier we spoke of the shells of mollusks, the cuticle of arthropods, and the bones of vertebrates. Mollusks and arthropods never figured out how to make a proper growing skeleton. Each adopted a scheme for coping with the problem of support, one with the advantage (whether incidental or not, one can't easily say) of providing outer protection. Vertebrates solved the problem of making a skeleton that could grow although in doing so, some of us—turtles and armadillos, at present—had to make additional structures to gain that outer protection.

Arthropods make an elastic protein, resilin, that has a higher resiliency (more of the energy used to stretch it is recovered when it's released) than anything found in a mollusk or vertebrate. Mollusks like clams and scallops use an alternative protein, abductin, to open their paired half-shells, as when a scallop swims by clapping them together. Vertebrates use yet another protein, elastin, in ligaments and blood vessel walls: See Figure 2.3. One of resilin's main roles is storing the energy that decelerates an insect wing at the end of one stroke to accelerate the wing at the start of the next. Is making insect wings beat efficiently more crucial than either making a good hinge

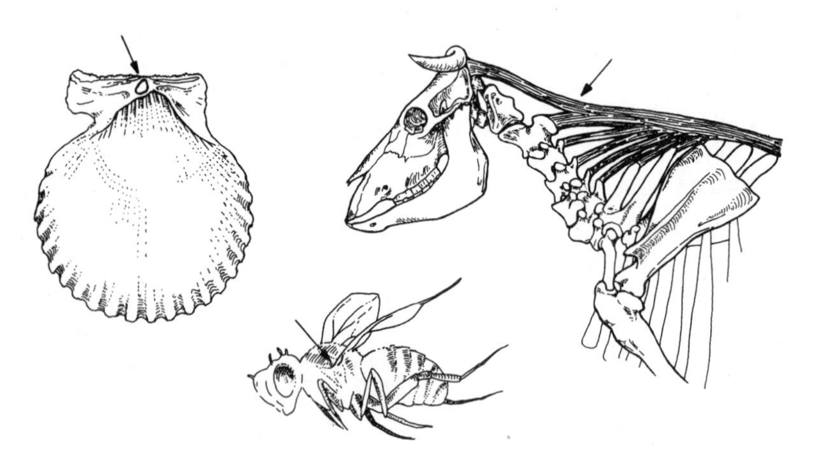

FIGURE 2.3. *Three different elastic proteins, characteristic of three different organisms. The scallop uses abductin in the hinge of its shell; the fly has a pad of resilin in the hinge of each wing; the ligament connecting the head and thoracic vertebrae of a cow is largely elastin.*

ligament to connect the half shells of a swimming scallop or making a low-loss elastic to support the head of a grazing sheep? Probably not; most likely scallop and sheep would be better off if they could trade proteins with a fly.

Mollusks aren't inevitably losers in these comparisons. They alone have mastered the trick of keeping a muscle contracted without expending energy. Ideally, you don't need energy to produce a force that just supports a load; you need it only to move a load. Exerting force against an immovable load—Atlas holding up the sky or a chain supporting a chandelier—ought to use no energy. In practice, though, our own muscles take energy to do anything, even if nothing moves. But the sedentary clam can stay clammed up at virtually no cost.

So nature must follow an inherited plan. The human designer, on the other hand, can borrow devices from other designers. If the device is covered by patents, then royalties or litigation form part of the process, but most useful items are common knowledge in the public domain. For that matter, two competing manufacturers can buy identical parts from the same supplier. Nature has trouble doing anything analogous. Still, cross-lineage transfer of technology is not absolutely ruled out, and a pair of examples illustrates its nicely offbeat flavor.

1. Termites harbor in their digestive systems protozoa called mixotrichs that are critical to their ability to get energy by digesting cellulose. Colonial life in termites may even have originated as a device to ensure adequate opportunities for passing along "infections" of mixotrichs.[11] These protozoa were long assumed to be propelled by organelles called undulipodia on their surfaces. Oddly enough, the undulipodia turn out to be a bunch of bacteria, spirochetes in particular, arranged as shown in Figure 2.4. The protozoa have adopted bacteria as engines the way a human might use a team of horses. The symbiotic association of course is much closer; it's obligatory for both kinds of organism.[12] Even more curious, this kind of symbiosis has occurred on at least one other occasion, with a different bacterium in a different intestinal protozoan in another termite. In this latter case the engines are the flagella of rod-shaped bacteria. Each of several thousand bacteria on each protozoan has about a dozen flagella, oriented on its surface so they all contribute to moving the protozoan around.[13]
2. Only coelenterates, such as jellyfish, know how to make certain special stinging cells, their nematocysts. Contact with a big coelenterate

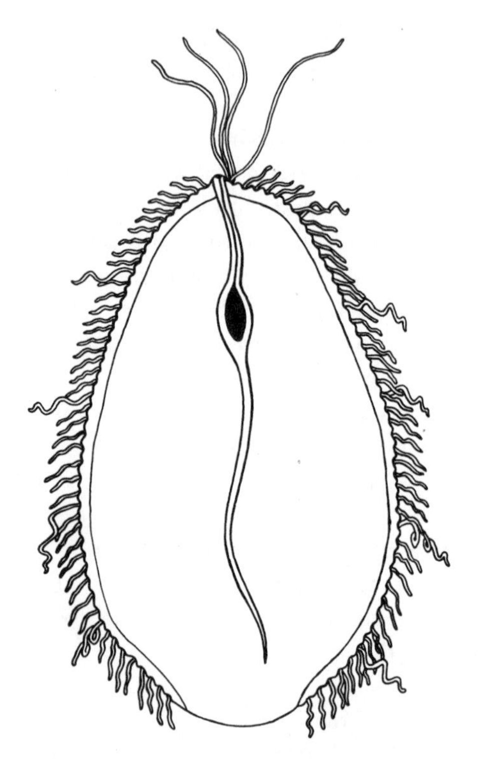

FIGURE 2.4. *A mixotrich with a pox of spirochetes. While the mixotrich has several flagella of its own (at top), for some reason it doesn't use them for locomotion.*

(the Portuguese man-of-war is especially vicious) is extremely unpleasant for a person and often fatal for a fish. A few creatures somehow manage to contact coelenterates without triggering the nematocysts. These, including a few fish, then use the coelenterates for shelter, protection, and even food. Certain nudibranch mollusks (marine snails without shells) take the trick a giant step farther. Not only can they contact coelenterates without damage, but they can incorporate the nematocysts into their own skins quite undischarged and still defensively potent.[14] Here no symbiosis occurs; rather a jellyfish technology is appropriated and redeployed: They steal loaded guns from the army.

In both these cases something mechanical crosses from one lineage to another. More frequent are chemical transfers, in which, for instance, an animal becomes poisonous to potential predators by ingesting a normally toxic plant (as when monarch butterflies eat milkweeds). The general point is that recognizing such cases shows that we know what to look for. So we can confidently declare such transfers uncommon and assert their difficulty for natural selection.

As a designer, then, nature is not only glacial in speed but lacking in versatility and erratic in performance. Fundamental innovation comes hard, and once achieved, it disseminates almost entirely within a lineage. To a remarkable extent the dazzling diversity in nature represents superficial features of systems of an exceedingly conservative and stereotyped character. No patents exist to be licensed; infringement by copying is impossible (although fortuitous infringement—convergence—carries no penalty); and the bottom line is immediate profit—surer reproduction. Trial, error, patience.

DESIGN FROM A COMPARATIVE PERSPECTIVE

Design in human technology is far less constrained. All the complications of life histories are swept away. Fundamental change may be hard to bring about, but it faces no fundamental barrier. Of course, for the most part we build on past accomplishments, and we do so within limits set by human ingenuity and modes of thinking, by materials at hand, and by the social support at any time and place for innovation. The inventor, after all, must eat, and society must be willing to adopt an invention. Without a doubt, human societies have run the gamut from hostility to hospitality toward technological innovation. The Roman Empire, given its size and duration, must be rated quite low; ships, building materials, even weaponry changed little for hundreds of years. By contrast northern Europe and North America during the nineteenth century were obviously high; innovations included railroads, steamships, telecommunications, synthetic fibers, and electric motors and lighting, just to name a few.

Design in human technology generates change and progress in a way that's a lot easier on the intuition than design in nature. On the other hand, it's a lot harder to encapsulate succinctly. An inventor, a concept, a model; testing, approval, dissemination, improvement—the elements come easily to mind. Alexander Bell thinks of a simple way to convert sounds to electrical signals and back again. After a lot of tinkering he builds a successful model, he patents the model, and then he and others commercialize the telephone. Thomas Edison's more efficient mouthpiece then supersedes this element of the Bell system. The design process clearly involves the planning, anticipation, and deliberation of which natural selection is incapable. But the really basic items of human design are so old that we know little or nothing of their beginnings. Who, after all, invented the right angle? Who first fabricated things from metals? Good histories of technology have been written, but they can't see back more than a few millennia.

While the processes of design differ dramatically between the two technologies, important elements are common to both—perhaps more than are commonly realized.

Cultural dissemination. Nothing forbids invention and cultural transmission among animals (plants are a different matter). Monkeys and apes invent copiously, but my favorite case involves birds, not known as such a brainy bunch. Some years ago individuals of four species of tits in Britain found that if they pecked just right at the caps of milk bottles left on doorsteps, they'd be rewarded with fine meals of cream. Cream might seem an unusual food for birds, but if they can get it, it's a grand source of fat, their fuel for flight. Those cream-fattened fliers might have enjoyed greater reproductive success; their offspring might have preferred pecking at objects in some class that included bottle caps; natural selection might have fine-tuned the behavior in succeeding generations. Dissemination among British tits in fact proved far more rapid. Milk bottle pecking spread quickly as birds learned from one another—until a change in the design of the caps foiled birds and some fascinated biologists. As our bio-bard John Burns quipped, the case "smacks of the loosely preadaptive inasmuch as tits were the birds to try it."[15]

Natural selection. Nor does anything prevent accident and selection by percipient humans. Lucky accidents must have been fairly important in early human history. Domestication of plants and animals, the origin of cooking, the use of a strangely shaped stone followed by deliberate shaping of others—all need only minimal foresight or initial deliberation. Certainly folk medicine is mostly selection from among nearly random ministrations. We learned that chronic aspirin therapy lowers the risk of clot-associated cardiovascular diseases from observations incidental to the use of aspirin for arthritis. Many a toothsome recipe must derive from accidental alteration of a routine procedure.

Both natural and human technologies bow to economics. Improved reproductive success—"fitness"—and corporate advantage are much alike. Corporate advantage is especially close to what the biologist means by fitness according to the economist John Kenneth Galbraith.[16] He points out that corporate managers typically favor expansion—equivalent to population growth—over payout of profits to anonymous stockholders. Furthermore, one hears complaints that the corporate culture takes an unduly short-term view of what's beneficial; immediate utility is thus the main measure of success in both technologies.

The role of isolation. A lot of evolutionary change apparently occurs in small, genetically isolated populations, in which competition with other members of a species is limited by geographic or ecological barriers. In general, organisms that don't do a lot of moving around generate more different species than those that do. When a barrier falls, a now well-tuned form may come into renewed competition and displace another—part of why biogeography has become such an important part of evolutionary biology.

With a globally uniform technology, we've less opportunity for such temporary sheltering, and continuous stirring replaces the rare mixing of a Marco Polo. But we still set up isolated habitats of reduced competitive pressures. For better or worse, a lot of secret work done under military auspices ultimately appears in open, nonmilitary circumstances. The space program of the United States sheltered and subsidized the microelectronics industry. Surely government-supported biomedical research (with results available in the public domain) and the profitability of the pharmaceutical companies are joined at the hip.

The conservative bias. No dominant technology displaces easily in a purely competitive encounter. Cars long ago standardized on the Otto cycle internal-combustion engine. Its nearly universal use doesn't prove it the most economically appropriate way to power passenger vehicles. A hundred years of refinement and easy access to fuel and maintenance are advantages not easily overcome. If that engine is ever displaced by another, it won't be through pure competition in an open, global marketplace. Present standards for the resolution of broadcast television are clearly anachronistic, but with hundreds of millions of sets in use, they're proving hard to shake off. Similarly, insects constitute the majority of the world's animal species despite the awkwardness of periodic molting and the resulting difficulty of making large terrestrial forms. In neither technology does fundamental superiority of a newcomer give assurance of success. Their respective histories hold each a captive.

The time course of change. In times past we believed in the steady advance of human culture; everything went "onward and upward." Nowadays we recognize that our history is bumpier, more episodic. In biology we now generally accept the idea that evolutionary change may occur as intermittent bursts, something first brought to our attention by Niles Eldredge and Stephen Jay Gould.[17] The issue remaining is just the relative importance of their "punctuated equilibrium" and steadier change through time (phyletic gradualism). Most of the examples on which arguments turn are fairly arcane. But consider the explosive radiation of an already old group of small creatures called mammals following the demise of the dinosaurs sixty-five million years ago. Millions of years of stagnation and then, wow! Another such episode happened more recently, a mere twenty million years ago. Grasses evolved, and with them came extensive grasslands. Grass may be easy to get to, but it's poor fodder for the unprepared, abrasive stuff with a low energy content for its bulk. Within a relatively short time, teeth in many lineages of mammals evolved into a form (Figure 2.5) that could manage to munch this miserable material.[18] Less dramatically, punctuated equilibrium implies that most of the time natural selection just maintains forms well arranged for what they do; once an organism is well established, random alterations are especially unlikely to improve it.

Looking for equivalent bumpiness in human technology requires that

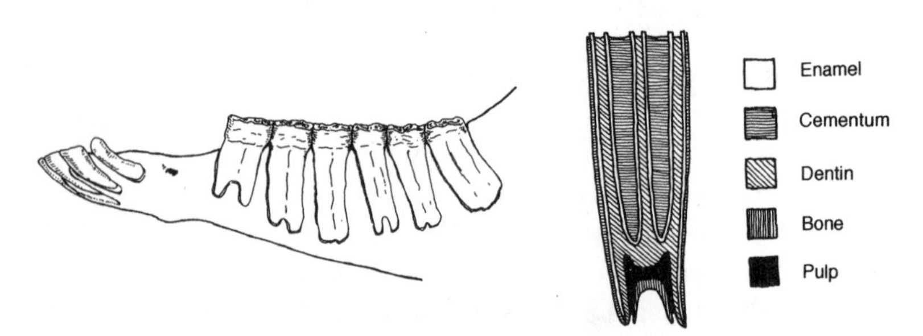

FIGURE 2.5. *A horse's teeth are typical of those of large, grazing mammals. Vertical columns and layers of materials (enamel, cementum, and dentin) differ in hardness. In use the harder material will always protrude farthest, so the teeth do not get smooth as they wear down.*

we speed up our time scale. Doing so reveals the same variation in how fast change occurs. The small electric fan you buy today has changed little in eighty years,[19] but the laptop computer was almost unimaginable forty years ago. Fans, toasters, and other small appliances came hard on the heels of light bulbs and the wiring of households. Even for a stand-alone device, change is irregular. The body of a single-lens reflex camera looks and works like one of twenty or fifty years ago, but between skin and skeleton a layer of novel electronics has recently been interposed. Probably the development of semiconductor amplifiers (transistors) in the late 1940s and the related creation of digital integrated circuits (chips) in the 1960s have been the main instigators of the most explosive changes in contemporary human technology. By contrast, our homes, our vehicles, our household appliances, and our clothing have changed far less than we might have predicted forty or fifty years ago. Besides radically new electronics and a few pieces of soft plastic, I see precious little novelty when I look around my house.

Incremental progress. Natural selection, as emphasized a few pages ago, is an evolutionary rather than revolutionary process, one building on small changes. Of course we humans can make major technological revolutions. But what do we really do? Early education gives the impression that James Watt invented the steam engine and Henry Ford the automobile. When examined in detail, though, almost every recent human invention is part of an incremental sequence. Key individuals certainly play important roles, but almost every recent history of technology views the development of specific technologies as more gradual than what most of us learned.[20] That long-term progress is unsteady doesn't mean that short-term change isn't incremental; just as in nature, geologically episodic change isn't revolutionary when viewed generation by generation.

New uses for old devices. Evolutionary biologists speak of preadaptations, preexisting features that make organisms suitable for new situations. For instance, amphibians and the rest of us legged vertebrates appear to have evolved not from familiar ray-finned fishes but from a group of lobe-finned fishes. (One representative of this group survives in deep water off the east coast of Africa. Until its discovery in 1938 we thought that coelacanths had been out of the picture for the past hundred million years.) Living in oxygen-poor swamps, these lobe-finned fishes had evolved air-breathing lungs and muscular fins that allowed them to

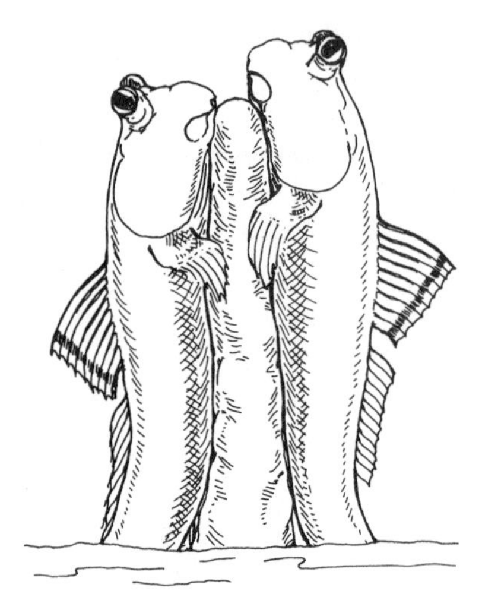

FIGURE 2.6. *Not prehistoric lobe-finned fish but contemporary ray-finned ones, a pair of mudskippers. They prop themselves up in the manner suggested for creatures in the transition from fishes to tetrapods. No proposal of descent is involved; the mudskippers just show that fish can do it.*

keep their heads above water, perhaps as in Figure 2.6. The lungs and lobe fins that permitted life in ancient swamps amounted to preadaptations for our kind of ambulatory, air-breathing, terrestrial life.[21]

The sudden expansion of human culture has brought to light all kinds of preadaptations among organisms. For instance, a weed with seeds that are hard to separate from those of a crop plant can suddenly find itself superbly well suited to a vast new habitat. And some small flies (chironomids), whose larvae normally live attached to the rocks in rapid streams, now thrive as a nuisance in the aerators of sewage treatment plants.[22]

Preadaptation may be so common in human technology that no one pays it much attention. Computers that displayed programs as text and were instructed from a keyboard were obviously preadapted to become word processors. The waterwheels long used as power sources provided a way to apply rather than extract power in the first generation of steamboats. Indeed, using preexisting things in novel ways enjoys a special countercultural mystique, giving particular gratification if the items have outlived their original applications. For instance, an old oil drum can be sliced lengthwise to make a rotor for one kind of windmill. Some of us do it all the time. I recently pressed a plastic water pipe into service as a curtain rod and used an ordinary metal lathe as a torsion-testing machine.

Parallel developments. Nature often makes the same thing in several different lineages. Such convergence includes some truly remarkable cases.[23]

Fleshy, spiny, leafless plants evolved in the deserts of the Old World within one family, the euphorbs (such as the crown of thorns plant); and similar plants evolved in the American deserts within another family, the cacti. The most recent common ancestor of cacti and euphorbs was not fleshy, spiny, and leafless; they're truly convergent. Marsupial and placental mammals are in many cases amazingly similar. Marsupial mice, moles, flying squirrels, dogs, and others look like their placental equivalents but (on overwhelming evidence) represent an independent bush of mammalian evolution. The human (vertebrate) eye and the octopus (cephalopod) eye look alike and work alike but form another case of independent and convergent evolution. A good design is a good design, and that different lineages are driven by natural selection in similar directions shouldn't be surprising. Convergence tells us a lot about functionally important characters, since anything that converges must make a difference to reproductive success. It also directs attention to what is relatively easy (in some sense) for the evolutionary process.

We pay less specific attention to convergence in human cultures and technology, but it's there. Some parallels, such as the invention of the calculus by Leibniz and Newton in the seventeenth century or the suggestion of evolution by natural selection by Darwin and Wallace in the nineteenth, are purely intellectual. In each instance the intellectual climate must have been right. Technological parallel development is probably commoner. Did Marc Brunel or Eli Whitney invent interchangeable parts, Howe or Singer the sewing machine, Swan or Edison the light bulb? Or, in each case, both? Bell's first telephone patent beat a competitor by a few hours. Others were so close to flight that the Wright brothers had to strike a careful balance between immediate public disclosure and proper patent protection for their aircraft. When technology is changing, obvious next steps (need and ease, as in nature) will occur to more than one person.

In early human history, when contacts between cultures were much more limited, convergence must have occurred. Whether it's common or uncommon is debated between anthropologists of isolationist (independent origin) and diffusionist (spread from a single origin) predilections. Did such things as weaving, archery, and metallurgy arise more than once or only on single occasions? Impressed with the extreme commonness of convergence in nature, I ally more easily with the isolationists.

Extinction. We usually think of extinctions in Darwinian terms, but they result as much from general change in habitat or circumstance as from directly competitive inferiority. Characteristics beneficial during normal times are likely to work against an organism faced with catastrophic change.[24] After all, those normally nice characteristics were selected when the future mirrored the past; a plant that does well in the shade will have a hard time when the forest disappears! At the least, when the world changes rapidly, extreme specialization is likely to be counterproductive. One wonders whether the role of habitat change in large-scale extinctions and the disadvantage of specialization hold practical lessons for human technology. Certainly, the causes of extinction in human technology must be comparably complex. Saddlemakers and farriers declined not because of automation but because of automobilization. Perhaps the internal-combustion engine will fade too, in the face of more expensive fuel or unacceptable emissions (in effect, habitat change), not a competitor that's superior under current conditions.

Much has been written about evolutionary and revolutionary change. But these terms acquire special connotative hazards from their biological associations and political analogs. Maybe we ought instead to stick with less burdened words, such as "gradual" and "jumpy" or (more pedantically) "incremental" and "saltatory" for changes in small steps and great leaps. We're biased toward a jumpy view of history since we like to focus on heroes and to divide things up into discrete events, time periods, and categories. An incomplete fossil record makes us think that nature is similarly jumpy. Perhaps recognizing that we're prone to one bias will alert us to the pitfalls of the other.

The larger point is our recurrent theme that looking at both natural and human technologies forces us to think about each in novel ways. Here we've seen surprising similarities in practice despite the vast difference in underlying mechanism. Similarities have had center stage; farther along the differences will loom larger.

Chapter 3

THE MATTER OF MAGNITUDE

Size matters, and like evolution, it will pervade all that follows. For one thing, an effective design for large things often works poorly for small things, and vice versa.[1] For another, our two mechanical technologies span an enormous range, from a virtual macromolecule to the largest of human structures. For yet another, nature's products are generally smaller than ours, although the ranges overlap extensively. Since the two technologies share the same planet, they experience the same pressures, temperatures, gravitational accelerations, winds, and water currents. In many ways, though, the influence of such physical factors on the two often proves profoundly different; practical reality depends very much on how big something is.

How much size matters has long been recognized. Galileo gave it his full attention, correctly calculating that (in the absence of air resistance) an animal of any size ought to be able to jump as high as any other. This means that relative to body length, the small ones win hands down. (Even in the real world of draggy air, fleas are truly impressive, clearing the bar at several hundred times their own length.)[2] The great seventeenth-century French polymath Descartes put the matter this way: "The only difference I can see between machines and natural objects is that the workings of machines are mostly carried out by apparatus large enough to be readily perceptible by the senses (as is required to make their manufacture humanly possible), whereas natural processes almost always depend on parts so small that they utterly elude our senses."[3]

What confuses our intuitions—but didn't mislead Descartes—is our own atypical size. The smallest fully competent organism (thus excluding viruses), a bacterium that gives you a mild pneumonia, is about 0.2 micrometers long, about a fifth of a thousandth of a millimeter and just visible as a dot in a good light microscope. The largest in volume is a large whale, a little more than 20 meters or 60 feet long. That's roughly a hundred million–fold range. On an appropriate geometric scale, as in Figure 3.1, we humans hug the upper end. At about 2 meters long, we're about ten times shorter than the biggest whale but ten million times longer than the smallest bacterium. We use, for these comparisons, a geometric rather than an arithmetic scale, counting each additional zero,

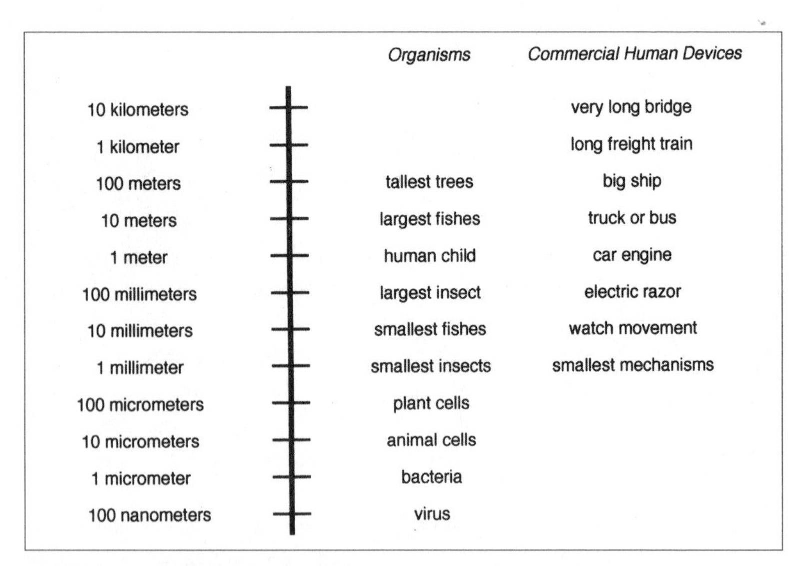

FIGURE 3.1. *The size ranges of organisms and our mechanical devices. Some arbitrary judgments beg forgiveness. The longest (if very thin) organisms are probably some multinucleate fungi, and the Great Wall of China is vastly longer than the longest bridge. Nor are subcellular items, such as microtubules and bacterial flagella, included.*

or order of magnitude, as an equivalent increment. As a pioneer of biomathematics, D'Arcy Thompson, puts it, "It is a remarkable thing, worth pausing to reflect upon, that we can pass so easily and in a dozen lines from molecular magnitudes to the dimensions of a Sequoia or a whale. Addition and subtraction, the old arithmetic of the Egyptians, are not powerful enough for such an operation."[4] (D'Arcy Thompson needs a few words. He's almost exclusively known for *On Growth and Form*, a large book written in 1917 and again in 1942, unquestionably the best-known work on mechanical aspects of biology. Part of the book's continuing impact—it's still in print—comes from its shear linguistic splendor; more, perhaps, reflects its accessibility, startling breadth, and creative insight. While certainly worth reading, as biology it's strange and anachronistic, a search for a kind of geometrical perfection in nature to which evolution by natural selection is largely irrelevant. Nonbiologists such as architects often assume that *On Growth and Form* is in the mainstream of biology or biomechanics. So I hasten to explain that Thompson is a much-beloved godfather rather than someone whose intellectual genes we proudly carry.)

Not only are most organisms smaller than we, but in most groups smallness is the ancestral condition and largeness the specialization.[5] Big fossils are impressive, but little ones are more likely to lead somewhere. Nature starts small. Organisms are basically built up from cells rather than divided into cells; the earliest fossils are microscopic. Human technology goes the other way. Our ships, buildings, and bridges may be larger than ever, but the factor of increase has been small and the times involved have been long. More impressive is the way our systems (or their parts) have gotten smaller. The first steam engines were enormous, operating slowly and at low pressures. Jet turbines are small, fast, high-pressure devices. Most extreme of course are electronic devices; compare today's microscopic semiconductor junctions within large-scale integrated circuits with the huge vacuum tubes of the 1930s.

LENGTH, SURFACE AREA, AND VOLUME

Length, surface, and volume aren't at all the same kind of thing. Consider a pair of cubic boxes, as in Figure 3.2. If an edge of one is twice as long as that of the other, then the larger one will have not twice but four times the surface area of the smaller. At the same time the larger will have fully eight times the volume of the smaller. Similarly, if one of the cubes has

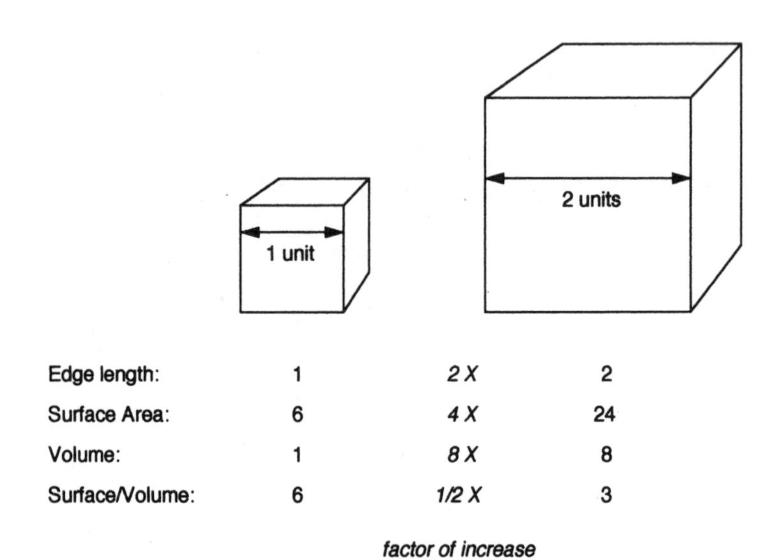

		factor of increase	
Edge length:	1	*2 X*	2
Surface Area:	6	*4 X*	24
Volume:	1	*8 X*	8
Surface/Volume:	6	*1/2 X*	3

FIGURE 3.2. *Two cubes, one with sides twice as long as those of the other. The bigger has four times the area, eight times the volume, but only half the surface relative to volume of the smaller.*

edges ten times longer than the other, it will have a hundred times as much surface area and no less than a thousand times as much volume. Put as a general rule, area increases as the square of length ($2^2 = 4$; $10^2 = 100$), while volume increases as the cube of length ($2^3 = 8$; $10^3 = 1,000$). The rule works for any set of similarly shaped objects, such as spheres or (at least roughly) salmon. When things grow big, volume increases more drastically than does surface area. Therefore, being big means having lots of inside relative to your outside; being small means having lots of outside relative to your inside.

Biological objects, whether trees, people, or bacteria, don't vary much in density—all are about as dense as water—so mass and weight follow the rules for volume. A fish twice as long as another of the same shape will weigh about eight times as much; let cooks take notice.

Thus variables that follow volume, such as weight, will increase faster than variables that follow surface or length. The consequences aren't trivial. For instance, heat is generated throughout an animal's insides but is lost across its surface. If two animals, a large one and a small one, produced heat at the same rate (relative to their volumes), the larger one, volume rich and surface poor, would be warmer. But body temperature varies little among mammals and birds of all sizes. We larger creatures simply produce less heat (relative to our volumes). We need proportionately less food. We can get by with a thinner shell of insulating fur, fat, or feathers. We can also walk or swim about in a cooler climate. Being warm-blooded would have been no enormous accomplishment for a large dinosaur but a lot more remarkable for the small mammals contemporary with the dinosaurs. Indeed, warm-bloodedness occurs only in animals above a few grams in mass; as a fine convergence, the smallest birds (hummingbirds) are about the same size as the smallest mammals (shrews). Both hummingbirds and shrews are voracious eaters, and at night both let their temperatures drop, essentially hibernating, lest they starve before morning. For warm-blooded aquatic animals the minimum size is still greater. Warm-bloodedness is no small trick for tiny animals; all that surface makes trouble.

Our technology makes elaborate use of heat—for instance, in fabricating materials. But we use large ovens and fabricate in large batches, so the actual energy requirements aren't that bad. For an organism only a millimeter or centimeter across, making a hot spot either internally or externally would be far more costly, relative to its volume. Keeping a large building heated is cheaper, relative to its volume, than is heating a small house. Colonial bees can heat their nests communally; no solitary insect can do so. Nor is this business of heat exchange the only consequence of how the relationships among length, surface area, and volume depend on size. All processes that involve exchange of material with the surroundings are ruled by those relationships—for both better and worse. If you're small, getting oxygen in and out is relatively easy, even without resort to lungs or gills, but at the same time you're more vulnerable to chemical assault by predator or pollution since no part of your inside is very far from your surface.

As noted, organisms have to grow in size without serious interruption of their functioning. That raises peculiar complications for size-dependent variables. An adult can't be, and in fact isn't, just an enlarged child, as you can see from Figure 3.3. Consider two people of ordinary corpulence. One is tall, and the other short. (It doesn't matter if the short one is a small adult rather than a youngster.) The weight of each ought to follow the cube of height, at least roughly, and the soles of the feet of each will have to bear that weight. But soles—now we're talking about an area. If

FIGURE 3.3.
An adult and an infant of about five months drawn to the same apparent height, each with head, limbs, and torso in correct proportion. To emphasize the change in shape, the baby has been given adult posture.

the tall person were just an enlarged version of the short one, then the tall one would impose more pounds (weight) on each square inch (area) of those soles. As it happens, though, we're more subtly made. As investigators at the Nike Research Center found out, tall people have disproportionately large feet, just disproportionate enough to fix the weight per area of sole.[6] Of course weight here is a kind of anticipated weight based on the cube of height; if you get fat, your feet don't enlarge in compensation, even if they flatten a bit in response.

Another example: A falling body falls faster and faster, until its increasing drag just equals its weight and it gets to a final, steady velocity. The surface of a falling object determines its drag while the volume of the object sets the weight that draws it earthward. Since a larger body has more volume for its surface and since weight equals drag at its final velocity, the larger body will fall faster, as you can see from Table 3.1. Falling is mechanically hazardous for a human; it's a serious danger for a nestling bird only because predators may lurk below. As a great biologist, J. B. S.

TABLE 3.1. *Final falling speeds for spheres of water's density at sea level. (Over this huge range one can't simply assume that drag varies with the square of speed. I've ignored transonic phenomena in the calculations.)*

DIAMETER	FALLING SPEED	
1 meter	330 m/s or	738 mph
10 centimeter	104	233
1 centimeter	15	33.6
1 millimeter	3.6	8.1
0.1 millimeter	0.27	0.6

Haldane, put it in an essay entitled "On Being the Right Size," "a mouse is uninjured, a man is broken, a horse splashes."[7] (In a bleak book about coal mining in England, *The Road to Wigan Pier,* George Orwell—who later wrote *Animal Farm* and *1984*—wondered about how mice get into the mines: ". . . possibly by falling down the shaft—for they say that a mouse can fall any distance uninjured, owing to its surface area being so large relative to its weight."[8] Common knowledge to Orwell and other British Socialists at least—Haldane's essay first appeared in their newspaper.)

In short, we big, terrestrial animals live in a gravity-dominated world. For an animal of more ordinary size, gravity matters a lot less. For aquatic animals, it's of no great gravity at all.

SIZE AND FLIGHT

No mechanical feat is more impressive than flight. Nature showed us that flight was possible, something that could not have been self-evident. But while flying animals pointed the way to airplanes, they misled us badly on the particulars—mainly because the practical problems of flight are strongly size-dependent. Put another way, a flying machine's size controls both its design and performance.[9]

First, a flying machine has two separate size-dependent missions, staying aloft and making headway. For a tiny flying insect, staying aloft against gravity's pull is easy. But making headway against the drag of the air is tougher than for a bird or airplane. At issue, again, is how drag and weight scale with size. To stay aloft takes an upward force that just counterbalances weight; going forward requires a force equal to drag at that flying speed. Weight depends on volume, while drag depends on surface area. Halving body length reduces weight fully eightfold while reducing drag about fourfold. Thus the smaller creature finds weight less troublesome but drag more so. Thus it flies more slowly and finds itself more severely affected by any wind—for better (taking advantage of air currents) or worse (dealing with headwinds and navigational complications).

Second, a wing's lift varies with its area, just as its drag does. Therefore, doubling length (while keeping shape unaltered) gives a craft four times the lift but eight times the weight, which doesn't sound auspicious. One solution is having really large wings on the larger craft; another is to fly somewhat faster: Like drag, lift goes up with speed through the air. So here again, larger ought to mean faster, roughly what we see among flying animals, from tiny insects to large birds and larger planes. It needn't (and doesn't) mean very much faster, though, since a doubling of speed increases both lift and drag, not twice but about four times. A fruit fly might hit three miles per hour, a bumblebee can do twelve or so, only large birds can exceed forty or fifty miles per hour, while for airplanes that's about the slowest they can go.

A third size-dependent factor complicates design and performance. The best wings produce a lot of lift while suffering little drag. This relationship between lift and drag depends slightly but significantly on size and speed, especially for small wings traveling slowly. Lift relative to drag gets worse as the craft gets smaller. The culprit is a property of fluids, whether liquids or gases, known as viscosity. Put simply, viscosity is a fluid's resistance to flowing, its internal stickiness. Its influence gets steadily more pernicious as systems get smaller and slower. The wing of a tiny insect moves through air a bit like a rod pushed through thick syrup; its shape, upon which its lift depends, is considerably obscured by the air carried with it. Very small and slow wings have more drag relative to their lift; here nature is the designer with the harder assignment.

Consider gliders, whether living or not. The angle at which a glider descends in still air depends almost entirely on that ratio of lift to drag; maximization of the ratio gives the flattest, most nearly horizontal glide. Birds can't make glides as flat as sailplanes, simply because they're smaller, as in Table 3.2. In practical terms, a bird can't glide as far from a given height. Is human technology better since our wings have less drag for their lift than those of birds? The comparison is so clouded by size that no simple judgment is fair or useful. As gliders, insects are still worse off than birds, and only large ones, such as locusts and butterflies, do much gliding at all.[10]

Gliding through moving air makes matters still murkier. Both human gliders and gliding animals ride air currents to prolong and direct their

TABLE 3.2. *Least (and thus best) angles of descent for a variety of gliders.*

FLIER	MINIMUM GLIDE ANGLE
Sailplane	1.5°
Small airplane, engine off	3.0°
Albatross	3.0°
Falcon	5.5°
Pigeon	9.5°
Monarch butterfly	12 °
Flies, etc. (calculated)	30 °

flights; it's called soaring. So time aloft may be quite as important as the distance that might be covered in a simple still-air glide. Time aloft depends equally on descent angle and descent speed. Smaller generally means both steeper and slower, so while the gliding bird may descend more steeply than the sailplane, it descends more slowly. Thus time aloft is about the same for an eagle and a sailplane. Gliding insects descend still more steeply but more slowly yet, and the monarch butterfly is little worse than bird or plane.

Size affects the design of flying machines in a fourth way, a subtle one that misled most of our early attempts to fly. Flying animals use beating wings to produce both lift and thrust. Efficient human aircraft (which helicopters and harrier jets are not) divide lift and thrust production between fixed wings and propellers. Should birds emulate our separation of the two functions? Or why can we achieve decent efficiency only by disentangling the two while birds needn't bother?

An aircraft must push air rearward faster than its flying speed in order to keep going forward. At the same time it has to push air downward faster than its ascending to keep up its ascent. But ascent speeds are tiny compared with forward speeds; indeed, for most of any flight of bird or airplane the ascent speed is zero. For level flight, *any* downward push will make lift, and the greatest efficiency (for a reason best avoided at this point) is realized when the largest amount of air is given the least downward speed. If the craft is going forward, though, it must push air that's already moving fast. Long, fixed wings deflect a lot of air downward; short propellers spin rapidly, so pushing less air but giving it that necessarily greater speed.[11] The separation of functions buys efficiency—smaller engines and less fuel—for airplanes. But flying animals, being smaller, proceed more slowly and don't encounter such rapid, oncoming wind.

Thus separating propeller and wings buys little advantage, and flapping wings make both lift and thrust quite nicely. We'll return to this comparison in Chapter 10.

SURFACE TENSION AND DIFFUSION

If we humans, so large and lumbering, care little about viscosity, we care even less about surface tension and still less about diffusion. But for the tiny insect that accidentally touches a wing to a puddle of water, surface tension can be a matter of life or death. Nature gives it close attention, making surfaces that either wet easily or vigorously repel water, depending on the roles they play.

Surface tension comes from mutual attraction among the molecules of a liquid such as water. If molecules attract each other, they prefer to cluster—which is to say that they prefer to form the least amount of outer surface. Surface tension works much like a bunch of people trying to get close together, perhaps to lessen heat loss. A droplet of water tends to take the shape that gives it the least surface for its volume. If the droplet of water rests on a surface to which its molecules are less attracted than they are to each other, it will round up into a flattened sphere, as, for instance, a raindrop on a well-waxed car. In an orbiting spacecraft the droplet will be almost perfectly spherical. If, by contrast, the water molecules are more strongly attracted to the surface than to each other, the droplet will spread as a thin film over the surface.

We see that water rises in a thin glass tube since water and clean glass attract each other strongly; we note that mercury drops in the same tube because of its low attraction to glass. We commonly add a little detergent to water to reduce its surface tension and make it give a more cleansing wash. We notice (usually without knowing why) that an absorbent, say, a cotton ball or a cloth of natural fiber, compresses as it dries; "absorbent" means that water adheres to it, and the surface-minimizing tendency of water pulls the wet fibers inward as the water evaporates and loses volume. One can use the phenomenon to make a self-starting, small-scale siphon; a glass of water will empty if you drape a piece of cotton fabric from the bottom of the glass up and over its rim and down into an adjacent, lower sink. Still, such things are far from critical to our daily lives.

But for a water strider, as in Figure 3.4, surface tension is life itself. Its legs have a waxy coating and don't attract water; its weight depresses the

FIGURE 3.4. *A water strider standing on the surface of a pond. Notice the dimples in the water's surface under each leg.*

water's surface under each leg; the water pushes upward as it tries to flatten its surface and thus minimize its area; the water's surface ends up depressed just enough so that its upward force offsets the water strider's weight. If the insect lifts two legs off the water, the other four depress their surface dimples slightly more. In its world surface tension is a major player; some surface insects move forward by squirting detergent behind. What about us? Why can't we walk on water? A downward force, weight, has to be balanced by an upward force, surface tension times the length of the foot-water-air contact line. The downward force is a matter of a volume, the upward one of a length. We're so big that we weigh too much for feet of any manageable edge length. My sixty-kilogram mass would require feet with eight thousand meters (five miles) of edge, but a ten-milligram mosquito-size insect needs a mere millimeter of total foot edge.

The downside of being small enough to walk on water is being unable to get through the water's surface. That surface has about the same behavior for an insect about a millimeter long as the canvas wall of a tent has for us. We can dive in, we can row a boat by repeatedly dipping the oars, and we can swim using crawl, butterfly, or backstroke; the tiny insect, though, must remain above or below the surface.

Nor is the surface of pond or puddle the only place surface tension matters. Water rises in the conduits of trees and evaporates from leaves. If you stop sucking on a straw, air enters the top, and the soda goes back down, so why doesn't air go back into the leaves? Surface tension turns out to be critical for keeping the air out; the pores in a leaf's cell walls will let water evaporate out, but they're just too small for air to get in. Around thirty atmospheres of pressure would be needed to pull air through the air-water boundary in pores a ten-thousandth of a millimeter across. Again, the relevant size (here pore diameter) is small, so surface tension can be enlisted for a major role.[12]

Diffusion comes into play on an even smaller scale. It's a consequence of the endless random and independent wandering of every molecule of every gas or liquid; left alone, substances such as oxygen and nitrogen or fresh and salt water mix together. The usual demonstration of molecular diffusion involves opening a bottle of perfume in a classroom; after a short while everyone smells the aroma. The odorant is spread, it is claimed, by that random wandering of molecules—to diffusion.[13] Not so. Except for the tiniest bit next to one's nasal epithelium, the perfume has been carried around by the irregular and turbulent motion of air in the room, something totally different from random molecular motion. So ubiquitous is such convective motion of air and water that diffusion is almost impossible to demonstrate on a perceptually relevant scale.

But as we go from small to smaller, to subcellular dimensions, diffusion becomes a potent agency of transport and mixing—within ourselves and all other organisms. Impulses most often go from one nerve cell to another by the diffusive spread of a transmitter substance. With a gap between cells of only about one fifty-thousandth of a millimeter, the diffusional delay is about one ten-thousandth of a second. For subcellular distances, diffusion is certainly speedy. Indeed, almost all transport of material within animal cells takes advantage of diffusion. But the usefulness of diffusion depends drastically on size; a tenfold increase in distance slows diffusive transport a full hundredfold. Animals made up of more than one or a few cells can't ordinarily rely just on diffusion for moving material within themselves. They have to augment it with hearts, blood vessels, pumped lungs, digestive tubes, and other devices that force fluids to move.[14]

The machines of human technology, much larger than cells, only occasionally make use of diffusion. One blood-cleansing machine used for dialysis of people with kidney failure relies on diffusion in and out of very tiny pipes with a huge aggregate surface area. One famous (or infamous) process takes advantage of the different rates of diffusion of molecules of different sizes. The rare but fissionable uranium 235 moves faster than the common, bigger, and nonfissionable uranium 238 as they diffuse (as gases) through a porous barrier. Being big and impatient, we ordinarily resort to stirring and pumping to get around the slowness of diffusion over appreciable distances, doing just what animals do when they make circulatory systems. A cell-size creature might well wonder why we bother.

GRAVITY AND INERTIA

We've now noted three phenomena—viscosity, surface tension, and diffusion—especially important in small systems. Others—in particular, gravity and inertia—dominate large ones. Gravity has already raised problems here. It causes large objects to fall faster than small ones. It doesn't let large creatures support themselves with surface tension on water's surface. And it makes it necessary for large aircraft to fly faster if they're to stay up with decent economy.

Gravity's size-related mischief takes more subtle forms as well. Consider the moving waves made by wind blowing across a body of water. What keeps them wavy is the water's inertia. What makes the water flatten out are the water's surface tension and weight. For ripples two-thirds of an inch or less between crests, surface tension is the more important thing flattening the water; its molecules pull together and minimize its surface area. For larger waves, weight—gravity—predominates, and water's preference for flowing downward is what flattens the water. The shift makes big waves and small waves behave differently. In particular, the relationship between the size of waves and the speed at which they roll along depends on whether they're big or small. For big waves, bigger is faster; increase their crest-to-crest distance, or wavelength, fourfold, and waves travel twice as fast. An ordinary boat can't easily exceed the speed of waves as long as its hull, so a boat four times as long can go twice as fast before the cost of propulsion starts to rise disproportionately: Large ships go faster than small ones, and even small ones go faster than ducks and muskrats. But for tiny ripples, ones less than two-thirds of an inch apart, the rule is just the opposite: Smaller is faster. The world of a minute surface boat, such as a whirligig beetle, must be something like a freeway with small, fast sports cars and large, slow vans.

Inertia, a property of both solids and fluids, is the tendency for something either to remain at rest or to keep moving unless persuaded otherwise by some external force. Put another way, to get an object going takes force, and a moving object exerts a force when it stops. More specifically the force equals the mass of the object times the acceleration or deceleration that alters its motion. What matters here is that the force associated with inertia follows an object's mass. A massive system can exert a lot of force by stopping suddenly. Conversely, bullets must have enormous speeds to offset their limited masses, and the lower speeds of short-muzzle handguns are commonly offset by using heavier projectiles. Humans have long used stone-ended clubs and metal-headed hammers, sledges, mauls, picks, and axes—heavy things that stop abruptly. Large pieces of metal can be shaped by dropping even larger pieces on them, something of industrial importance for well over a century. A large animal can inflict substantial damage on another by kicking; even a human can injure another by punching. But inertial aggression without weaponry has little value for creatures much smaller than we are; the most pugnacious ants don't kick their antagonists. Even for us, the effectiveness of a punch depends on the inertia—the mass—of its mark. Kicking your cat is nasty; kicking a mouse is ineffectual.

Put the other way around, small things, with less mass, are easier to start and stop—to accelerate and decelerate. As noted earlier, all of nature's jumpers could, if air resistance didn't matter, achieve about the same height. That implies that their takeoff speeds must be the same. But the short-legged flea achieves that speed in a vastly shorter distance than does the long-legged kangaroo; the flea's acceleration is much greater. Bigger may mean higher speed, but bigger also means lower acceleration, a rule of thumb that works for both living and nonliving systems. Try catching a resting housefly in your hand! The jackrabbit starts faster than the best racehorse and the most violent drag racer. Once started, though, the large mover can coast better than the small one. Stopping a large ship takes miles; a ferry must reverse its engines or it will smash the dock; automobiles must be equipped with brakes. But a swimming microorganism will halt almost instantly—typically in less than its body length. Its surface-dependent drag is relatively huge; its mass-dependent inertia is trivial.

Inertia also affects how fluids flow. At small sizes, viscosity predominates, and flows are orderly, laminar affairs. Each bit of fluid does nearly the same thing as its neighbors. At larger sizes, inertia increasingly offsets viscosity, so the bits of fluid tend to keep doing whatever they have been, despite any different motion of their now more temporary associates; we call such flows, with their chaotic eddying, turbulent. (Figure 3.5 illustrates the difference.) Flow immediately around an aircraft or ship is inevitably turbulent; flow around a microorganism is as assuredly laminar. (In between some tuning of transition points may be achieved by changes of shape or surface texture.) Turbulent flow really stirs things up; laminar flow is surprisingly ineffectual at mixing. A microperson's spoon wouldn't easily stir milk into coffee. Blood flow in all but the largest vessels of large mammals is laminar; almost all flows in industrial and household plumbing are turbulent.

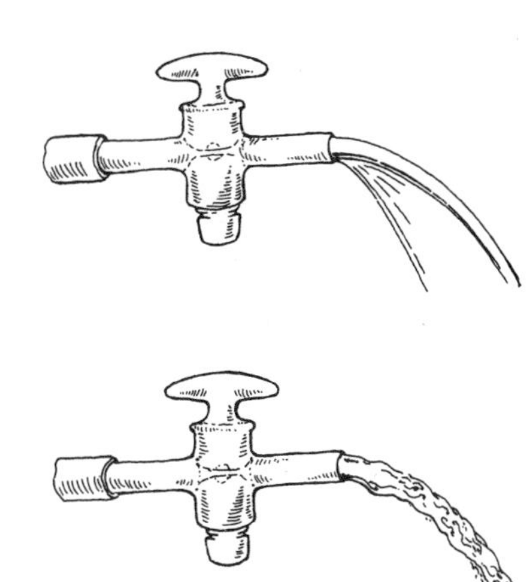

FIGURE 3.5. *As speed increases, the flow of a liquid within and from a pipe shifts from being laminar (above) to turbulent (below). The bigger the pipe, the lower the speed at which the transition occurs.*

The two regimes don't just differ in self-stirring; almost every rule for fluid flow comes in two versions. Viscosity is a kind of internal stickiness, and it causes small objects moving slowly through fluids to carry along a lot of the fluid. Seeds of such plants as dandelions and milkweeds can thus descend slowly by using a bunch of fine hairs as an analog of a parachute, as in Figure 3.6; the hairs carry along enough air so the bunch behaves like a balloon. But the device scales up badly since for larger sizes and speeds, viscosity becomes less important than weight and inertia. So neither technology can use the fluffy seed solution for slowing the descent of larger items. For slowing these, rather curiously, the two have gone separate ways. The parachutes used by humans or our bundles of baggage have only fairly crude analogs among terrestrial or arboreal organisms. Nature prefers another design, that of spinning, autogyrating seeds (fruits, strictly) of maples and other trees. While these passive autogyros scale up satisfactorily and have been considered for use by humans, parachutes have consistently proved handier.[15] Physical reality precludes using tufts of fluff if one is large, but it imposes no rigid choice between autogyro and parachute.

COLUMN AND BEAM

If you double the length of a column that supports a roof, how much fatter must you make it? Even a casual look at some of the rules used by mechanical engineers reveals still another role of size. These rules must

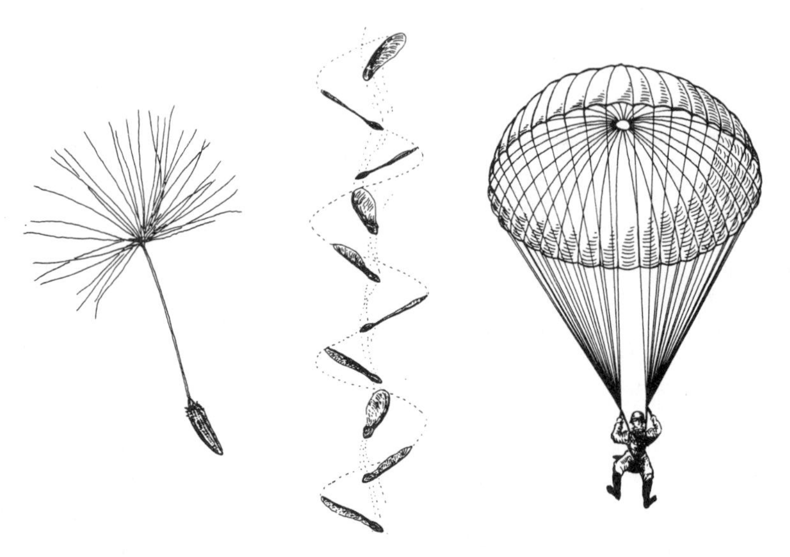

FIGURE 3.6. *Three ways to descend more slowly in air: the drag-increasing fibers of a dandelion seed, the lift-producing autogyrating samara of a maple, and a drag-increasing conventional parachute.*

apply equally to designs in nature. For a simple example, we'll consider two circular cylinders (Figure 3.7) made of an ordinary material and carrying loads that don't vary over time.

Look first at the upright column, a cylinder supporting its own weight and the weight of some load on top. As you might guess, failure by crushing will happen only in a short, fat column. We'll worry about one long and thin enough so it fails by sudden buckling to one side or another, as when the ends of a piece of dry spaghetti are pushed together. What does it take to start such a collapse? The critical force varies with the fourth power of the column's diameter divided by the square of the column's height. That's a combination sufficiently hard on the unaided intuition to demand specific numbers. What, then, would happen if we make the column twice as big, doubling both diameter and height? The force that starts buckling then goes up by 2^4 divided by 2^2, or 16/4, or fourfold. Swell, a twofold size increase gives a fourfold increase in resistance to buckling.

But it's really not at all good. If we're being completely consistent about doubling size, we end up increasing fully eightfold the weight of both the column and whatever loads it. So scaling the entire system up by

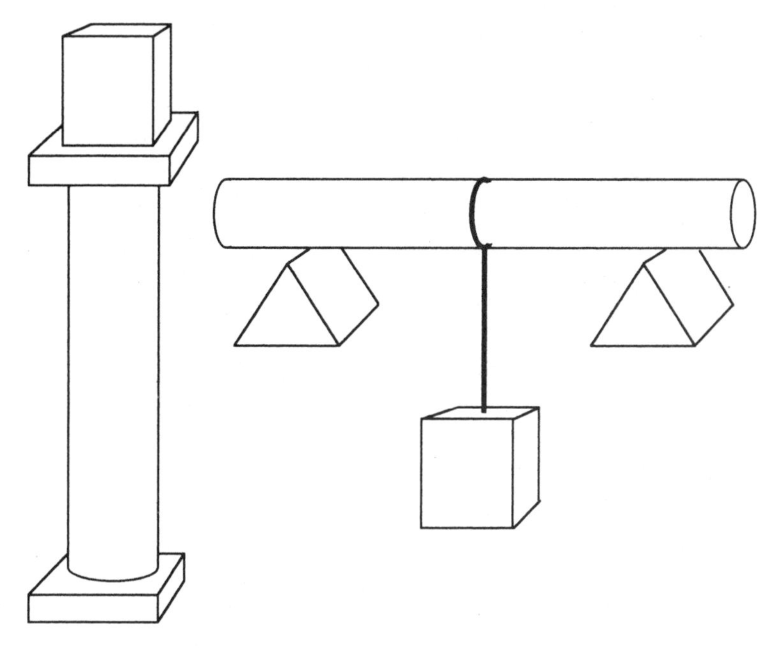

FIGURE 3.7. *A cylindrical column with a load on top; a similar circular cylinder supported near its ends and loaded in the middle.*

a factor of two gives a column that is four times stronger, to be sure, but one that must bear eight times the load! At best, the safety factor is halved; at worst, the column breaks. For the larger column to serve as well as the smaller one, it must be fatter. Worse, being fatter, it suffers still greater self-loading, which requires it to be fatter still. At the least, the larger structure will need different proportions, and if the differences in size are very great, it may need a stiffer material or have to be designed differently. Large mammals have stiffer (and thus more fracture-prone) bones than small ones. The daddy longlegs (or harvestman) walks easily on spindly, multiply flexed legs, while the legs of the elephant are straight, substantial columns.

The same rule applies to a cylinder serving as a horizontal beam between two supports. And it works whether the load acts at a single point in the middle or uniformly over its length. For the cylinder to bend downward in the same proportion after a doubling of size, something must be altered; its material must be stiffer or its thickness must be more than doubled. Put another way, if the distance between the beam's supports, the beam's diameter, and the length, width, and height of the load all are doubled, then the beam will sag downward not twice as far but four times as far. Once again, larger is weaker, relatively, whichever technology is in charge. Thus, if the same design is used, the larger bridge will incur a greater penalty from self-loading; scaled up sufficiently, it will collapse from self-loading alone. Elephants are bonier than cats but still must tread more carefully.

These examples are just that—examples, picked from a rich diversity of size-related phenomena. But they show how severely size affects design, both imposing constraints and affording opportunities. Gravity is important if you're big, diffusion if you're small. And so on. Of particular relevance here is the fact that the technology of people must be different from that of nature simply because the two span different size scales. A rule for design may apply to both, but if the rule has in it a size-related factor, the particular way it applies will differ between the two.

Chapter 4

SURFACES, ANGLES, AND CORNERS

Turning from size to shape, we start with things of such everyday commonality that they normally pass unnoticed. That's of course one of this book's main purposes: drawing attention to what we might otherwise overlook.

FLAT VERSUS CURVED SURFACES

We humans have a great affection for flatness. We make floors flat. And walls and roofs and steps and desktops and paper and books. We make our roads as flat as possible; the fancier the highway, the flatter the paving. Beneficence in the Book of Isaiah includes "making the rough places plane"—not domed or trough-shaped, but flat. Of course we're not fanatically flat. Our automobiles and airplanes have few flat surfaces. Jars, cans, and pipes are almost always cylindrical.

Nature, on the other hand, makes very few flat surfaces. The only ones both common and fairly large are photosynthetic structures: the leaves of many plants and the fronds of large algae. Flatter is better for these latter, since surface area facing skyward matters most for gathering light. Looking for flat surfaces beyond leaves and fronds, one scrapes around among such small fry as fish scales, bat wings, and duck feet.

First, the advantages of being flat. A floor that's easy to walk on at any point and in any direction will be flat—uniformly horizontal. So flatness has definite utility in a world dominated by gravitation, which is the world of large, terrestrial creatures like us. But the gravitational imperative can't be the only virtue. A wall of minimal area that separates two compartments will ordinarily be a flat one. If you want a set of surfaces that pile smoothly on top of one another in any alignment, flat surfaces are the easiest to make and the most versatile in use. How would one read a hemispherical page? How would one shelve a library of conical or dome-shaped books?

Roofs, whether horizontal or sloping, don't have to be flat; domes and arches have long and distinguished histories. But flat roofs are especially easy to make and handy to use, at least if they're tipped so water runs off. Identical, straight beams can be laid parallel to each other, flat boards to fit on top need cutting in only two rather than three dimensions, roofing paper can be unrolled and fastened down with minimal fitting, and shingle or slate can be applied as a strictly two-dimensional operation. Intersections between roof elements form straight lines, so junction devices are simple. By contrast, intersections of elements of vaulted or domed roofs (as in Figure 4.1) are usually curved, often complexly so.

In short, flat is easy and convenient, and our technology capitalizes on that. We stockpile material in a limited number of sizes and shapes; with only simple cuts we then make a wide range of structures. The prac-

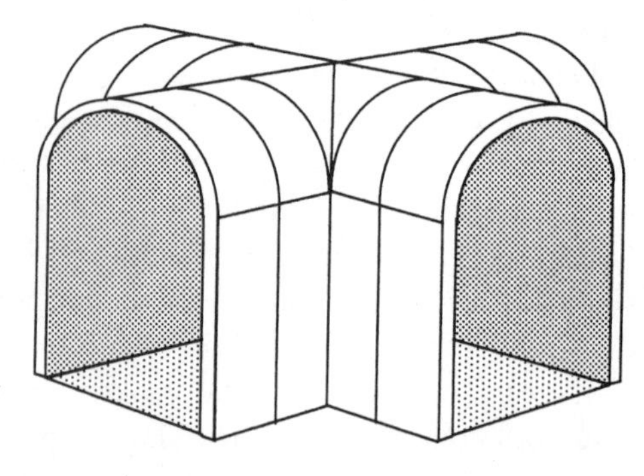

FIGURE 4.1. *Each of these intersecting barrel vaults supports the outward thrust of the other's walls. The resulting groin vault, used first by the Romans, makes efficient use of stone construction, but at the cost of awkwardly curved roof intersections.*

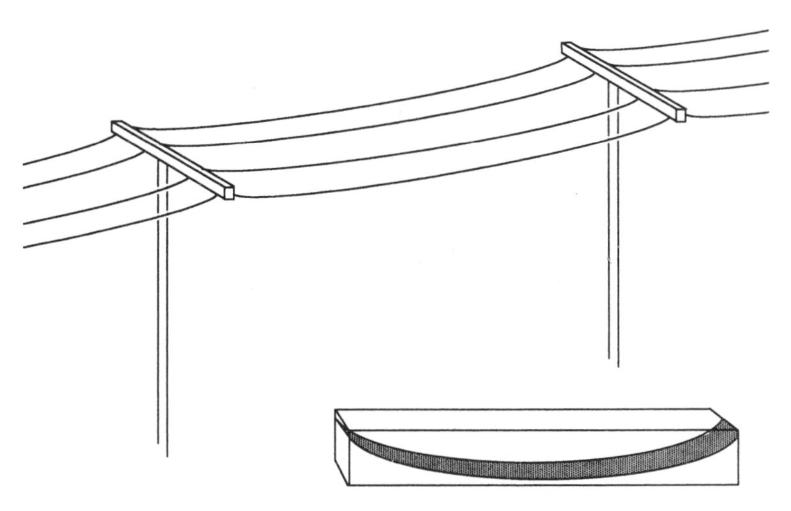

FIGURE 4.2. *Wires, unless absolutely weightless, must sag between supports. A beam that's supported at its ends may look uncurved, but only because its thickness conceals the effective curvature; it has a virtual hammock inside.*

ticality of both sawmill and paper mill rests on the flatness—and consequent versatility—of their products.

At the same time flatness has its disabilities. In the real world extending a straight line indefinitely is no trivial task, unless the line is exactly vertical. Neither a person nor a spider has trouble stretching a thread between two points. But neither can avoid gravitational sag in the middle. For a long rope the sag is obvious. The longer the distance to be spanned, the harder the surveyor must pull on the measuring tape to get an accurate reading. To achieve perfect straightness, to thwart the malevolence of gravity and self-loading, takes an infinitely forceful pull. That of course will break thread, rope, or tape. Hang something in the middle, and the situation grows worse, the sag more obvious. So, as in Figure 4.2, telephone wires always sag between poles, and hammocks droop earthward in use even if minimally swaybacked on their own.

We now see that a flat floor is something special. How might we make one? Stretched sheets simply won't work, and in practice we use beams—structural members with some thickness so they can resist bending. In a very crude sense (as the figure shows) a beam's thickness hides a curve, a concealed sag. The greater the load (including, of course, self-loading), the thicker must be the floor or the horizontal beams that support it.[1] The flat surface exacts a considerable price, paid as thickness in flat roofs, sagless bookshelves, and so forth.

Bookshelves—mention of them brings up a concrete example that could easily repay the cost of this book. A bookshelf is a beam supported at its ends like the one in Figure 3.7. How does the sag of the shelf depend on its load and dimensions? First, the distance the center sags follows the load. (We'll ignore self-loading since books greatly outweigh shelves.) Doubling the load doubles the sag. Second, the sag varies with the *cube* (third power) of the length of the shelf. With the same load, a thirty-six-inch shelf sags 73 percent farther down than a thirty-inch-long shelf. But ordinarily the load itself increases with the length of the shelf since the longer shelf carries more books. So as a practical rule the sag increases with the *fourth power* of length, and a thirty-six-inch shelf actually sags just over twice as far as a thirty-inch one.[2]

So stick with short shelves! If, however, you insist on long ones, bear in mind that even a little extra thickness helps a lot. Sag varies inversely with the *cube* of beam thickness. So the thirty-six-inch shelf need be only a little thicker—about 40 percent—to sag no more than the thirty-inch shelf. A thickness of one inch instead of three-quarter inch will do nicely.[3] That's not an off-the-shelf piece of lumber, but glue and clamps don't deter the determined bibliophile. In any case, assiduously avoid furniture suitable solely for bric-a-brac, and load a sample shelf before booking delivery.

(The general point—longer means weaker when something is bent—came up in the last chapter and will reappear later. In recent months it has become a personal matter. To illustrate the principle for a television program on size, I first put a board between two sawhorses three feet apart and loaded it with a 140-pound weight: me. Midpoint sag was half an inch. I then separated the sawhorses by six feet to get the four-inch sag expected in theory and by my test the day before. But I had inadvertently switched to a board with an oblique grain, and I crossed the line between treading the boards and walking the plank. Fortunately the board was the only real casualty.)

How does nature manage flatness? Dealing with the awkward problem of sag underlies the design of leaves quite as much as that of bookshelves. Leaves commonly stick out from branches more or less horizontally. Those veins on the undersurfaces of many leaves, as in Figure 4.3, may look trivial, but they're a way to increase the functional thickness of leaves with only a little extra investment in material. They are in fact beams, and with them, leaves sag a lot less than they would otherwise.

A second approach effectively thickens and thus stiffens a flat surface

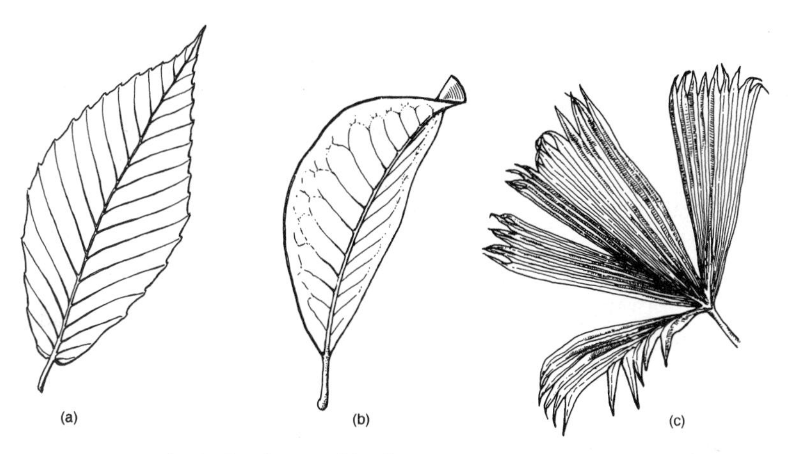

FIGURE 4.3. *Thin leaf surfaces avoid bending in various ways. Veins may provide supportive trusses (a), the whole leaf may be cambered lengthwise (b), or pleats can make a ridge and valley self-trussing system (c).*

with a little curvature. You can't make a flat sheet of paper stay upright when you hold it at the bottom—unless you curve it a little, as you do every time you hold up a page to read it. Quite a few biological surfaces get stiffness with just a little bit of curvature. A lot of leaves have a pair of downwardly concave surfaces, one on each side of the midrib. The southern magnolia is a good example; besides their midribs, its leaves lack conspicuous venation. But with that little curvature (and some thickness of the blades themselves) they form decently stiff beams.

Feathers are similarly curved on either side of their lengthwise central axes. The curvature helps offset the pull of gravity, as with leaves. For feathers it also serves an aerodynamic function, one particularly important for the long feathers that form the fingerlike tips of many wings. About a century ago several people discovered that airfoils work best if they're curved, with their concave faces downward.[4] Early aircraft really did use curved plates for wings, and modern planes just hide the curvature a little by covering the lower concavity with a flat surface. Similarly, we make the blades of electric fans of thin, curved plates of plastic or metal, getting both stiffening and better aerodynamic performance from the curvature. Incidentally, stiffening increases whichever way a blade is curved, but the aerodynamics improve only if the concavity is on the surface that faces more nearly downstream. Check your fans; you'll occasionally find one with its blades installed backward, and it's a breeze to reverse the blades on the shaft.

A third way to stiffen a flat surface: Give it a set of pleats running in the direction in which bending is expected. A piece of paper extending between two supports will sag under self-loading alone, but if you fan-fold the paper a few times so the creases go from one support to the other, it will hold many times its own weight. By pleating, you're increasing effective thickness without going to the trouble of adding proper beams beneath the surface. We use the device in corrugated cardboard, as well as in grooved and rippled ceilings and roof panels. Nature sometimes uses the scheme in large leaves, particularly those with veins that radiate rather than branch.

When it comes to stiffening flat plates with a minimum of material, nothing touches insect wings. Insects commonly invest only about 1 percent of body mass in their wings. Yet the wings move at several meters per second through the air, and many reverse their movement several hundred times each second. To get sufficient stiffness for this demanding application, they combine curvature, veins, and lengthwise pleats. Curvature may be an aerodynamic necessity in all but very small insects, but it presents a special problem for beating wings: The direction of curvature that's right for a downstroke is wrong for an upstroke. Many insects have a fine solution: Their wings are built so they're curved by the force of the wind that strikes them. Since the wind on the wings reverses between upstroke and downstroke, their curvature reverses too.[5]

Thin, flat surfaces deflect with even small loads. The rounded shapes of our automobiles seem made for minimizing drag and wind noise and for maximizing sales appeal—the latter perhaps by tacit allusion to well-curved human bodies. In fact, the roundness of cars serves primarily to stiffen them, if more attractively than folds or ripples. Pressing a piece of metal into a curved shape is much simpler and uses less material than spot-welding a lot of stiffening strips to a plate. Sometimes we use a cruder fix; heating ducts are most often rectangular in cross section, so they form huge arrays of thin, flat, metal sheets. Minor changes in pressure or even temperature all too easily bow these sheets sideways and produce disconcerting noises. We minimize the problem (but don't fully fix it, at least in my house) by putting slight diagonal creases in the flat surfaces.

One can take a more general view of the problem of flatness. Consider a hollow sphere made of a thin material with a greater pressure inside than outside. What pressure can the sphere withstand before bursting? In effect, what you're asking (as in Figure 4.4) is how much stretch—or tension—the surface can take before you're left with two hemispheres. What's the relationship between the pressure difference across the wall of the sphere and the tension generated in that wall? Oddly enough, the amount of tension produced by a given pressure depends on the size of the sphere. The rule, often called Laplace's law,[6] is that the tension is equal to the pressure times one-half the radius of the sphere. For a given pressure, the bigger the sphere, the greater is the tension stretching *each bit* of its surface. Or, for a bigger sphere, less pressure is needed to get a given tension. (Starting to blow up a large balloon is easier than starting a small one; you need to blow less hard to get enough tension to do the job.)

One more piece of the argument: A bigger sphere has flatter, less sharply curved walls. Thus, in general, the flatter the wall, the more tension a given pressure difference produces. A fully flat wall would occur only on an infinitely large sphere; any pressure difference across such a flat wall would generate an infinite tension. That's just the same problem as drawing a string out straight between two supports: Any load (equiva-

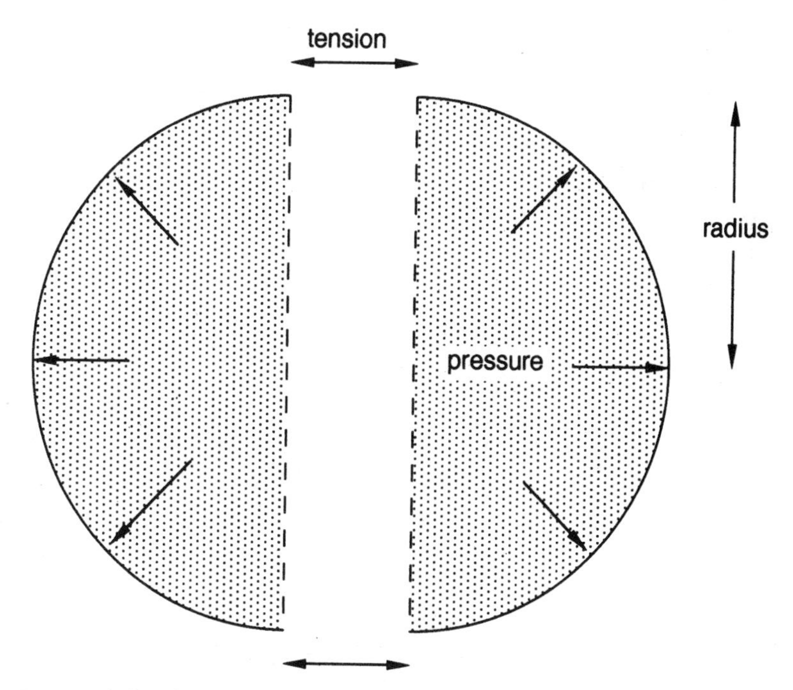

FIGURE 4.4. *The amount of tension in the walls of thin-walled spheres or cylinders produced by a given pressure inside depends on their size. The larger the sphere or cylinder, the greater is the resulting tension.*

lent to pressure) and the string can't be drawn straight unless pulled outward with infinite force (equivalent to tension).

Among its other implications, Laplace's law rules out making flat walls on balloons or any other internally pressurized structures. (One can of course fake flatness by using thick enough walls to conceal curved lines of tension, the way we use thick floors to internalize the curvature.) It instructs us, among other things, to make our pipes round rather than rectangular in cross section lest they split their seams with little provocation. It also explains why a rectangular carton of milk always (if filled) has bulgy sides. The sides must bulge, creating some curvature, or else they'll split. Only when packed together can full cartons have flat walls because only then do the pressures across their walls balance out. Aircraft fuselages are almost always cylindrical or elliptical. Grace has nothing to do with this, and drag minimization has only a little. Pressure is what counts; most aircraft in flight are pressurized to carry people at high altitudes, and that rules out thin, flat walls. One common commuter craft, a Shorts, has flat walls—it looks like a bus with wings—but since it can't be pressurized, it's limited to low altitudes.[7] Canisters for tea leaves can afford flat sides, but cans for beans or soup ought to be cylindrical. Their shape doesn't, as I once heard claimed, represent deliberate phallic symbolism, although can and phallus do have Laplace's law as common denominators.

Living things are usually made of flexible materials, and both they and their parts often have different pressures inside from outside. So Laplace's law tells us much about why nature abhors flat surfaces. Worms, guts, blood vessels, the alveoli of lungs, free-living cells—all must be cylindrical, elliptical, or spherical, indeed anything but flat-walled.

A similar rule relates the size of a dome and the load it can support. A bigger dome is less curved and therefore relatively weaker under load. Above some particular size, its weight alone becomes an unbearable load. Analogous rules apply to arches, to the main cables of suspension bridges, and so forth. In each case, the flatter the curve, the worse the tension generated by a given load, which limits how large we can make them. Nature may not use many rigid flat surfaces, but her stiff domes are legion; examples include eggshells, nutshells, clamshells, and our own heads.

RIGHT ANGLES

For humans, they're just right. In our world they're so ubiquitous that their roles defy short summarization: pages, desks, windows, floor and ceiling tiles, walls, shelves and drawers, boxes, shingles, any set of straight lines encompassing the wheels of a car or the legs of most furniture. Almost all the stock sizes of wood in the lumberyard are rectangular solids, as are bricks and cement blocks. Both Egyptians and Mayans made pyramids of rectangular blocks on square foundations. That twentieth-century style of art cubism (Matisse's derisive name) flaunts our obsessive rectilinearity. By contrast, from tropics to Arctic, simple dwellings are more often round: cones, domes, upright cylinders with conical or domed roofs, and so forth. I recently visited a museum that traced tens of thousands of years of archaeological and historical change; I noticed that millennium by millennium right angles become better established.[8] They're almost unfailing signatures of cultures of high technological complexity.

Nature has neither clear affection for right angles nor clear antipathy toward them. At least one bacterium is square,[9] and right angles occur between the edges of the skeletons of certain protozoa, the foraminiferans. Where a surface is covered with a single layer of cells, the walls or membranes that separate individual cells form right angles with that underlying surface. Where trees grow upward from level land, each trunk makes a right angle with the land; where the land slopes, there's still a right angle between trees and ultimate horizon. The pine trees in my front yard, for instance, have almost perfectly vertical trunks. We have to ask, then, not why nature avoids right angles but why we prefer them. Of particular interest in this regard are the semicircular canals of our inner ears. A set (Figure 4.5) consists of three canals, each at a right angle with the other two. They're important parts of the system with which we keep track of our orientation and acceleration. Do we like right angles just because our sensory equipment regards them as a simplest case in a geometrically complex world?[10]

In the last chapter size loomed large, and it plays a major role here as well. We're big creatures; between our size and terrestrial habitat, we're ruled by gravity. You don't tip over as long as you keep your center of gravity—the effective location of your weight—over your feet. So you stand vertically. You prefer to walk on horizontal surfaces, which much of the earth's surface is anyway. Vertical plus horizontal generates right angles, for you and the ground, for the walls and floor of a building, for a tree and the horizon. Stacks of things must stay close to vertical, or they'll tip. Walls of stacked blocks are far easier to make if upright, and the walls of large structures have over most of human history been made from stacked blocks.

Not so for smaller structures, for dwellings light enough to be packed up and moved, for ones easy to keep warm. For these, round and perhaps

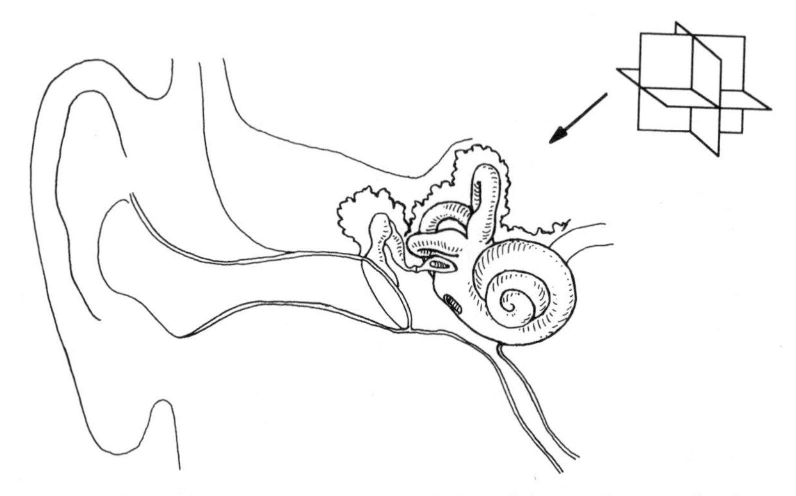

FIGURE 4.5. *We have three semicircular canals (beyond the arrow) associated with our inner ears. Each forms a plane at right angles with the others, so they occupy mutually perpendicular planes, as in the inset.*

inwardly tilting walls are better; tepees and similar conical and hemispherical houses have been invented by numerous cultures. Anthropologists and archaeologists[11] tell us that round houses typify nomadic or seminomadic societies; curvilinear buildings are more economical of material and easier to erect. By contrast, rectangular houses characterize sedentary societies. They permit more buildings in a small area—say, within a walled compound or city. Their interiors partition more easily, and since their outer walls can serve as common walls for adjacent structures, they're easier to add on to. On average, the round houses of primitive societies have less than half the floor area of the rectangular ones. A family may occupy several round houses, with still others serving as storage structures; a round building as a whole works like a single room of a rectangular one.

One's view of one's world is shaped by personal experience. For most of us that reflects our incorrigibly rectilinear culture. A peculiar test reveals an odd acquired bias. Back in the 1950s some psychologists tested the way two groups of Zulu reacted to a particular visual illusion. A rural group lived in traditional circular huts, while the other was urban, housed in conventional rectangular dwellings. The illusion, called the Ames window or rotating trapezoidal illusion (Figure 4.6), consists of a model of a window that's rotated about a vertical axis in front of a subject. Now windows are normally rectangular and not, as in the model, trapezoidal. When conditions for depth perception are marginal (one eye closed or the model quite distant, for instance), the urban Zulu (as do non-Zulu) commonly perceive the model as an oscillating rectangle slanted away from the viewer. In effect, we imagine a perspective that forces the model to fit our notion of a properly rectangular window. The rural Zulu are much less easily fooled. For that matter, their language lacks words for "square" or "rectangle."[12] Other cross-cultural comparisons yield similar

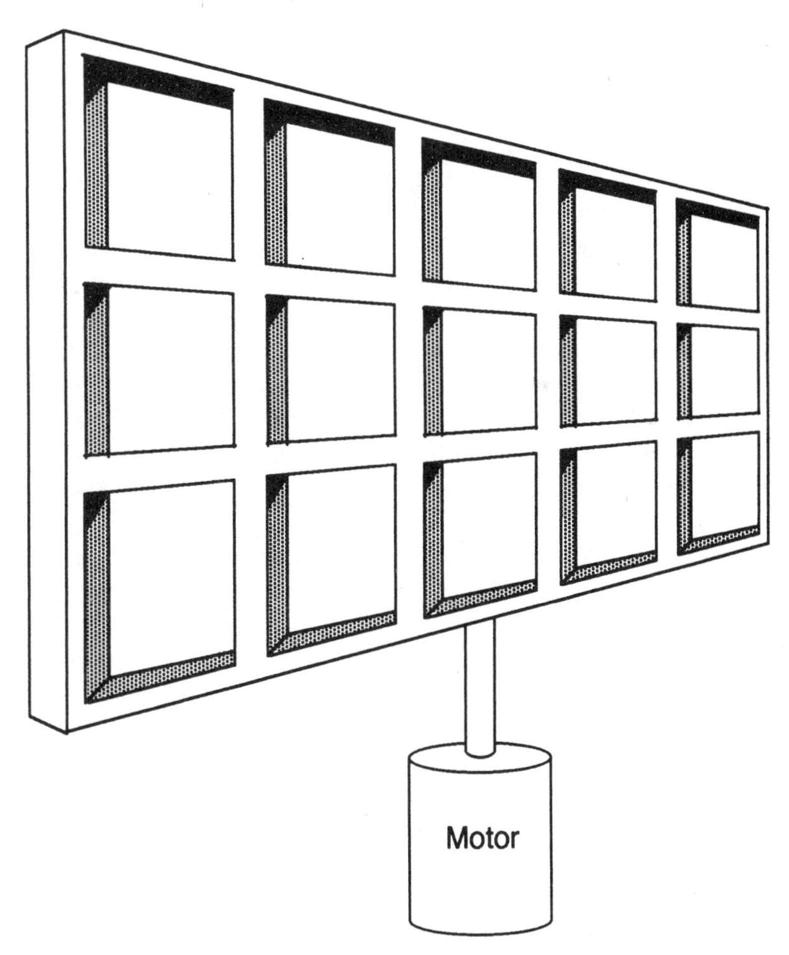

FIGURE 4.6. *An Ames window. This is not a perspective drawing.*

results. For instance, tepee-dwelling Cree Indians of northern Canada are more resistant to the illusion than are other Canadians.

Back to stacks. If you want to build with stacked blocks, rectangular solids are wonderfully versatile. As suggested in connection with flat surfaces, pieces of some shapes constrain designs more than pieces of other shapes. Curved blocks determine a specific size of room or house; foot-long blocks curving by ten degrees each will make a circular house 36 feet in circumference, about 11.5 feet in diameter. Uncurved rectangular blocks, by contrast, make rectangular houses of any size and internal partitioning. Rectangular houses can be roofed over with a set of identical beams, whether the roof is flat or sloped. Once you have a rectangular house, rectangular furniture is positively made to order. Rectangular books then fit nicely on rectangular bookshelves, and rectangular drawers into rectangular cabinets. So one right angle leads to another, starting with the decision to stack blocks.

Stacking blocks or bricks is particularly attractive for a technology that lacks good adhesives and cables. A stack is held together by compressive forces, blocks above pushing down on blocks below. Tensile—that is, pulling—forces matter little. Resisting tensile forces takes adhesives, cables, or tensile joinery, all more sophisticated than blocks or bricks. We put bricks together with mortar, but the brick-mortar joint isn't particularly resistant to tension. The mortar actually has two other roles: It keeps bricks from slipping sideways across one another, and it fills the spaces between them so bricks press evenly on one another without the separate contact points that might make these brittle materials crack. When building wooden houses, we join boards with nails. Like mortar, though, nails merely keep adjacent boards from slipping sideways under so-called shearing loads. For resisting tension, nails are hopeless; indeed, one uses just that weakness (and a claw hammer) to pull out misplaced or bent ones. So nailed structures are essentially stacked ones also, as you can see by examining the frame of a house (Figure 4.7). When wooden structures must be pulled on, we use screws and glue instead of nails, but that's too slow and expensive for making wooden houses. Boats face far less uniform and less predictably compressive loads than houses, so builders of wooden boats have had no choice but to master tensile joinery.

Furthermore, rectangular solids—such as bricks, beams, and boards—facilitate storage and delivery. Most piles of identical items—sugar crystals, dry seeds, gravel, nails—are unstable if their walls are steep-

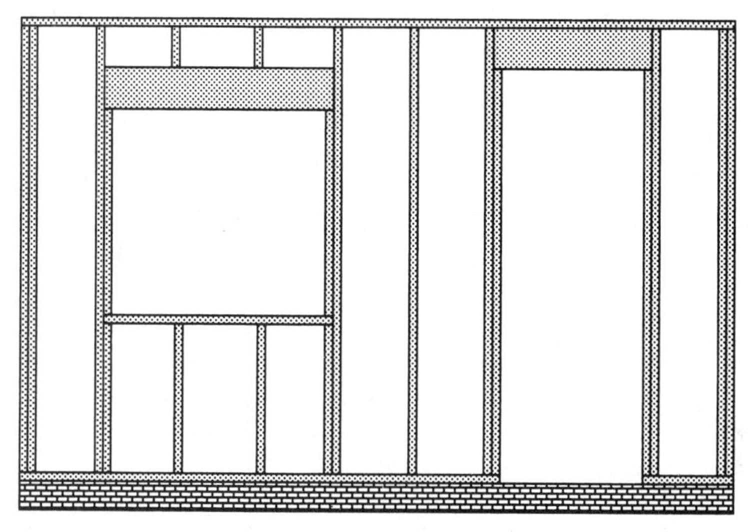

FIGURE 4.7. *The frame of the wall of a wooden house. With real care and straight up-and-down stacking (and no wind) it could be self-supporting without nails.*

er than some critical angle of repose, but rectangular solids evade the problem by fitting easily into regular, nonrandom piles. Stockpiles are simply larger rectangular solids made up of smaller rectangular solids, with vertical walls and no internal cavities. If one matches the larger solids with hollow boxes, also rectangular solids, one has a fine system for packing as well as stacking. The only awkwardness in the scheme is cultural. Having five fingers per hand, we persist in counting with a ten-based, or decimal, system. Rectangular solids persist in packing into larger solids in rather different arrays. Consider convenient packs or stacks made up of individual items. You might arrange eight items two across, two deep, and two high. Or you might arrange items in other arrays:

2 x 3 x 2 or 12 items	2 x 4 x 4 or 32 items
2 x 4 x 2 or 16	3 x 2 x 5 or 30
2 x 3 x 3 or 18	3 x 3 x 4 or 36
2 x 4 x 3 or 24	3 x 4 x 4 or 48

Multiples of ten aren't prominent. Hence the continued popularity of wholesale purchasing by the dozen and by the gross and the persistence of those words in our decimalized world.

Nature rarely builds by stacking but rather laces things together with cables of one kind or another. These include ligaments, muscles, and tendons in creatures like us; membranous outer stretched sheathing in worms, caterpillars, sea anemones, and such; and internal tensile fibers in plants. We have good access to nature's tension-resisting materials—we've long made our ropes of natural fibers—but nature more readily attaches tendons to bones than we attach cables to struts. Conversely, stacking isn't so good if your materials aren't very dense; weight is what gives a stack decent resistance to sideways forces, such as those that winds produce. Stacking loses its appeal if your materials are soft. Forget about stacking low-density material underwater, where buoyancy offsets weight and where the forces of flow get quite extreme. She may not stack much, but nature does sometimes pack things. Still, as we'll see, she doesn't pack them into rectangular boxes.

Orthogonality—right angleness—even if less useful for nature, sounds ideal for us. "Ideal," though, sweeps further relevant matters under the rug. If four rigid struts are joined by hinge pins to form a four-sided, two-dimensional structure, as in Figure 4.8, the result has no set shape and is unstable. Such an arrangement is called a mechanism—a poorly chosen term but one that at least implies mobility. As the figure shows, mechanisms make nice mechanical linkages. We use them for all manner of machines. Nature even uses them for complex movements, such as those of snake jaws, where animals swallow remarkably large and unmasticated mouthfuls,[13] and fish jaws (as in the figure), where a mouth can open and extend forward in a single motion.[14]

However good for eating, such sets of struts won't support much. By contrast, three rigid struts joined by hinge pins have a fixed and reliable

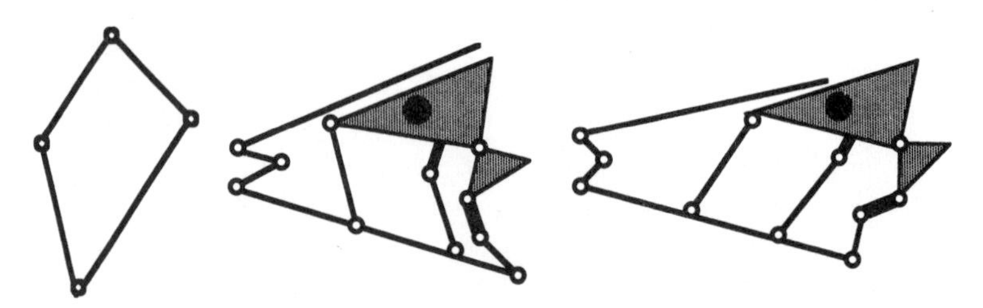

FIGURE 4.8. *Mechanisms. An especially simple one and two views of some (not all!) of the stiff elements of a complicated, biological one—the scheme by which a fish (a wrasse), on approaching prey, suddenly protrudes both upper and lower jaws.*

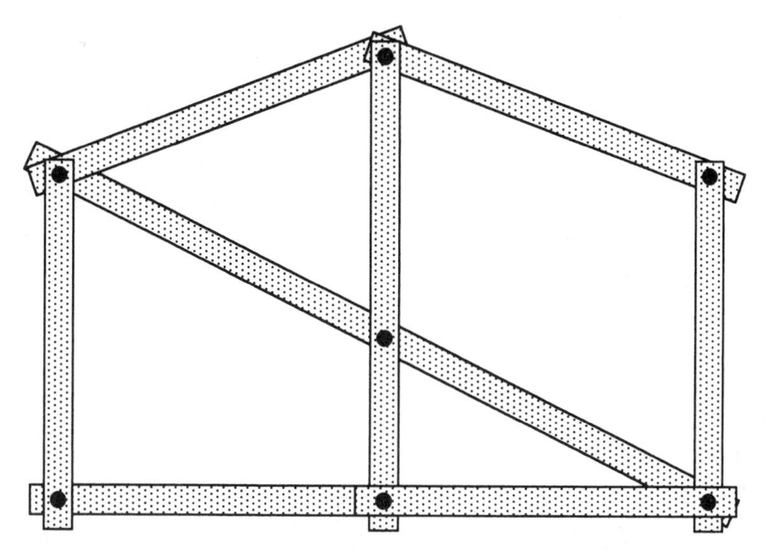

FIGURE 4.9. *A statically determined structure. This one is nonredundant. Remove any pin, and it becomes a mechanism, as you can check with cardboard and straight pins.*

shape; such arrangements (Figure 4.9 gives a more complex one) are called statically determined structures.[15] If you frame a wooden wall with vertical and horizontal pieces (such as the vertical studs running between the top and bottom horizontals of Figure 4.7), you've unfortunately made a mechanism. Some cross bracing must stabilize it against lateral loading; either a diagonal or two must be worked in, or else a sheet of plywood that will remain reliably rectangular must be nailed on. We make great use of such diagonals to create stable, rigid triangles out of the unstable, flexible rectangles that our overall designs keep generating. And we've been doing it for a long time. I was amused to see just this in a Bronze Age rock carving of some sort of sled or boat, near the west coast of Sweden, shown in Figure 4.10. A less aesthetic example is the scoreboard of the football stadium of my own campus, shown in the same figure.

Nature doesn't use triangles much—partly because she rarely goes in for real stiffness and partly because she gains what stiffness she uses in other ways. A few trusses formed of triangles, though, are recognizable—inside the wing bones of large birds, for instance. I know of no case in which rectangles are, as an afterthought, transmogrified into triangles with cross bracing. Not that natural afterthought is unthinkable. Nature is forever adding a little here and a little there to make an existing structure serve some new

FIGURE 4.10. *A contemporary structure in which the critical cross bracing looks like an afterthought, and a shallow carving (painted for contrast) of a cross-braced structure. The latter, from the Bronze Age, is on a rocky outcrop near the west coast of Sweden.*

function. That's the point of Steve Gould's fine essay about the thumb of the panda; the leading popular expositor of evolution describes how pandas make functional thumbs by dividing the first digit on each hand to give an apparent sixth finger, one unrelated to our own thumbs.[16]

Since four end-to-end struts are nicely flexible, one might imagine that a construction system with no predilection toward stiffness would prefer quadrilateral rather than trilateral arrangements of struts, but that's not quite the case. Most sponges have flexible skeletons made of stiff struts with their ends interconnected by flexible pads of protein, as in Figure 4.11. I once looked at a lot of sponge skeletons in search of triangles, geodesics, and other efficient strutting and trussing arrangements. None was obvious, and I realized that I was looking at sponges with my human bias toward stiff structures. I then looked for quadrilateral strutting but came up almost as dry; I had just brought to bear another human bias. While the tiny struts (spicules) never formed triangles, quadrilaterals were no more common than five-, six-, and seven-sided arrays. The greater the number of struts hitched end to end in each loop, the greater the overall flexibility that's possible.

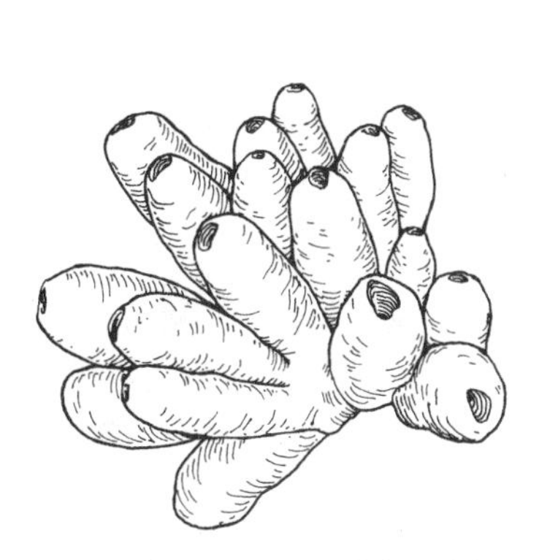

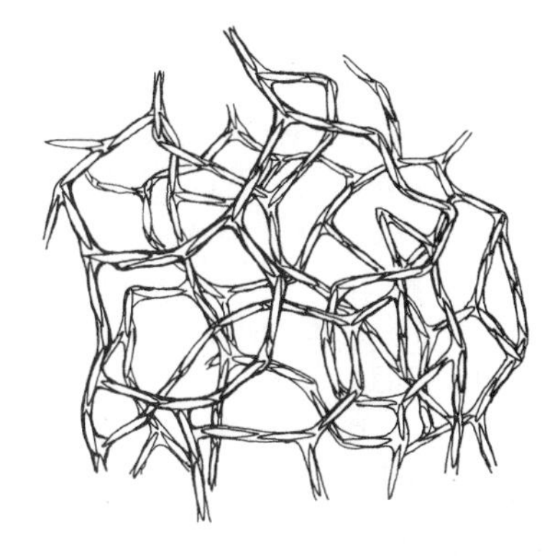

FIGURE 4.11. *A sponge (left) and the skeleton of a sponge (right). The skeleton is a meshwork of protein in which stiff struts are embedded. (Natural bath sponges are the supportive systems of animals that have the protein meshwork but, unusually, lack the stiff and scratchy struts.)*

Yet another problem of a world of right angles. For a chair or table that sits on the ground, stability minimally requires, first, three points of ground contact that form the corners of a triangle and, second, a center of gravity somewhere above that triangle. Four legs would appear better; the location of the center of gravity will be less constrained, so load shifting will be less likely to make the structure tip over. A four-legged chair tips less readily than a three-legged one. Furthermore, a little redundancy sounds attractive. The fourth leg, though, makes subtle trouble. With three, leg length isn't critical, and the structure will behave itself on almost any floor. That's why we mount cameras and telescopes on tripods. Making simultaneous contact with all four means that floor and legs must be carefully adjusted to fit each other or that one or the other must be a bit flexible. Our ordinary quadrupedal tables often come with height-tuning screws on their legs to minimize wobble on uneven floors. Tripedal tables need no such adjustability.

Cats, camels, and crocodiles use four legs, but analogy with tables isn't appropriate. For one thing, jointed legs are naturally adjustable in length. More important, a quadruped with one leg lifted forms a nicely stable tripod. Watch a cat slowly stalking; it keeps a leg off the ground for long periods. (That's a lot harder for people, who wobble a bit when standing monopedally.) But four legs may be too few for easiest walking since only one can be lifted at a time without losing stability. Most insects walk on six legs, so they can use their legs as alternating tripods—one on one side and two on the other—without loss of stability. That looks like a better arrangement.[17] Six looks like the optimal number of legs for a legged, walking biomimetic vehicle, about which more in Chapter 13.[18]

But we've strayed a bit from right angles, to which we now return. On reasonably flat terrain, we divide land into rectangular (and often square) plots, whether eighth-acre house sites or state and national boundaries. By far the largest number of arbitrary surveyed borders between states in the United States and provinces in Canada run east-west or north-south and thus intersect at right angles. Again, no rule requires it. Some diagonals have been used, and one interstate border (between Delaware and Pennsylvania) is a surveyed circular arc.

Simplicity suggests rectangular surveying. But it's not uniquely easy or automatically virtuous. Equilateral triangles divide a surface without gaps, and these can be laid out by drawing taut a set of strings of equal length. Since no angles need to be measured or divided, it may be the simplest surveying method of all. Its main drawback is the large amount of boundary of its plots relative to their enclosed areas. Regular hexagons, in which all six sides are of the same length, also divide a surface without gaps. As an amalgam of six equilateral triangles, a hexagon is no more difficult to lay out. For boundary relative to area, hexagons are as good as can be done. And they have another advantage if our whole spherical planet has to be divided. With the addition of twenty pentagons to make the system close on itself, the earth can be partitioned into hexagons without any distortion from its curvature. (Wyoming looks rectangular, but its northern border is shorter than its southern one.) But hexagonal division precludes running unidirectional roads along sequential boundaries. Neither triangles nor hexagons give the farmer fields that can be planted in rows of equal length. And neither facilitates arranging rectangular structures. Of course buildings don't have to be rectangular. A behavioral laboratory near where I live uses a hexagonal room surrounded by six others. An observer in the middle of the central room can keep watch on every bit of five others (the sixth provides access). Such reasonable schemes are merely unfamiliar and less than convenient to make from standard components. Figure 4.12 shows a street plan with the predominant elements hexagonal; to my eye it looks as functional as what we usually do.

We've clearly preferred to use rectangular plots of land for a very long time. While the ancient Egyptians used triangles of stretched ropes for surveying, the sides of their triangles weren't equally long but had a 3:4:5

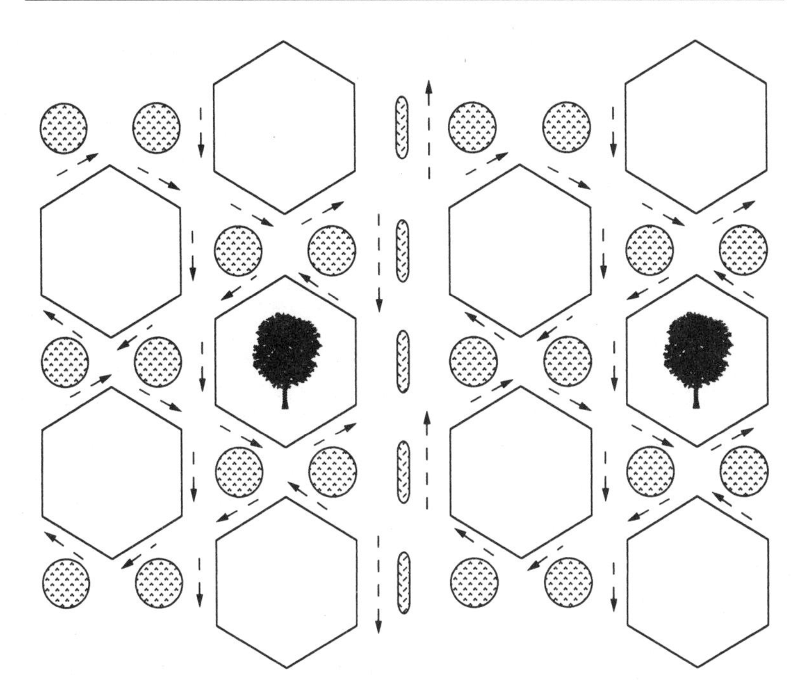

FIGURE 4.12. *A street plan that makes substantial use of hexagonal elements. Going "around the block" is a more literal matter in this arrangement.*

ratio. A 3:4:5 triangle has a right angle between the shorter sides, so they could keep their corners square. Nor were the Egyptians alone in that preference. The great Pythagorean theorem applies to a 3:4:5 triangle or any other triangle with a right angle: The squares of the short sides add up to the square of the long side, as, for instance, $3^2 + 4^2 = 5^2$. (Or 9 + 16 = 25.) Only people concerned with right angles should care about that subtle theorem. It was known (and quite likely was independently discovered) in ancient China, India, and the Middle East well before the formal proof that we attribute to the school of Pythagoras.[19] Some logical person once suggested that we mark a bare place on the earth (such as the Sahara) with a pattern that demonstrates the theorem, the one shown in Figure 4.13. Anyone looking at the earth would then know that it harbored intelligent life.

Back to hexagons. Consider how a group of animals might divide a habitat into individual territories. Our use of squares and rectangles certainly excels for plowing and road building. But other animals do little of

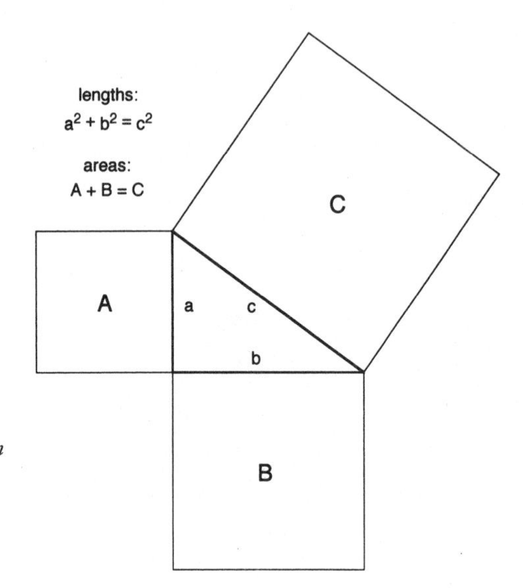

FIGURE 4.13. *A geometric representation of the Pythagorean theorem. If the angle between sides a and b is ninety degrees, then area C is exactly as big as A and B combined.*

either. If a border to be defended must be minimized, if all area is part of one or another territory, if all area is equally good to live in, and if individuals are equally effective in establishing and defending territories, then territories should be hexagonal. Even with such stringent requirements, natural selection occasionally finds the logic sufficient. Partitioning that closely approaches hexagonal has been found among sandpipers in the tundra, terns on the barrier islands off North Carolina, and bottom-living African cichlid fish in a breeding tank.[20] Of course the most famous cases of hexagonal partitioning in nature are the honeycombs and larval cells of bees and wasps. We make occasional use of the arrangement for internal stiffening partitions in hollow doors and for stiffening beamwork under the floors of airplanes. The latter is no minor matter. Flatness costs material, and aircraft designers are properly fanatical about weight economy. And the stress—force per unit area—of a stiletto heel on a floor is remarkably high. Thin floors braced with aluminum honeycomb puncture a lot less easily than if they were braced with well-spaced beams.

Just as squares don't give the best edge-minimizing partitioning of area, so cubes don't give the best surface-minimizing partitioning of volume. The best geometric solid for such close packing is distinctly arcane. It's a fourteen-sided figure, one made with eight triangular faces (a regular octahedron) that then has its six apexes cut off so each of those faces

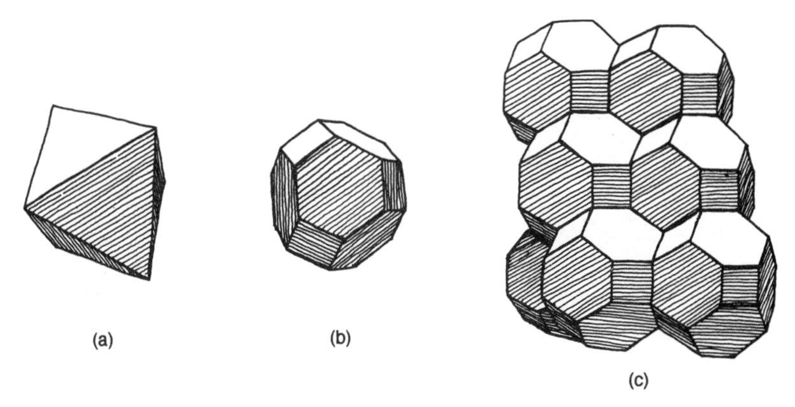

FIGURE 4.14. *An eight-sided regular solid called an octahedron (a). Cutting off each of its six apexes produces a fourteen-sided figure (b), a cuboctahedron or (gulp!) an orthotetrakaidecahedron. If the cuts are done just deep enough so all edges are again the same length, then (c) the cuboctahedrons will pack together without any spaces between. They turn out to do so with less surface area than any other volume-partitioning, close-packing shape.*

beomes a hexagon. For most of us that's beyond imagining, so Figure 4.14 is a necessity. Anyhow, such solids have eight hexagonal faces and six square faces (yes, right angles do appear), and mirabile dictu, they actually pack with no voids. Close approximations of them can be made by compressing a container filled with identical lead shot until the shot deforms enough to squeeze out all the air.[21] So where in nature might such a shape occur? The thin-walled cells that fill the middles of the stems of many herbaceous (nonwoody) plants approach that ideal shape, with an average of about fourteen faces on each.[22] I know of no specific use in human technology of this strange truncated octahedron, although it probably occurs fortuitously where large pieces of foam are made from small beads of the stuff.

Nor are right angles best for making shells of struts or flat panels; the geodesic domes of R. Buckminster Fuller are much more efficient ways to invest material. As in Figure 4.15, a series of struts (or intersections between panels) in such structures follows what, on a globe, we call great circle routes. Nowhere do they intersect at right angles. I once built several geodesic domes out of metal tubing. They worked well as jungle gyms for climbing—they were light in weight, were stable under load, and needed no anchorage—but when I tried to design a geodesic cabin—something more than bare metal strutting—I discovered their impracti-

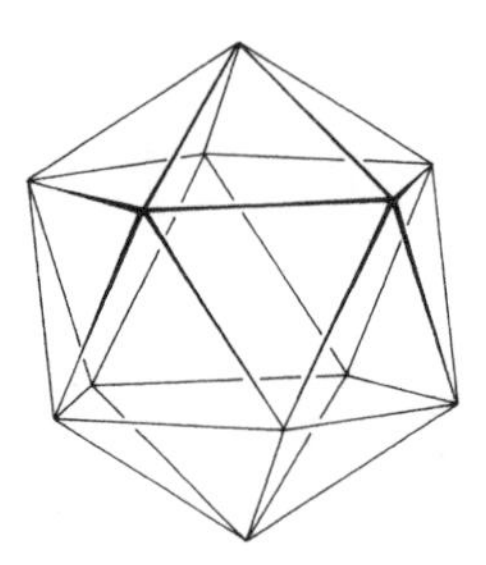

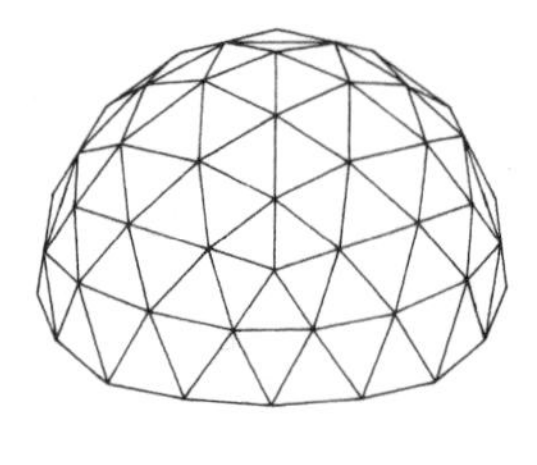

FIGURE 4.15. *Geodesic domes are derivatives of icosahedrons, figures with twenty triangular faces and twelve points at which five edges converge. The dome in the middle divides each face into nine smaller triangles (notice the five-strutted junctions retained from the original icosahedron). A viral shell of equivalent form (such as the cowpea chlorotic mottle virus on the right) has 180 protein molecules; the old icosahedral apexes are replaced by rings of five and the old faces by rings of six.*

cality. In a culture of rectangular lumber and plywood sheets, the labor needed to make a satisfactory geodesic structure is excessive, and wasted material ordinarily offsets its efficiency.

Organisms don't make much use of geodesic domes either. Perhaps that's because they make unstrutted, smooth domes when they want such regular things. Perhaps they rarely need hollow, regular, truncated spheres. One collection of geodesics is usually cited: the protein coats of spherical viruses. Identical protein molecules combine in such coats in just the same numbers—80, 180, and 320—as the number of faces in full-sphere geodesic structures. I think, though, that neither mechanical strength nor material economy explains their occurrence. More likely we're glimpsing the shortage of information brought up in Chapter 2. These structures permit a virus to cover itself with multiple copies of a single protein, with each molecule of that protein fitting into an equivalent position, something that can self-assemble with minimal complication.

CORNERS AND CRACKS

When two components come together, they usually form a corner. We've been asking, in effect, whether such a corner should form a right angle. Now let's look at the corners themselves. Here once again nature and human technology have different preferences. We build things with sharp corners—unless good reason dictates otherwise. Nature builds things with rounded (faired) corners—again unless some functional imperative supervenes. The underlying questions here are simple. First, what's wrong with sharp corners? Second, for human technology at least, where will sharp corners actually prove intolerable?

Answers to both questions flow from a third. In a stiff structure under load, where will the first crack appear? We all know the answer to that one: where there's the most force on the structure. Straighten out a bend or corner in a brittle bar, and any crack will start on the inside of bend or corner. Moreover, starting the crack takes a much lower force than you'd need to start one in a straight bar. The effect doesn't depend on some weakness caused when the bar was bent in the first place; a bar cut with a bend from a larger piece of material or one cast with a bend will behave in the same way. The inside of a bend is apparently a naturally weak place. What determines how much the bar is weakened by the bend is not so much the overall angle of the bend as, oddly enough, the sharpness of the inside of that angle.[23] Making a bend deliberately less sharp, fairing it, can gain a lot of strength.

To get a quick look at the effect of fairing the inside of a corner, cut out a flat strip of aluminum foil, about an inch wide, with an angle of about sixty degrees at its midpoint—a roughly boomerang-shaped piece. Pull on the ends with it flat on a table, and it tears in two, beginning at the inside of the angle. Make another strip, using a paper punch or a carefully rounded cut to get a less sharp inside angle, and pull again. The force required to tear the second is probably greater; fracture is foiled. Doing the same thing with plastic wrap may (depending on the kind of plastic) be even more dramatic; the rounded corner may prevent tearing altogether. Force concentration—that's what's wrong with sharp corners. The stiffer the material, the worse the problem becomes. We use the phenomenon to cut glass. Scratching the glass makes a kind of corner or at least a place where force is concentrated and from which a crack will predictably propagate. We can also get cloth to tear where we want it by making an initial slit, at least if the cloth isn't too stretchy. Tearing then perpetuates the slit, so the process easily continues.

Worse even than a sharp corner is a sharp corner at which two pieces of solid material are joined together, as at the corners of a picture frame. To the intrinsic fragility of the corner are added all the difficulties of applying fastenings that resist tensile loads; straightening a corner stretches its inside edges. But that's just how we build window and door frames, wooden boxes, drawers, furniture with legs, and so forth. You know from experience where such things fail—at the joints and usually at their insides. We know that making joined corners is a bad practice, but it's ever so convenient for construction!

Rounding corners helps, as you perhaps saw with the foil. Our technology does just that, but usually only where mandatory. I once broke an interior door handle of a car; examination showed a sharp inside corner in the bit of cast metal within its plastic trim. In the "exact factory replacement part" that corner was faired; someone had wised up. Long ago the teeth of gears were found less prone to break off if the bottoms of the gaps between the teeth were rounded rather than sharp-cornered. The portholes of a ship are always round to avoid initiating cracks when waves stress the ship's hull. We deliberately use windows with rounded corners on aircraft. An airplane's windows are gaps in the skin of the craft, and the skin is part of the mechanical structure, not just some covering that just keeps people and air inside and wind and weather outside. The claw of a decent hammer has a bit of fairing where it connects to the rest of the head. If we make two pieces come together at a corner, we often round the welded joint or attach a third piece that rounds or cross-braces the corner. And so on.

Nature's structures, more limber, must be less prone to cracking at corners. In addition, natural structures only rarely use separate pieces rigidly joined at corners. Most often we find single pieces of material, grown from a single site, with fairing incorporated into the basic element rather than added adventitiously. The key here, I think, is the growth process. Whatever the complications it entails, growth greatly simplifies the construction of faired corners on single structural elements. Figure 4.16 gives a pair of examples. The long ridge on a mammalian scapula (shoulder blade) is nicely faired with the rest of the bone. The branch of a tree grows smoothly out from the trunk with wood fibers running across the upper (functionally the inside) angle of the joint.[24] One of my less smart acts was to hit a wedge with a sledge into such a crotch in a piece of oak. Smashed eyeglasses a fraction of a second later gave real impact to a fair lesson about fairing, a lesson reinforced by a small scar on my brow. Where nature makes a sharply angled junction between pieces, the pieces usually form part of a mobile mechanism, as where the forearm attaches to the upper arm. There, of course, cracking isn't an issue.

We're gradually making more things from single pieces of material with nicely rounded internal and external corners. That we're deliberately copying nature is unlikely; rather we're taking advantage of plastics that can be cast or molded into more complex forms than we can get by

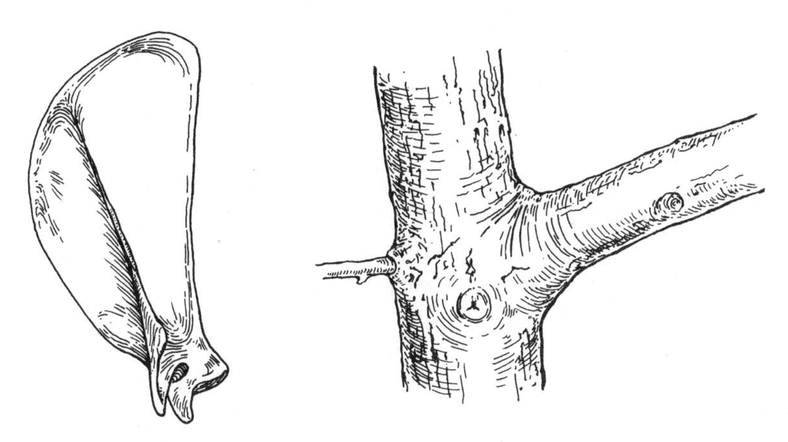

FIGURE 4.16. *Faired corners in nature: the scapula, or shoulder blade, of a cat and a branch forking from another in a tree.*

stamping things out of metal sheet. The curves not only make the products easier on shins and simpler to clean but, by reducing force concentrations and lessening the problem of crack initiation, also permit useful economies of weight and material. Garden carts made mainly from single plastic pieces can have far more intricate shapes than wheelbarrows with stamped pans. Fiberglass shower-baths share the same advantages over metal tubs and enclosures.

Three contrasts in mechanical design between nature and human technology are evident even at this geometric level. We build flat, nature builds curved; we hold right angles dear, nature is untouched by such affection; our corners are sharp, hers are rounded. Even in these most ordinary matters, we see pervasive differences between the two technologies. Of greater importance, we see repeated examples of the insights available if one avoids premature judgment of superiority and inferiority. Of still greater importance, we begin to see how each technology forms a separate, well-integrated entity, operating in an internally coherent context.

Chapter 5

THE STIFF AND THE SOFT

Natural and human technologies diverge as much in the materials they use as in their shapes. We prefer stiff and brittle materials while nature opts for pliancy. An airplane's shape at the gate differs little from its shape in flight, but a leaf in still air looks nothing like one in a storm. The difference takes us into materials science, a field both interesting and immediately relevant, one deserving more attention than it gets from us nonengineers. Blame it, if you wish, on how we're taught about the physical world. Elementary physics classes assume that masses act as if concentrated at points and that solid bodies are perfectly rigid. These fine abstractions, useful fictions, prove inadequate for the present story. If, horrors, an object is flexible, then forces acting on it may change its shape and the center of its mass, complicating things considerably. Worse yet, the changed shape may then incur different forces. Such analytical complications, though, represent our problem, not nature's. As we'll see, flexibility provides all manner of opportunities for clever design.

We need special tools with which to describe this more complex world. "Solid" will mean only that something isn't fluid, that it doesn't flow. Forget perfect rigidity; no solid is perfectly rigid. Even the formal test for telling a solid from a fluid presumes that solids give at least a little. Forcibly distort (without breaking) a solid object, and it snaps back when the force is removed; distort a fluid similarly, and it obligingly adjusts its shape. The jelled salad, a soft solid, retains the shape of its mold; by contrast, the coffee fits any cup.

Furthermore, what usually matters is not force but *stress*, force divided by the area over which it acts. Even a small force will push a needle into a hard material; the needle is sharp, so the effective area of action is minute, and the small force creates the large stress that does the job. Likewise, a sharp knife penetrates better than a dull one when both are pressed with the same force; the sharp one exerts a greater stress on steak or string bean.

MATERIAL EVIDENCE

A solid object stretches when you pull on it. If stone or bone, it won't stretch visibly, but it will certainly stretch measurably.[1] You might plot data for a gradual stretch on a graph where distance stretched runs horizontally and force runs vertically, as on the left in Figure 5.1. The graph, though, describes only your particular object. More useful is a graph

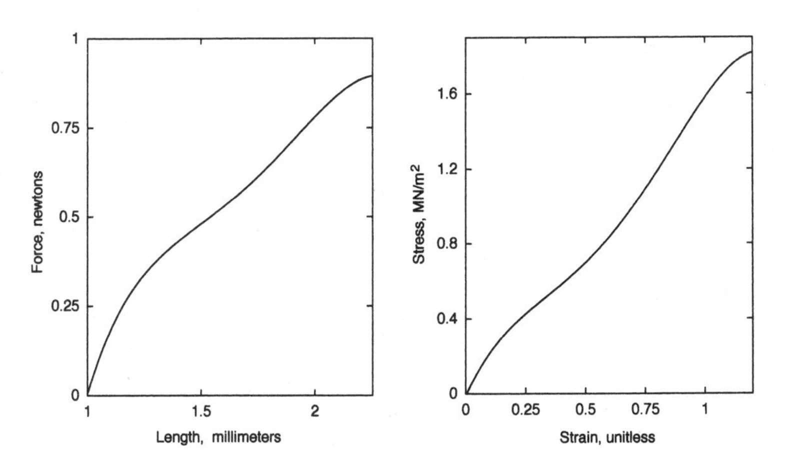

FIGURE 5.1. *A graph of force against stretch and the same converted to stress against strain. Strain is unitless; it's extension divided by original length or (multiplying by 100) percent extension. Stress is usually given in newtons per square meter or (as here) meganewtons per square meter. A newton is a force about equal to the weight of an apple.*

relating force and stretch, not for a specific object but for a kind of material—for stone or bone, Vermont granite or cow femur. To achieve that generality, instead of force we use stress—pulling force divided by the cross-sectional area of what we're pulling on, once again. Instead of distance stretched, we divide stretch by the original length of the object and call the result strain, or fractional elongation. Then, as on the right in Figure 5.1, we plot stress upward and strain across. In materials science, "stress" and "strain" are far from synonymous. Anyway, odious nomenclature aside, we now have a graph that applies *not just to our object but to a material.* A tensile stress of a million newtons per square meter (abbreviated MN/m^2, about 150 pounds per square inch) might give a strain of 0.1, or 10 percent; that's the value for one kind of rubber.[2]

With nonrigid solids, the world gets still more complicated; a rubber band and a lump of taffy lack rigidity in obviously different ways. The stress-strain graph of Figure 5.2 may help the reader as we make our way

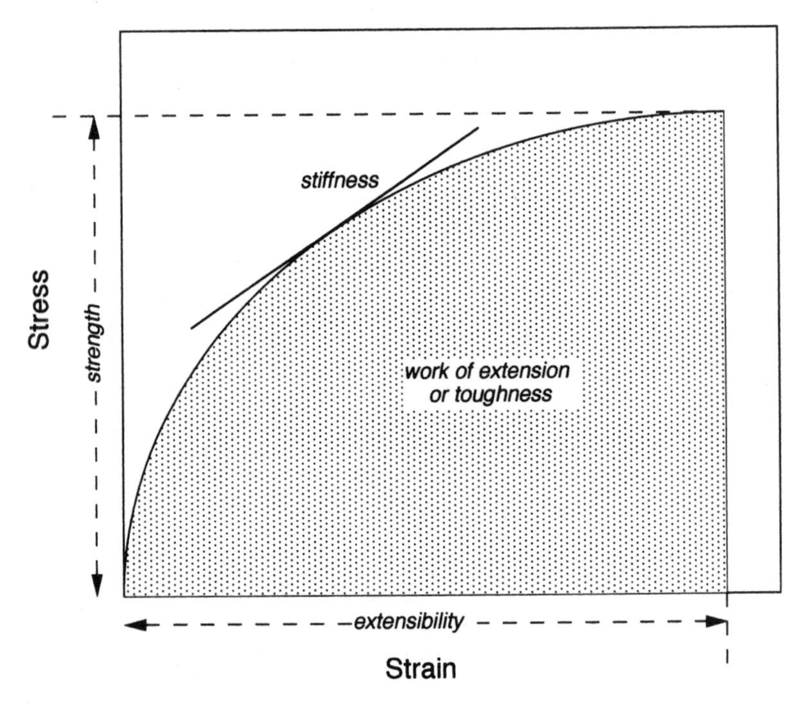

FIGURE 5.2. *Four material properties that can be extracted from a stress-strain graph: strength, extensibility, stiffness, and work of extension.*

through a ménage of relevant properties of materials. One such property is how *forcefully* a material can be stretched before it breaks—the maximum stress that it can withstand. That we call its strength. A second is how *far* the material can be stretched before it breaks—the maximum strain it can withstand. That's usually known as its extensibility. On the graph, these are the maximum height of the plotted line and the maximum distance it extends to the right. For a lot of materials (many metals, for instance) something else complicates extensibility. If extended up to a certain point, these materials snap back properly, but if extended farther, they stretch and stay stretched. Such materials have an elastic limit, and that may be the extensibility that matters in use. The elastic limit isn't limited to stretching; a bar bent beyond the elastic limit of its material stays bent.

Beyond strength and extensibility we encounter a third property: stiffness. Pull on something, and ask how much pull (stress) it takes to get a given stretch (strain). That's stiffness—stress divided by strain. Rubber is easy to stretch, which is to say that it has a low stiffness. For the rubber just mentioned, a stress of 1 MN/m^2 divided by a strain of 0.1 gives a stiffness of 10 MN/m^2. Steel, vastly stiffer, is harder to stretch by the same amount. One kind demands a stress of 20,000 MN/m^2 (3 million pounds per square inch) to stretch by the same 10 percent and so has a stiffness of 200,000 MN/m^2. Engineers call stiffness Young's modulus of elasticity after Thomas Young (1773–1829), one of the people who struggled to apply Newtonian mechanics to real materials and to whom, incidentally, we owe our notion of energy.[3] But "stiffness" will do for us since it corresponds quite closely to both our perceptions and our common speech. On the graph, stiffness appears as the slope (the steepness) of the plotted line.

Another slight complication: For many materials, stiffness depends on how hard or far they're stretched; a material may be hard to start or resistant to the last bit of stretch before it breaks. On the graph, its stress-strain line will be curved rather than straight. As it happens, the materials we humans most commonly use—metals, especially—give fairly straight lines, so we're accustomed to quoting specific values for the stiffnesses of specific materials. Almost all biological materials, though, have dramatically curved stress-strain lines, sometimes curved upward and sometimes curved downward. On the one hand, values for their stiffnesses either are only rough averages or else apply only to specific conditions. On the other hand, the complications permit great flexibility (to pun slightly) in how different materials can be used. For instance, as in Figure 5.3, the

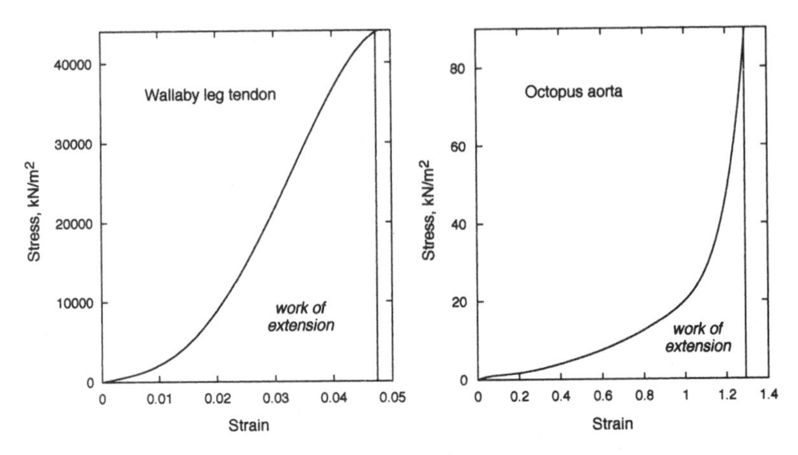

FIGURE 5.3. *Stress-strain curves for a tendon (data from Alexander, 1988) and for a large blood vessel (data from Shadwick, 1994).*

material of our tendons gives a curve that's concave downward with a lot of area underneath it. By contrast, the material of our arterial walls gives a curve concave upward with a relatively small area underneath.

Those areas underneath the curves represent a fourth property. Stretching something takes energy, and stretching something thicker or stretching something farther takes more energy; everyday experience does not mislead. We use "work of extension" as a measure of that energy (strictly, energy per unit volume) needed to stretch a material. On the graph that's the area under the stress-strain line, from zero out to the material's breaking point. Were all stress-strain lines straight, this wouldn't be a particularly interesting property. But as just noted, nature's materials give diverse curves whose shapes are fraught with functionality. "Work of extension" is sometimes called toughness,[4] an agreeably intuitive term that we'll find quite useful.

And another, property number five—resilience—what rubber has that taffy lacks. Say you keep track of the work of extension while stretching an object (such as a rubber band or steel spring) to something short of the breaking point. Then you release the object and keep track of the work or energy it returns. In our imperfect world the two won't be equal. The work going in as you stretch it will always be more than what comes out when it's released.[5] The work you get back divided by the work you've done is the material's resilience, a property alluded to several chapters ago when we considered the hinge pads of insect wings and the hinge ligaments of scallops. On the stress-strain graph it comes from two areas, the one representing the work obtained during release divided by the one representing the work done in the stretch.

At this point let's summarize in a list (with some nonbiological examples) for quick reference farther along:

MATERIAL PROPERTIES THAT MATTER	LOW	HIGH
Strength: stress (force per area) needed to break a material	jelly	steel cable
Extensibility: strain (relative stretch) needed to break a material	brick	rubber band
Stiffness: stress divided by strain as a material is stretched	soft plastic	brick
Work of extension, toughness: energy to stretch a material to breaking	cast iron	spring steel
Resilience: energy output upon release divided by input during stretch	taffy	rubber band

WHICH PROPERTIES MATTER WHEN?

With five properties of materials rattling around,[6] superiority or inferiority can't be judged on a single scale. A lot of talk goes on about the wondrousness of biological materials. Much of it is simplistic blather; one has to consider the suitability of materials for specific applications. Even here, though, a little mental extensibility is in order since almost any task can be done in a variety of ways. Some examples ought to flesh out these points.

We fashion tension-resisting things out of many materials: steel and other metals; natural fibers, such as jute, sisal, manila, and hemp; synthetic polymers, such as nylon, polyethylene, and polypropylene. We also make them in many forms: chains, bands, and bars as well as twisted and braided ropes and cables. Their task seems specific enough: to resist pulling (tensile) forces. Making them best should be simply a matter of maximizing strength relative to either weight or cost. But no, the business is more subtle and multifaceted.

Consider stiffness. If you anchor a boat with a line made of unstretchy stuff, you'll get into trouble. If the boat moves about its anchorage, sooner or later the line will come taut and try to stop it. How much force will that take? According to the best authority, Sir Isaac Newton, the force will equal the large mass of the boat times the deceleration of the boat. If the line comes taut at a very specific length—that is, if it has a high stiffness—the boat will stop abruptly, with a lot of deceleration and thus a lot of force. That might break even a hefty anchor line, or worse, it might rip off part of the boat or dock. Much better to use a stretchier, less stiff rope, one that stops the boat less abruptly and thus with a lower force. Not only is damage less likely, but a weaker rope will do; using a material of lower stiffness allows using one of lower strength! (Similarly, a longer mooring line may be better than a short one. Extensibility may be the same for long and short, but the actual extension will increase with the length of the line.) In other applications, stiffness is clearly a virtue. Both bicycle chains and the belts on your car's engine work by resisting tension, and the less they stretch in the process, the better. Flex, yes; stretch, no.

The same argument applies to living systems. A large, grazing mammal has a ligament that runs from a ridge on the back of its skull, beneath the skin on the back of the neck, to the bony upward extensions of the vertebrae of its neck and chest (Figure 5.4). That ligament has fairly low stiffness—it's stretchy—which helps keep a heavy head from being badly shaken or the ligament from getting torn loose when the animal trots along. If you buy an unsliced lamb's neck (a good source of chunks of meat for skewering), you can dissect out that ligament and feel its elasticity. By contrast, tendon, connecting muscle to bone, has to be stiffer stuff. A muscle works by shortening, usually pulling two bones closer together. Any stretch of its tendons means that the bones move that much less. The shank end of a leg of lamb has a nice long tendon that you can remove and pull on to get a feel for its resistance to stretch. Its response contrasts sharply with that of a neck ligament.

Through the 1800s natural technology far outdid humans in making unstiff tensile elements—that is, stretchy ropes. We humans did play some tricks with our stiff ones, creating what might be called virtual elasticity. One can, for instance, use a rule mentioned earlier about not being able to draw a string out to a perfectly horizontal line. Try pulling on a heavy chain that sags in the middle; you have to pull harder and harder as the slack is diminished. The chain doesn't stretch, but its weight makes the two ends move apart as if it did. If you must moor your boat with a chain, use a heavy one, or else hang a weight from the chain's midpoint. These days, of course, we routinely make stretchy stuff, such as rubber[7]

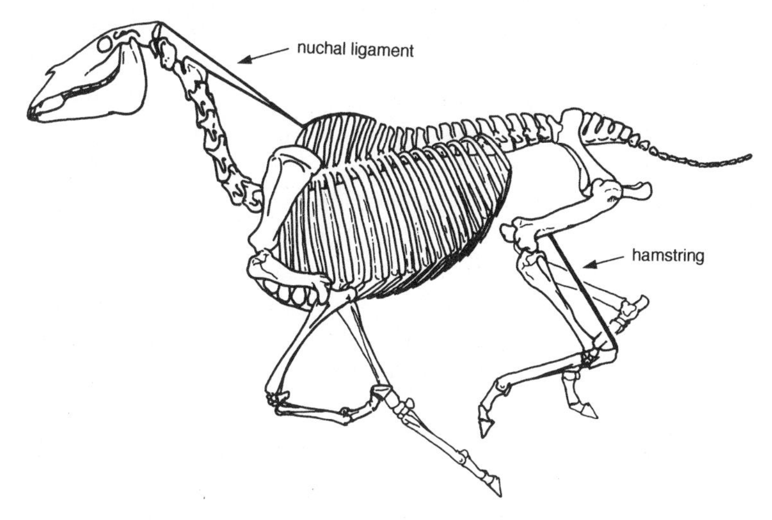

FIGURE 5.4. *The approximate location of the nuchal ligament that supports the head of a grazing animal, such as this horse, and the hamstring on the rear of the hind leg, roughly equivalent to our Achilles tendon, which connects the big muscle on the rear of the calf to the heel.*

and other polymeric materials. We've converged with nature not only in making such materials but in what we make them from; even our fully synthetic ones are mostly the products of carbon-based organic chemistry.

For resilience as well as stiffness, what's best depends on the application. If one wants a good hinge ligament for a swimming scallop, then high resilience is important. In clapping its half shells together—to expel water in a pair of jets—a scallop contracts a large (and tasty) muscle. What reopens the shell is the elastic recoil of the hinge ligament; in shortening, the muscle not only has closed the half shells but has strained the ligament, giving it the energy it will then use for reopening. The higher the resilience, the less of that energy is wasted. Conversely, high resilience may be quite undesirable. For a spider to catch insects that fly into its web, the prey must remain attached to the silk. A web of high resilience would, like a trampoline, tend to fling the prey back out. Best would be a web made of silk with high strength, extensibility, and toughness but low stiffness and resilience. Spider silk[8] fits this peculiar bill superbly, compared with the other materials of both technologies. But the requirements are unusual, and what's noteworthy is the match of material and application.

Resilience management highlights the dichotomy between nature and human technologies. We don't do well at making materials that extend a long ways and then return to their original lengths with low resilience—materials like spider silk. To do so, a material must turn into heat most of the energy put in, but it mustn't flow (as a fluid) or get self-destructively hot in the process. Our springs, whether of steel or rubber, are just that: nicely springy. Our normal way to get lower resilience, as in Figure 5.5, is to use additional mechanical elements in parallel with springs, so-called shock absorbers, as found alongside or within the springs of our cars and as parts of door closers. These devices, properly called dampers, have no preferred length and no resilience at all. Like springs, they resist any stretch or crunch; unlike springs, they don't spring back, converting mechanical energy into heat instead of storing it elastically.

Keeping resilience and rebound low is equally important in nature; the orb web of a spider is only an extreme example. For instance, for a tree to sway in the wind with too much resilience—too little damping—wouldn't be at all handy. The magnitude of sway would increase, and the roots would gradually work loose. Nonetheless, nature doesn't limit resilience with anything quite like our shock absorbers. Instead, she commonly provides damping right in her basic materials, such as the wood of trees. Materials of low resilience predominate in nature. High resilience is a sign of something special going on, such as reversible energy storage (as in hinge pads and ligaments) or deliberate flow-induced shaking (as in our vocal cords and the seed-flinging equipment of some plants).

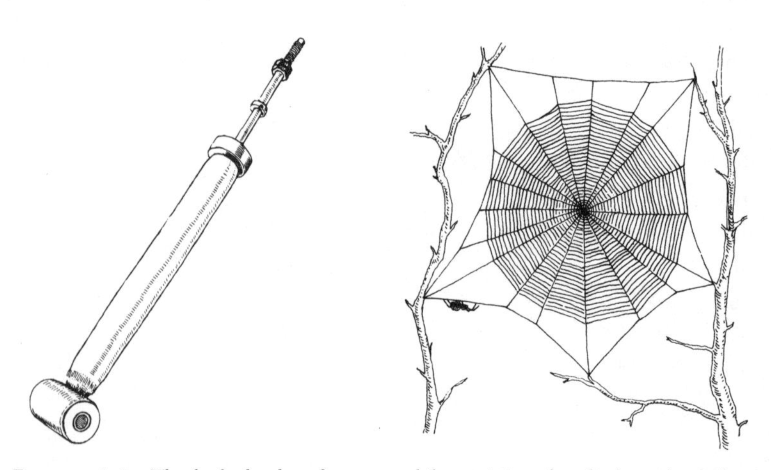

FIGURE 5.5. *The shock absorber of an automobile, consisting of a cylinder with a leaky piston inside, and a spider web of silk in which a single material provides both elasticity and (through low resilience) damping.*

Animals often offset unwanted motion with schemes much fancier than simple damping: with sensors, feedback circuits, and contracting muscles. You're more stable when standing upright than any department store mannequin. Why? Your base of support is equally narrow, and your tendons are highly resilient. But if you shift your center of gravity slightly outward from your base of support, sensors in your tendons and muscles detect the shift, and your brain then tells a host of muscles to make slight adjustments in their lengths. ("Postural reflexes" is the general subject in textbooks of physiology and neurobiology.) Only occasionally does human technology achieve such sophistication. Large airplanes, for instance, use the same kind of active controls to reduce the bumpiness of flying through irregularly moving air. You may notice flaps on the rear of the wings that seem to wander slightly but disconcertingly up and down as the plane flies along. They're computer-controlled and not under the pilot's direct command; no pilot could react fast enough as the plane whizzes along. One hears talk of active automobile suspensions, systems that would detect a bump just before a wheel hit it and then extend or retract the wheel as needed.

No material better illustrates the complex diversity of mechanical properties than wood. To most of us it's ordinary stuff, with nothing more than superior and inferior varieties. But specifying quality demands that we look at the match of its wide range of mechanical properties with an equally wide range of applications. To identify our local trees, I use a book that mentions the traditional uses of the wood of each species.[9] The specificity may represent tradition, but the tradition comes from experience, not arbitrary ritual. White pine makes matchsticks, red spruce makes canoe paddles and the sounding boards of pianos, black willow makes artificial limbs, black walnut makes airplane propellers, white oak makes whiskey barrels, American beech makes clothespins and spools, Osage orange makes bows, yellow poplar makes barrel bungs, lignum vitae makes pulleys, and basswood makes drawing boards. That's in addition to structural timbers, for which yellow pine is notable among the softwoods and red oak among the hardwoods. These things were sorted out by generations of craftspeople, ignorant of the arcana of stress-strain graphs but finely attuned to the practical behavior of what they had available.

INTERRELATIONSHIPS AND CONSEQUENCES

Spider silk is extensible, strong, and tough but not particularly stiff or resilient. Collagenous tendon is stiff, tough, and resilient but not very extensible or strong. The mechanical properties we're worrying about may be defined quite distinctly, but they don't necessarily go their separate ways in the real materials available to the two technologies. In short, the full range of properties in all combinations isn't available to either technology. Put another way, seeking a material with a desirable value of one property limits one's choice of values of the other properties. So a design that capitalizes on advantageous characteristics of a particular material must also deal with that material's less handy properties. More generally, if two technologies use different basic materials—say, concrete versus fresh wood—many other differences will follow.

Consider stiffness, strength, and toughness. Stiff materials are usually not very tough, but strong materials (if not too stiff) are commonly quite tough. Bricks, for instance, while stiff are notably nontough. The problem with such stiff materials is that cracks all too readily extend through them. A sharp blow by a hard object cracks a brick, especially if the brick is supported at its ends and the blow is delivered in the middle. Mortar keeps bricks supported along most of their surfaces lest the bricks crack under force concentrations that they can't redistribute (as does wood) by sagging a little. By contrast, mild steel, nylon, spider silk, and fresh wood are strong materials; none is especially stiff; all are nicely tough. Objects made of them can't easily be cracked in two but must usually be cut completely across. You don't hear of people splitting blocks of freshly cut wood with karate chops.

James Gordon, retired naval engineer, student of crack propagation, and biology watcher, has pointed out that humans usually build to a criterion of adequate stiffness while nature most often builds to a criterion of adequate strength. Perhaps it implies too much judgment by both technologies to call their contrasting preferences for stiffness and strength a fundamental difference in philosophy of design. For both the preference is essentially accidental, since no one's really in charge, and incidental, a matter of the materials most readily available. While far from absolute—we're really viewing tendencies or alternative defaults—the difference certainly represents a pervasive divergence between human and natural technologies.

Bricks, cement blocks, cast concrete, stone, ceramics, and glass are the most extreme of our stiff materials. Cast iron and high-tensile steel are notably stiff as well, and wood in the form of cut and dried timber is much stiffer than the stuff in trees even if less stiff than steel or ceramics.[10] Making value judgments, cracks about being stiff and uptight, are all too easy. They're also unfair. The materials we've found plentiful, versatile, and durable happen to have been stiff ones. If your tool is a hammer, you'll prefer nails over screws; if you discover a good mortar, then stones suddenly look attractive; given a power saw, you might build cabins of squared-off timbers rather than logs. With an ample supply of rot-resistant cypress or red cedar, you might choose wooden pilings over stone piers. Furthermore, tall structures on land—and thus fully vulnerable to gravity—are simpler to build from stiffer than from less stiff material.

What's important are the consequences of our predilection for stiff materials, the strictures it puts on our structures. First, it makes them peculiarly vulnerable to any accidents or unusual loads that might start cracks. These include both unanticipated loads imposed by our uses of the structures and loads from extreme but rare environmental forces, such as hurricanes, heavy snow or ice, and earthquakes. Second, fractures are more perilous than deformations; a structure can't so easily bounce back or grow back from a complete break. So our stiff materials court disaster more than nature's flexible materials. These disasters have stimulated both soul-searching and useful analysis by engineers as well as some very readable literature.[11]

Gordon also pointed out a third and more subtle consequence. Most amply stiff structures will be sufficiently strong as well, but adequately strong structures may not be especially stiff. Put another way, more material is usually needed to build something that's stiff enough than to build something that's strong enough. Should we conclude that our technology is wasteful because it puts such a high value on stiffness? We certainly find unstiff floors, ones with too much "give," disconcerting, even when they're strong enough to avoid breakage quite safely. I rebuilt a deck on my house a few years ago using thicker beams; the original deck had given long service even under heavy partying, but it felt just a bit too lively. The low stiffness of modern skyscrapers, suspension bridges, and airplane wings bothers quite a few of us. But any value judgment between strength and stiffness as design criteria is unfair unless we examine the practical and historical options and alternatives. Still, earthquakes are undeniably harder on stiff houses than on limber trees.

The bones of large animals are fairly stiff; they support the animals against gravity and serve as the levers and attachments that allow muscular engines to drive body movements. Just as a stretchy tendon would defeat the action of a muscle, so a leg bone that bends would lack standing. While nature does use stiff materials, she ordinarily reserves them for applications where stiffness is crucial. Much stiffer than bone are the real biological ceramics, the ones with less protein and more inorganic material than bone. Hard coral is a particularly stiff material, but coral forms new and more widely dispersed colonies after a reef breaks up in a storm, so failure of stiffness doesn't mean failure of fitness. The shells of marine mollusks are stiff too. But the shells provide enviable protection. Anyway, both mollusk shell and hard coral skeleton are made of calcium compounds that may cost very little. The ocean isn't at all short of calcium. Quite the opposite, it's supersaturated with the stuff, and very little energy need be invested to extract calcium from seawater. The enamel of our teeth is even stiffer than mollusk shell, but soft teeth would lack bite. For each instance we can recognize a specific rationale for making stiff materials and structures.

Like engineers and architects, nature must deal with the lack of toughness of stiff materials. Teeth and bones and hard coral, after all, do crack. Significantly, both technologies commonly use their stiffest and thus most crack-vulnerable materials in small pieces. Tooth enamel forms only thin layers or outer surfaces; teeth contain more dentin than enamel. Similarly, if you want to cut hard materials or get very long service from the blade of your power saw, you choose carbide-tipped blades. "Tipped"—we limit the stiff ceramic to tiny pieces at the tips of the saw's teeth.

THE VIRTUES OF FLEXIBILITY

Using unstiff materials can do far more than circumvent a few cracks on the cheap, and here nature wins hands down. I've picked examples from a wide range of possibilities to emphasize two general points. First, flexibility pays off both under tough external conditions—winds, waves, impacts, and the like—and in connection with internal operations—the flow of blood and the tugs and twists of muscles and tendons. Second, flexibility can do more than improve toughness and impact resistance. It provides a way to make structures that can change their shapes when loaded in highly specific and useful ways.[12]

How might one design a long structure that needs a lot of surface area and is exposed to rapid water currents? Something stiff, like the hull of a ship, requires substantial bracing. It may break suddenly, and it's in real trouble if it hits another stiff structure. A kind of marine alga or seaweed—a kelp—that lives on a wave-swept rocky coast (as in Figure 5.6) may be as much as a 150 feet long, as long as a big ship. The attachment

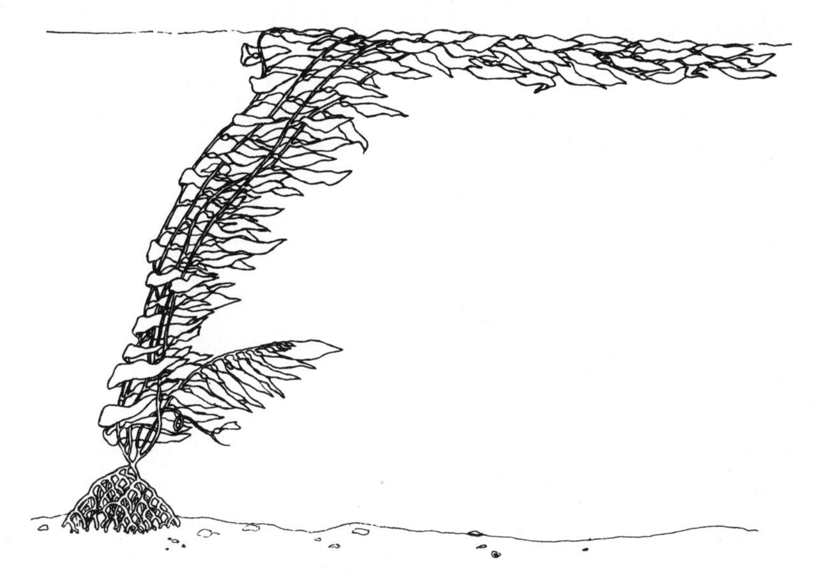

FIGURE 5.6. *An especially long marine alga,* Macrocystis, *from the Pacific coast of North America.*

to a rock at one end seems remarkably weak for a long structure subjected to the drag of storm-generated waves, and the whole plant seems not just flexible but positively flimsy.

Mimi Koehl, who works on all kinds of flexible marine organisms, identified the kelp's trick. She's a flexible thinker and recognized what earlier observers hadn't considered.[13] Waves produce flows that reverse direction every few seconds. When a nicely flabby kelp grows longer than the distance the surrounding water travels between reversals, its drag stops increasing. Any additional length simply goes with the flow. Its speed is zero relative to the water around it, so it has no drag! If water flows three feet per second and reverses every four seconds, then only the first dozen feet of kelp pull on the attachment. Koehl's seaweeds grow to great lengths to avoid drag. But the trick works only if you're really flexible: One part has to extend in a different direction from another. Long kelps flex like ropes.

Our flags and pennants have a distressing tendency to tatter in strong winds, distressing unless, as do some plant ecologists, you use the tattering as a measure of exposure to wind.[14] Flags in flows are draggy things; for the same shape and area, a flag has around ten times as much drag as a rigid weather vane. So flexibility clearly confers no automatic advantage. The kelp's evasion is impractical on land because winds are too fast and don't reverse often enough. With a wind speed of forty miles per hour (fifty-nine feet per second) and a reversal even as often as every ten seconds, a limp object would have to be almost six hundred feet long to reach the dragless zone. Nature certainly flies flexible flags on long poles; hers are called leaves, and they aren't hauled down during storms. To do its business, capturing energy from sunlight, a tree needs a big area of leaf, which could make big trouble. Most of the drag of a tree comes from its leaves; the trunk contributes little. Transfer of that force, drag, from leaves to trunk to roots makes trees go down in storms.

So what's the drag of a leaf? A few years ago I did some measurements in a highly turbulent wind tunnel at speeds a leaf might meet in a storm.[15] A leaf may be flexible, but leaves experience more nearly the low drag of a rigid weather vane than the high drag of a flag. Their strategy involves a clever use of flexibility. For instance, as a wind pulls a single maple or tulip poplar leaf away from the tree, the wind catches the lobes on either side of the stem end of the leaf blade. Those lobes bend upward, and the blade curls into a cone, as in Figure 5.7. As the wind increases, the cone rolls ever tighter. Even in highly turbulent and fluctuating winds, the cone is stable and experiences relatively low drag—around a

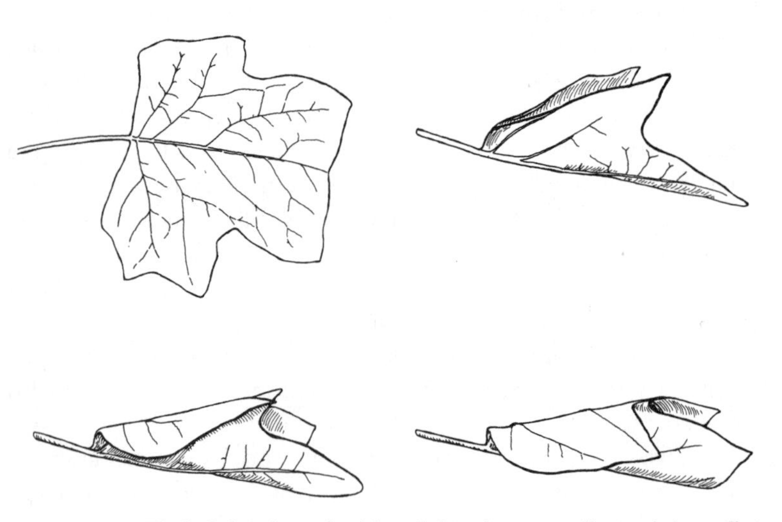

FIGURE 5.7. *The leaf of a tulip poplar (also called a tulip tree or yellow poplar) in still air and winds of 11, 33, and 44 miles per hour (5, 15, and 20 ms).*

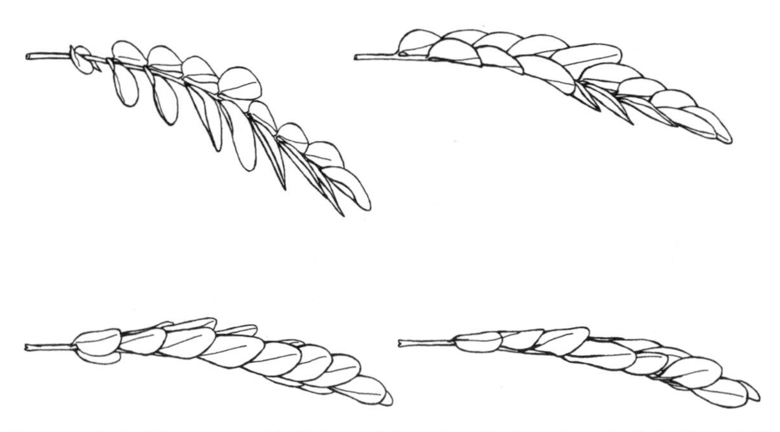

FIGURE 5.8. *The compound leaf of a black locust in still air and winds of 11, 33, and 44 miles per hour (5, 15, and 20 ms).*

quarter of that of a square flag of the same area as the leaf. One can imitate the behavior with simple models of paper or plastic, but they work only crudely; mere flexibility isn't enough. It takes just the right amount in just the right places. The leaf's solution, though, isn't without its disadvantage. Curled up, a leaf exposes less area to the sky, and exposure to light, after all, is central to its business. So a maple tree curls its leaves the way a sailing ship furls its sails, to lower its drag temporarily.

Curling into a cone is only one scheme leaves use to reduce drag in windy weather. Leaves made up of pinnate—that is, feather-arranged—leaflets, such as those of black locust and black walnut, reconfigure into cylinders that tighten as the wind rises (Figure 5.8). Besides individual tricks, leaves play communal ones. In rough weather the needles of pines splay out less and get more like the hairs on the tail of a horse, once again achieving relatively lower drag. The stiff leaves of the American holly swing together over the stem, packing like a multilayer sandwich (Figure 5.9). Leaves that curl into individual cones can also curl into communal cones when they grow close together. Other strategies for leaf reconfiguration undoubtedly exist, but no one has yet done a systematic survey.

Not that we've never done what leaves do; in high winds the blades of many old windmills would furl, if made of fabric, or pivot, if rigid. Perhaps we rarely face the tree's problem since we don't often need to expose a lot of surface to the sky while that surface is well off the ground.

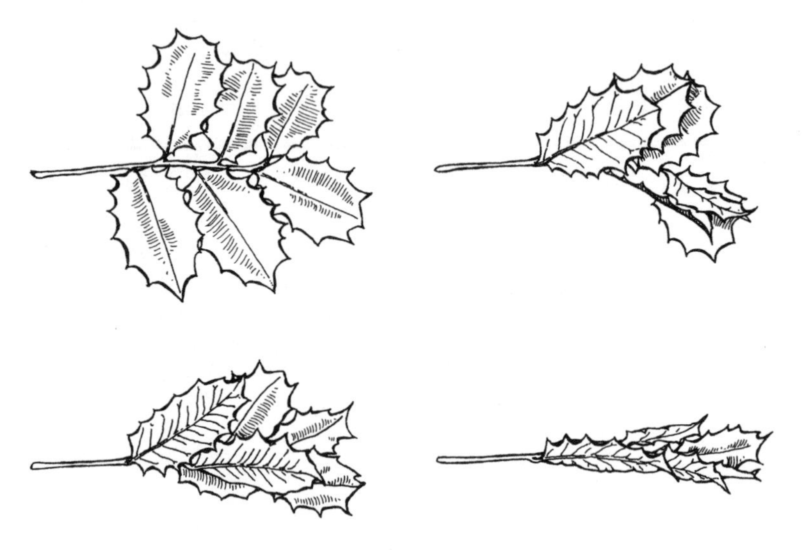

FIGURE 5.9. *A group of leaves on a branch of an American holly in still air and winds of 22, 33, and 44 miles per hour (10, 15, and 20 ms).*

Would some slight saving in material make us tolerate utility poles and antennas that bent over in storms?

The mechanical bases of some of these reconfigurations interest us here. A cluster of leaves usually feels even less drag than do individual leaves. To cluster, the flexibility of the stems of the individual leaves is as critical as that of their blades since the stems have to twist readily. That presents a queer problem in design. To hold the blade outward so it can absorb sunlight, the stem of a leaf must resist bending. Thus it must be resistant to bending loads while being compliant to twisting loads; it has to twist more easily than it bends. "Twisting in the wind" isn't just a slogan from the last days of the Nixon presidency.

One way to aid twisting while preventing bending turns on a simple device that's easily demonstrated. Try first to bend and then to twist either the cardboard core of a roll of foil, plastic wrap, or toilet paper or else a plastic soda straw. Then make a lengthwise slit in your cylinder, and try again. The slit weakens it, but not equally so in both respects; the cylinder's resistance to twisting is reduced far more than its resistance to bending. In fact, any lengthwise groove or line of weakness or even any deviation from a round cross section will do the same, if less dramatically. Many leaf stems have lengthwise grooves along their tops (Figure 5.10).

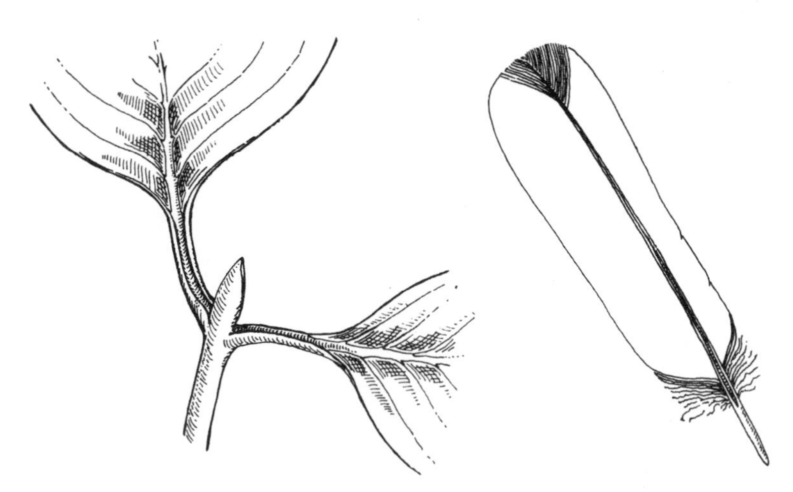

FIGURE 5.10. *Lengthwise grooves on the upper side of the leaf stem (petiole) of a dogwood and on the underside of the wing feather of a bird.*

Together with some peculiarities of the cells and fibers inside, the grooves ease twisting relative to bending, which reaches four or five times that of cylinders of plastic or metal.

The feathers that form the tips of a bird's wings face an equivalent problem. Here again, ease of twisting is functionally important. Wings propel a bird the way a propeller moves an airplane; they just beat up and down instead of spinning around and around. A propeller blade has to be twisted lengthwise to give a decent push to the air passing across it. If it were to spin in the other direction, the twist would have to be reversed. But twice per stroke the feathers of a beating wing reverse direction and so must reverse their twist. At the same time the feathers must resist bending; after all, the wings lift the bird, so in flight its body truly hangs from its wings. Again a structure must twist but not bend, and as on many leaf stems, a lengthwise groove runs along each feather's shaft. Melina Hale, now an accomplished biologist but at the time a first-year undergraduate, drew my attention to the possibility, which measurement then confirmed.

One difference, though, between wing feather and leaf stem: The groove on the feather is on the bottom, while that of the leaf stem is on the top. That too makes some sense, as you can see with your cardboard core or plastic soda straw. In bending, one side is stretched while the other is compressed, and the slit makes least trouble for bending when it's on the side that's stretched. For the leaf stem that's the top since the weight of the blade bends it downward. For the wing feather it's the bottom that's stretched; its own lift makes the feather bend upward.

A lengthwise groove, then, indicates twistiness. Where else and for what else might the device be used? Consider a starfish. Using the suction cups on the tiny feet that protrude beneath its five arms, it can (very slowly) grab a clam. It then pulls steadily until the clam tires and gives enough so the starfish can insert its stomach between the half shells and digest its dinner. But a starfish bears arms that must adjust to fit all clams. Patricia O'Neill looked at how starfish, sand dollars, and other echinoderms work. She found that the groove that runs outward along the bottom of each arm permits a starfish to twist each arm to perfect its grip—as it must, since living clams clam up forcefully and persistently.[16]

In this look at bending and twisting, we again encounter something possible but rarely used by human technology. We mainly put twist-prone structures where no forces will twist them. A road sign neither bends nor twists much if it's mounted on a round pole. But it twists readily on a cheaper pole, one merely creased lengthwise (Figure 5.11). We just avoid any wind-caused twist of creased poles by putting only well-centered signs on such poles. Similarly, an ordinary I-beam girder is very prone to twisting, but we hide that flexibility. We always use at least two such beams adjacent to each other, so each keeps the other in line.

Another aspect of how flexible structures bend matters to nature more than to us, or at least nature makes a virtue of what for us is mainly a nuisance. The way a structure bends depends not only on the stiffness

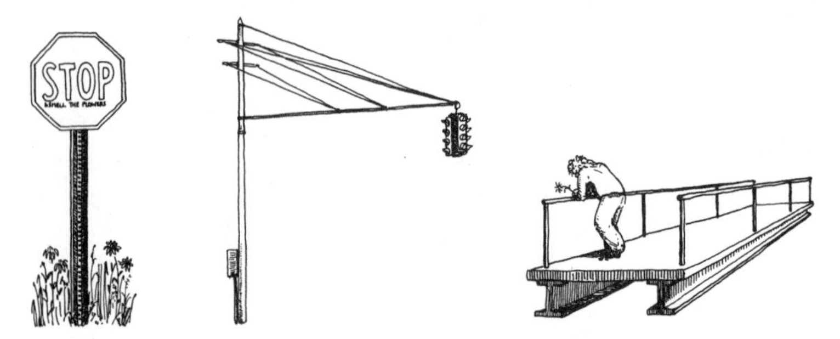

FIGURE 5.11. *The creased poles often used for road signs twist easily, but we mount the signs symmetrically so winds cause little twisting force. For less well-centered loads, we use more twist-resistant cylindrical poles. I-beams twist easily, but we circumvent that weakness by using them in pairs or groups.*

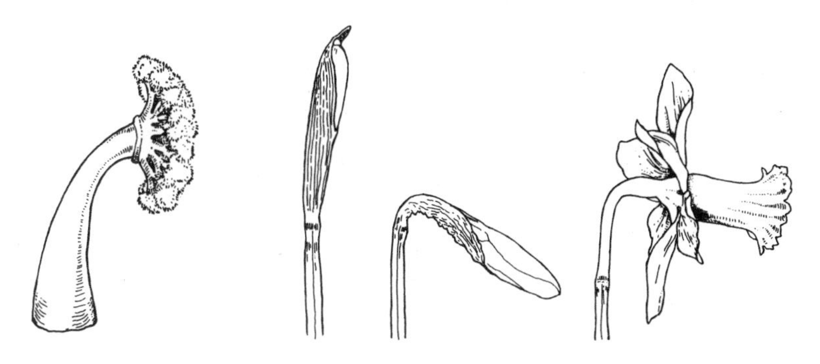

FIGURE 5.12. *Both a large sea anemone,* Metridium, *and the flower stem of a daffodil bend at specific, predetermined places.*

of its material but on the amount of material and the way it's arranged. Altering the amount of material even a little can have a large effect; recall how slightly thicker bookshelves sagged much less. For a cylinder, resistance to bending follows the fourth power of its radius; doubling the radius gives a fully sixteenfold greater rigidity. Or, looked at the other way, a simple kind of bending joint can be made by judicious local thinning. Mimi Koehl showed how a tall sea anemone (see Figure 5.12) lets a current bend its crown of tentacles just enough to position it best for capturing the small edibles that the current carries. Normally a sideways force on the top has its greatest effect lowest down; an anemone of uniform stiffness and diameter would bend over at the bottom. But the trunk of the anemone is a little skinnier at the top, and that's enough to make an adequate bending joint about which the crown rotates.[17]

Something like this happens when a daffodil flower emerges. The main difference is that it happens only once. Initially the bud points skyward, but it bends at a specific point just before opening. What appears to be happening is creation of a temporary joint, more likely by temporarily decreasing stiffness than by thinning, and then letting gravity do the work. I've made strange-looking daffodil flowers by inserting the stems upside down in siphons so the buds point down instead of up. No bending then happens, and the resulting flower, turned upright again, looks expectantly skyward rather than shyly downward.

Not that we never use localized flexibility. On my desk I have a small file box made of a single piece of soft plastic. The hinge is just a horizontal thin zone between trough and lid that makes bending happen where it's wanted. The box has now opened and closed when prompted for about a decade, and it's not obviously worse for the experience.

SWELL PIPES

But enough twisting and bending of protruding parts. Let's go inside. Contraction of your left ventricle generates a hearty pressure that forces blood out into your arteries. Then that ventricle relaxes and gets refilled. Blood pressure at the heart varies in each stroke from about 0 to 120 millimeters of mercury. Zero? Why then do we measure a range of about 80 to 120 with a cuff on the arm? Simply because our arteries are stretchy enough to damp the pressure fluctuations of the heart. When the heart contracts, the blood it pumps stretches the arterial walls. When the heart relaxes, the arteries deflate and in the process do a little passive pumping on their own. The synchrony is automatic, and it's why you can count heartbeats by feeling arterial diameter changes at your wrist or neck. The damping is a good thing too. Blood entering the small vessels moves a lot more smoothly, and lower peak pressures produce sufficient flow of blood. Atherosclerosis—stiffening of the arterial walls—means trouble. One signal of its presence is a wider spread between the maximum and minimum pressures measured on your arm.

Many ordinary pumps (for instance, the piston pumps we use to inflate bicycle tires) are as pulsatile as any heart. We might smooth the flow with arterylike elastic pipes, but we ordinarily take a different tack. Our pumps often have several chambers that work at different points in the cycle. Thus each chamber produces its peak pressure at a slightly different time, and the overall pressure never drops to zero. For the same problem we use a different solution, one that doesn't need materials that are especially flexible and, as I'll now explain, most peculiarly flexible.

Inflate a cylindrical balloon or a rubber condom. What happens isn't like stretching a rubber band. Starting to stretch a rubber band is easy, while it takes more and more force to stretch it farther and farther. Starting to inflate any balloon takes as much pressure as (and often more than) further expansion; after the start a constant pressure does the rest of the job. Furthermore, one part of a cylindrical balloon inevitably expands almost to the bursting point before the remainder does much at all. This odd (if familiar) behavior reflects something that came up in the last chapter, Laplace's law. In expanding, the wall of a balloon gets flatter. The rubber may be stretched farther, but pressure stretches a flatter wall more effectively. So the rubber gets harder and harder to stretch while the pressure gets better and better at stretching. The two just about balance. (Only the fact that the stress-strain line for rubber has an extra steep bit just before the breaking point permits cylindrical balloons to work at all. If the line were fully straight, one part would actually break before the rest stretched at all.)

In an artery, such ordinary elastic stretchiness would produce a local bulge, an aneurysm, something even worse than atherosclerosis. Fortunately, normal arteries expand uniformly, unlike balloons. How do they manage this crucial trick? To achieve uniformity, a stretchy pipe has to be very flexible when inflation starts but get stiffer and stiffer as the process continues; expansion must start easily but then become more and more difficult, disproportionately so. In their walls arteries have fibers of collagen, the same unstretchy material that makes up most of a tendon. But the fibers kink up when the walls aren't expanded, so they're mechanically irrelevant; a slack rope withstands no pull. As in Figure 5.13, expansion of the wall extends more and more of the fibers to the point where they provide tensile stiffening. This very unordinary stretchiness produces the upwardly curved stress-strain plot for arterial wall of Figure 5.3. Again we see how flexibility in nature is a subtle and multidimensional business. We (our arteries, at least) are really special.

Still, however nicely designed, we're not unique. Robert Shadwick, of the Scripps Institution of Oceanography, and his associates looked at some creatures whose circulatory systems work much like ours but that are about as distantly related to us as animals can be.[18] They found almost precisely the same variable flexibility in the arteries of squid and octopus. The main difference was that these creatures tune their arterial

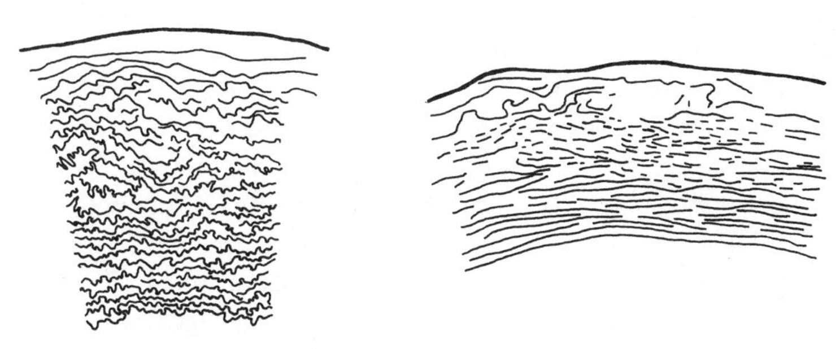

FIGURE 5.13. *Arterial wall unstretched and stretched, showing how initially kinked fibers straighten out, making the material gradually get stiffer.*

stretchiness to work at their lower blood pressures, as do our low-pressure closer kin, toads and lizards. Most startling, octopus and squid achieve their variable flexibility with a different elastic protein from the one we vertebrates use. So their arterial flexibility must have a substantially different genetic basis and must represent an independent evolutionary innovation. Wherever a pulsating heart pushes blood through flexible vessels, the vessels simply must be designed to avoid aneurysms.

As if to emphasize the point, a third group of animals has evolved aneurysm-resistant vessels. Shadwick found variably flexible arteries in crabs and lobsters, with yet another material basis and once again tuned to work at the appropriate blood pressures. He also recognized that this tuning of flexibility permits portentous predictions. Given a sample of its arterial wall, one can pretty well guess an animal's blood pressure. By this test, giant squid (which we've never kept alive) apparently have blood pressures as high as our own.

An upwardly curved stress-strain plot means that little energy is stored up when the material is stretched; the line has no great area underneath it. That can be a safety feature. A balloon, with a less curved plot, absorbs a lot of energy when you inflate it (whew!), and most of the energy gets liberated violently if the balloon bursts. Upwardly curved plots are common among biological materials such as skin, including the stretched skin of bat wings and duck feet; such as cartilage, as in our external ears; and such as ligaments. A cut or puncture doesn't release much energy, and (here we anticipate the next chapter) release of energy is what keeps cracks propagating. So a slight injury to such a material doesn't have the catastrophic consequences of a pinprick in a balloon.

But upwardly curved plots aren't universal. The plot for tendon (as in Figure 5.3) has what looks like a dangerously high area underneath. But on that area hangs functional significance of another kind. Tendon is stiff and doesn't stretch very far—10 percent beyond original length is about the limit—nor could it and still do its job. R. McNeill Alexander, probably the foremost investigator of bioelasticity, found that its energy storage even at these low strains permits a kangaroo to jump more cheaply. On landing, tendon is stretched; tendon and animal then rebound, with about 40 percent of the absorbed energy reappearing. We do likewise when we run. An Achilles (heel) tendon absorbs a lot of the energy released when a leg decelerates at the end of one step. That energy then helps reaccelerate it at the start of the next.[19] Energy storage makes getting about on legs much more efficient. Legs with energy-storing tendons don't reach the efficiency of wheels, but they're better than legs without step-to-step storage. These tendons, then, represent still another way to put nonrigidity to practical purpose.

If you want specific imagery for the present comparisons, consider a cat's ear and a door hinge. In one technology, orientation is changed by making things bend; in the other, by making them slide or roll. My plastic file box has bending hinges; when my joints move, my bones slide along one another. Once again the distinction between the technologies is one of degree and default. And once again such things as historical and evolutionary continuity, the availability of materials, and the modes of manufacture underlie the distinction, in short, the factors that underlie the different ways objects get designed and built.

If you want a context for the present comparisons, consider whether without the contrasting world of natural design, you would have wondered about the consequences of living in structures and using devices that are built of stiff stuff. Or would you have guessed how multifaceted is flexibility? What's most familiar biases our thinking, and what's most familiar is mostly what we ourselves make.

Chapter 6

TWO ROUTES TO RIGIDITY

Nature teaches us the virtues of flexibility. Leaves, large algae, and feathers show us how to economize on material, how to change shape as environmental forces change, how to enlist the environmental forces themselves to produce those changes. Still, nature doesn't entirely avoid stiffer stuff. Shell, coral, tooth, bone, and wood are relatively stiff. So too are the chitinous exoskeletons of lobsters and big beetles. Stiff materials may not be necessary for building large creatures, but they certainly help. But while both technologies use stiff materials, they differ in what kinds they use. One difference is startlingly absolute: no "relative to this" or "preponderance of that."

THE NONMETALLIC WORLD OF ORGANISMS

No organism that we know uses any piece of metal for any mechanical purpose; for that matter, none biosynthesizes a piece of metal at all. Here "piece of metal" means something fairly specific. Alloys of different metals are included; we use precious little in the way of unalloyed metals ourselves. But excluded from the definition are both organic and inorganic materials in which chemical compounds bind metal atoms with nonmetal ones. No legalistic subterfuge is meant since such metal-containing compounds don't behave mechanically like metals. Steel and bronze are proper metals because the metal atoms bind directly to each other, while iron oxide or copper sulfate are simply metal-containing compounds.

Quite peculiar, this complete absence of metallic materials in nature. To start with, most (perhaps all) organisms contain atoms of the kinds of metals that might do mechanical tasks. These atoms aren't there by accident; creatures require them, and they have enzymes that can synthesize large metal-containing molecules. The most familiar of course is hemoglobin, which contains iron. Other iron-containing compounds help cells transfer energy from compounds that store it to ones that fuel biosyntheses, that work muscles, and so forth. Hemoglobin itself has evolved on numerous occasions in both animal and plant kingdoms. To these iron-containing compounds within us we add ones containing copper, zinc, chromium, tin, and possibly nickel. One group of animals, the ascidians or tunicates (called sea squirts or sea pork; one is shown in Figure 6.1), has blood cells mysteriously loaded with vanadium. A few other metals

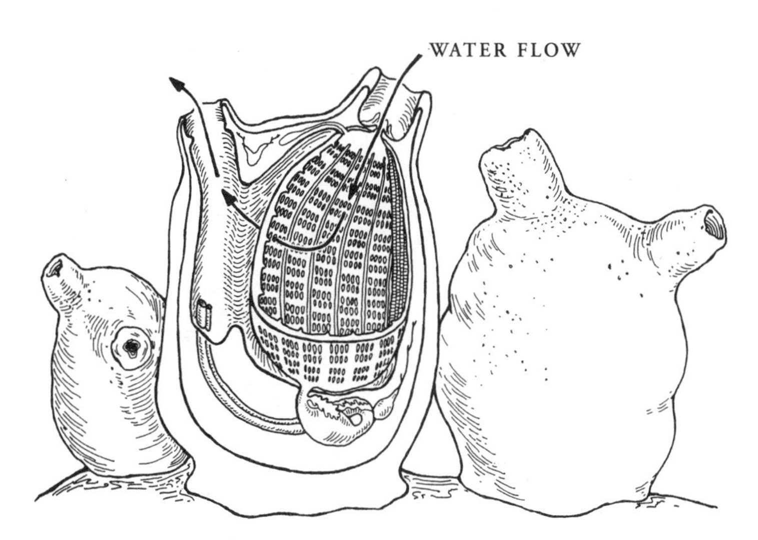

FIGURE 6.1. *Several adult ascidians, one cut away to show how water passes through its filtration apparatus. These animals are reasonably common on rocks and wharf pilings, usually below the low-tide line. They're in the same phylum, the Chordata, as are vertebrates such as we.*

are needed in small amounts by other organisms.[1] Small quantities, to be sure; our own most abundant metal is magnesium, still only a twentieth of 1 percent of our weight. As a pure metal magnesium reacts too readily to make safe structures; burning magnesium made the flash of old-fashioned photographic flash powder. Of iron, an adult has only about four grams (mostly in hemoglobin), ten times less than our magnesium.[2] Still, we contain metal, we use metal, we require metal.

Some organisms even build serious mechanical devices out of metal-containing compounds. Many mollusks (snails in particular) feed with organs called radulae (Figure 6.2). A radula works like a cross between a cat's tongue and a chain saw. The flexible, horny structure goes in and out, rasping food from a surface and carrying it mouthward. Hard denticles on the radula do the rasping, and some of these contain a lot of metal. But the metal turns out to be in the form of metal salts—of iron or copper—suitably hard, but minerals rather than metallic materials. Nor are radulae unique in this respect. Chewing poses a peculiar problem for animals that feed on leaves and wood—cows, caterpillars, and such. For their volume, these parts of plants don't provide much nourishment, so the animals have to eat large amounts. Worse yet, the plants seem to have taken considerable trouble (evolutionarily) to make themselves difficult to bite off and grind up. Grasses and tropical woods are full of sand (silicon dioxide), most likely just to hobble herbivory. So tooth wear poses a worse problem for herbivores than for carnivores; recall the horses' teeth

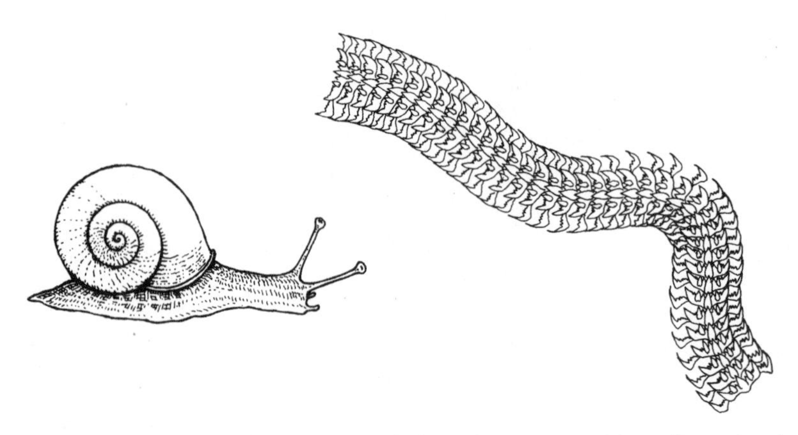

FIGURE 6.2. *A snail and its radula, a toothed strap that comes out of the snail's mouth and moves in and out, scraping like a rasp.*

of Figure 2.5. Metal salts can harden nonmetallic materials; herbivorous insects, for instance, have zinc or manganese in their mandibles, but again as salts rather than as metallic materials.[3]

About twenty years ago a short paper announced the presence of "iron-rich particles" in bacteria that oriented and moved directionally in magnetic fields.[4] Superficially the magnets sounded metallic. But the original report made no such claim and gave no indication of the form of the iron. It proved to be not metallic iron but a compound of iron and oxygen called, for obvious reasons, magnetite. Nothing obscure or radical—we use compounds such as magnetite to coat magnetic tape and computer disks. Magnetite has now turned up just about everywhere that magnetic sensitivity is known and in some places where it's only suspected: in bird brains, in honeybees (in the abdominal segments, as it happens), in salmon, in some rodent brains. For that matter, it occurs in our own brains.[5] Our sensitivity to magnetic fields remains uncertain.

WHY DOESN'T NATURE USE METALS?

So, ironically, no chunks of metal in organisms. In a sense this absence of metallic materials tests one of our main points: that natural design is severely constrained. If you take the opposite viewpoint—that nature will do all that's best and brightest—then you have to make the case for metals being unavailable, inappropriate, or at least no better than naturally produced nonmetallic materials. How good is that case?

Supply limitations. Perhaps living things find metals scarce because their distribution on the earth's surface is too spotty. After all, humans long ago found that ores or crudely refined metals were worth moving long distances despite their weight and bulk. In classical antiquity, Spain, Brittany, and Cornwall exported tin, the crucial additive that made soft copper into hard bronze. Organisms build themselves of more widely distributed elements: carbon, from the carbon dioxide in the atmosphere by way of photosynthesis in plants; hydrogen, from water, also through photosynthesis; oxygen, from the atmosphere, where it is a by-product of photosynthesis;[6] nitrogen, from the atmosphere through the action of nitrogen-fixing bacteria and a few other agencies. Calcium and phosphorus complete the list of elements in us in amounts over 1 percent. Calcium occurs in many rocks and in most natural waters. Phosphorus makes up less than a tenth of 1 percent of the earth's crust and a still smaller fraction of seawater, but it's well spread around. So far so good with the argument for effective scarcity of metals.

Still, the list of major ingredients in organisms corresponds poorly to the composition of the surface of the earth. While none of the elements just mentioned is rare, some elements present in abundance aren't used in any quantity. Aluminum, a splendid structural material, is the most notable; it makes up about 8 percent of the earth's crust. Iron, long our favorite nonprecious metal, makes up fully 5 percent. While rich ores are spotty, decent supplies of aluminum and iron are widespread at least on land. Besides, organisms do well at acquiring elements from dilute sources. The ascidians that concentrate vanadium get it from seawater, where it constitutes only two parts per billion. Upon closer scrutiny, therefore, ill distribution weakens as a basis for arguing the effective scarcity of metals. Metals are better spread than one might think, and organisms get many elements from sparse sources.

Nonetheless, for at least two reasons the scarcity argument shouldn't be casually dismissed. First, using a small amount of metal as a cofactor for some enzyme isn't the same as using enough to build a respectable amount of mechanical equipment. Our bodies use iron, but we use far less of it than, for instance, calcium. Calcium, a major component of bones, makes up two hundred times as much of our weight as does iron. Maybe organisms can't afford the cost of extracting large amounts of iron. Second, abundance in the earth's solid crust may be a misleading criterion if the earliest and most biochemically innovative episodes of evolution occurred in the sea. Seawater doesn't just contain calcium; it's supersaturated with the stuff. So the sea gives up its calcium for free. Iron, copper, and aluminum are vastly scarcer in the sea—one part in twenty-five hundred for calcium against one part or less in one hundred million for these latter.

Chemical trouble. Perhaps chemistry limits the use of metals. Organisms may be versatile chemists, but not all reactions are equally easy for them. Mechanically useful metals occur in nature almost entirely as compounds rather than as pure materials; they're combined with such other elements as oxygen, chlorine, silicon, and sulfur. Getting metal out of ore demands a lot of energy; worse, it requires a concentrated energy input, such as high heat or high voltage. In chemical terms, metals have to be reduced from the oxidized compounds in which we find them. A scale called an electrochemical series gives a good view of the relative difficulty of purifying different metals; it looks at the barrier to sticking enough electrons on to the oxidized form to get the pure stuff:

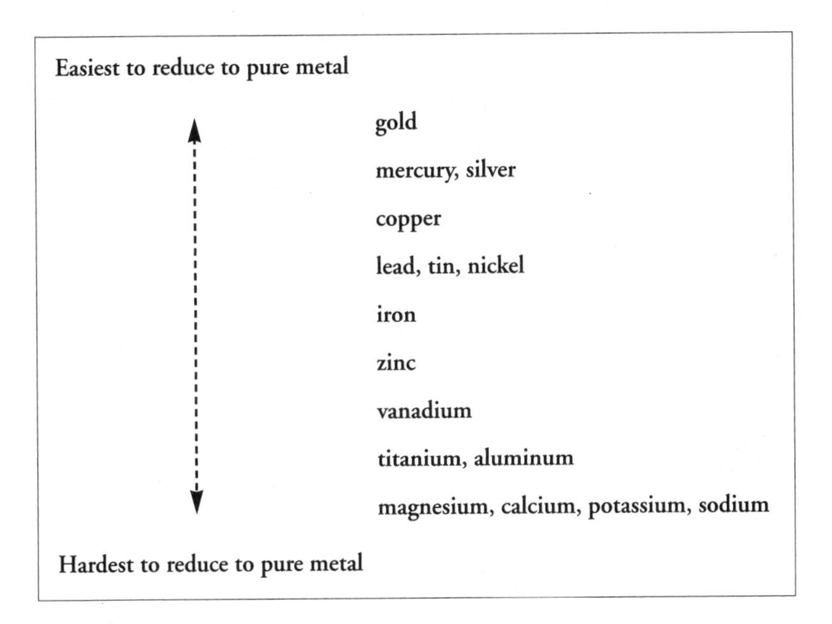

The scale runs from very stable, inactive gold, so indifferent to oxygen that it doesn't even need periodic polishing, to highly reactive sodium, which combines with just about anything it touches and spontaneously burns when exposed to air. The reactivity of the chemical compounds of these elements runs in the other direction. Compounds of mercury and gold are weakly bound and unstable. The top three elements occur in free, uncombined form in nature, copper ores sometimes contain metallic copper, and metallic lead is very rare. Their ores easily yield copper and lead. Reducing iron, farther down, from compounds to metallic form is tougher. Reducing aluminum, still farther down, is even harder.

Metals harder to isolate are also harder to maintain. The issue here is spontaneous oxidation, rusting. Silver tarnishes a little; sodium reacts violently even with water. We can use nearly pure aluminum only because it forms a white rust, aluminum oxide, a nicely impervious coating that protects the underlying metal. Rusting is rapid, but it's also self-limiting. (We also use various coatings and other processes to offset aluminum's reactivity in the presence of oxygen.) Iron oxidizes less eagerly, but the ordinary oxide has the inconvenient habit of flaking and peeling to expose fresh metal. Most organisms are aerobic; they use oxygen to oxidize carbon-containing compounds as energy sources. Under oxidizing conditions, perhaps metallic structures are impractical or less desirable than alternatives. In other words, preventing rust may be more trouble than it's worth.

An indication that these chemical limitations—the cost and difficulty of reduction (blast furnaces and so forth) and the subsequent cost and difficulty of preventing reoxidation (periodic painting of steel ships and bridges)—may be important comes from a dog that didn't bark. Aluminum is the third most abundant element on the surface of the earth, after oxygen and silicon. Every clay contains the element. Yet as far as I know, no organism does anything with it, even as some trace nutrient or biochemical cofactor. Exposure to aluminum leads to internal accumulation of the element but to no notorious pathology; it's much less toxic than the heavier metals.[7] The fact that aluminum (and titanium, for that matter), far down the electrochemical series, isn't used at all suggests that iron, a little higher, can be used only with difficulty.

Excessive density. Perhaps organisms don't use metals because life evolved in water. Of the metals that might make decent structures, only chemically troublesome aluminum isn't particularly dense. It's only 2.7 times as dense as water. The equivalent figure for copper is 8.9, for iron 7.9, for tin 5.8; even titanium is fully 4.5 times water's density. Just a little structural iron, for instance, would make an organism a lot denser than water. Thus gravity would impose a serious load if an organism grew up from the bottom of a body of water, and it would just about rule out swimming. The minerals with which nature builds bones, shells, and the like are less dense. Silicon dioxide has a density 2.2 to 2.6 times that of water, calcium carbonate 2.7 to 2.9, and calcium phosphate 2.2 to 3.1; the ranges reflect different crystalline forms. Organisms that don't have air inside (as do trees) are mostly denser than seawater, but only slightly. Even a thick-shelled clam sinks more slowly than a chunk of iron. A small deposit of fat or a tiny bladder of gas can offset slightly denser minerals and keep an organism from sinking at all. With substantial use of iron or copper, density compensation would take much greater investments of space, material, and energy.

This difference in density between metals and these minerals may also be important for animals that live in sediments beneath bodies of water. If agitated, particles sort themselves out by density (and, of course, size). Particles of most sands, silts, and clays have densities between two and three times that of water, so animals made of lightweight organic compounds,[8] even with a lot of mineral, are less dense than the surrounding sediments. By contrast, an organism with substantial amounts of metal might be self-burying. The habitat is far from unusual; most organisms are small, and an extremely diverse fauna of tiny creatures live in the spaces between the grains of lake bottoms, coastal beaches, and continental shelves. Yet despite these consistent and suggestive differences in density, the heaviness of metals doesn't seem a weighty enough factor to account for their complete absence in nature.

Inability to grow. I once encountered the argument that metallic structures couldn't grow the way the nonmetallic components of organisms grow. I'm not persuaded that they couldn't grow. Furthermore, neither mollusk shell nor arthropod cuticle grows, but both grace eminently successful phyla of animals. Similarly, trees make their wood once and for all, increasing their girth by adding new wood peripherally as annual rings. In short, the argument is so much vertebrocentricism.

When we ask why organisms don't use metals, we need to consider another possibility: Maybe metals lack mettle and aren't all they're cracked up to be. To explore that possibility, we need to look both at what makes metals special and at how human technology uses them.

THE UTILITY OF METALLIC CONSTRUCTION

Humans have long been willing to expend enormous effort to obtain metals. Ancient Middle Eastern and Mediterranean metallurgy is well known. Iron smelting was widespread in Africa. The Indians of North America strove to obtain metallic copper where they had any access to it, as in northern Michigan and adjacent areas of Canada. The Inuit of Cape York in northern Greenland laboriously pounded bits of iron from a meteor deposit to use as edges for their tools.[9] Did human technology have easy access to some better alternative to metals? The possibility seems remote. Why, then, are metals so useful? In particular, they're malleable and ductile; with minimal risk of breakage, they can be pressed or hammered into shape, or they can be drawn out into thin wires and sheets. Good malleability and ductility imply as well an agreeably high level of toughness, defined in the last chapter as requiring a lot of energy to be broken.

What's special about metals is easiest to see by stretching a sample of one and plotting the result on a now-familiar stress-strain graph. Figure 6.3 gives a plot for steel: for the lower edge of an I-beam that's being bent downward or the bowed-out side of a round column that's pushed sideways. The stretch initially meets a high resilience response; removal of the weight on the beam or the push on the column restores the original length immediately and forcefully. In this region the plot ascends along a fairly straight line. Double the stress, and you double the strain. Used like this, a piece of metal makes a handy spring for a scale to weigh things; the numbers on the scale will be equally spaced, and the scale will rezero itself between uses.[10] Most biological materials such as tendons give curved rather than straight lines in equivalent plots of stress versus strain. Organisms, as we saw, put the curvature to good use. That metals give straight lines indicates intrinsic superiority only for making simple spring scales. What matters is their high resilience.

Things get special beyond this initial upward line, beyond what's called the elastic limit. Typically, the plot levels (or nearly levels) abruptly as the metal is stretched farther; it enters its plastic region. Without any (or much) additional loading, the metal can be deformed a lot further, which is what ductility is all about. This horizontal extension of the stress-strain line generates a lot of functionally important behavior.

If you use a metal in its resilient, elastic region, you've got something that can be loaded repeatedly without getting permanently bent out of shape. The piece of metal tolerates abuse. Overload it, and it deforms

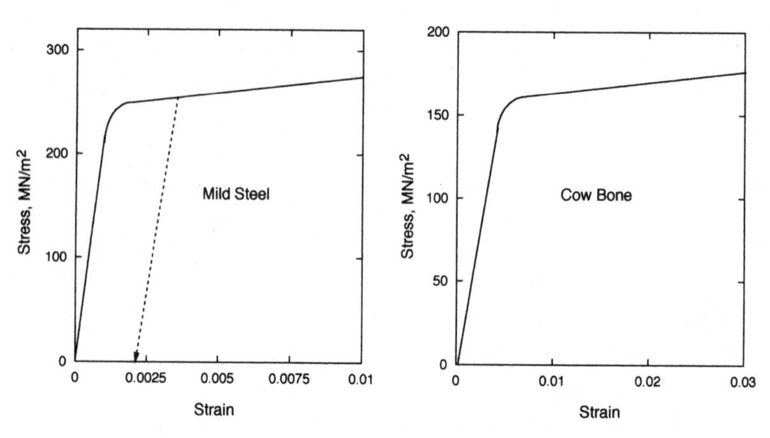

FIGURE 6.3. *Stress-strain curves for mild steel and for cow bone (data from Currey, 1984).*

plastically. That may spoil it for further use, but metal hasn't actually broken, and with good design, catastrophe needn't follow. The structure just bears its load in a slightly different position. The bicycle frame may be bent, but you haven't been dumped off, and you may even be able to ride home. The bed and side walls of my old pickup truck got pretty dented, but only the resulting rust demanded attention. Recall that the area under the stress-strain curve represents the energy absorbed in straining something. For a metal accidentally distorted beyond its elastic limit, all that area under its curve enhances its safety; straining the material soaks up energy that might otherwise make trouble. A well-strained resilient material, such as rubber, will snap back dangerously when unloaded again. But metals in the plastic region aren't at all resilient; they're plastic, not elastic, and the work of deformation gets benignly converted into heat rather than stored as elastic energy. Repeatedly stretch or bend a piece of rubber, and it warms almost imperceptibly (tires are a little warmer after rapid driving). Repeatedly bend a strip of metal beyond its elastic limit, and it warms noticeably.

Another view of breakage comes from a look at how cracks propagate. Cracks extend because of force concentrations at their tips. Cracking relieves the force, releasing energy that powers further progress of the crack. Force concentration means that the stress at a crack's tip is much higher than just the force applied to the object divided by its overall cross-sectional area. In their plastic region, ductile metals respond to higher local stresses by stretching rather than cracking. Stretching distributes the force over a greater area and thereby reduces the stress—or at least ensures that an increase in force doesn't cause a corresponding increase in stress. By contrast, cracking merely moves the location of the high stress from one place to another as the crack extends. That can increase the stress further since less material then remains intact to bear the load. Put one way, metals are fairly safe considering how stiff they are. Put another way, ductile failure is better than brittle failure; a dented cup holds water better than a shattered one.

We make good use of the ductility and particular stress-strain curves of metals when we make things of them. Push or pull on a metal piece with enough violence, and it will bend to your will as you force it into the plastic region. Remove the load, and the metal retains the new shape, but it now has a new elastic region; it now follows the dashed line of Figure 6.3. The new shape can then give long, dependable service, snapping back elastically when modestly deformed. Neither the stiffness nor the strength of the material has suffered much by the treatment. That's what its maker did to my pickup truck before I bought it, and that's how almost every curved panel on almost every automobile is made. With the aid of huge presses and hammers, that's how we make all kinds of ordinary objects, such as nails with heads. For large objects we extend the plastic region by working with hot metal; the process is called forging. To make wire and hollow pipes, we take similar advantage of the ductility and malleability of metals.

The stress-strain plot of bone, to take a well-studied biological material, looks superficially like that of mild steel. It has an initial upward-sloping elastic portion and then behaves plastically at greater extensions (Figure 6.3, again). Loading is most often in the elastic region, and the plastic portion provides a margin of safety. But as John Currey, who has loaded all kinds of bones in all kinds of ways (picking a bone with Currey is a serious matter), points out,[11] what happens in bone differs from what happens in metals. Bone isn't ductile but is instead viscoelastic. That is, in its plastic region it's viscous; it flows. Worse, it develops tiny cracks in all directions and softens from their interactions. A bone so loaded then has to be rebuilt, even if it hasn't actually fractured. Rebuilding depends on its being a living, growing material. As we use it, bone works about as well as does a decent metal, perhaps even a little better in terms of the mass needed to support a given load. But it does so only by virtue of continuous micromaintenance.[12] Without such rebuilding, bone loses much of its value. Humans have long had access to bone but have mainly used it—as do the Inuit in particular—when little else, such as wood, stone, or metal, has been available.

Any proclamation of the magnificence of metals must be tempered by a few cautions. Metals are even more mechanically diverse than are woods, and no one metal proves perfect in all properties at once. Our cultural biases, stemming from both personal experience and the way we teach history, may give metalworking undeserved credit for permitting technological complexity. The use of metal for more than ornamentation was a glory of Mediterranean and Middle Eastern civilizations. We mustn't forget that sophisticated but largely nonmetallic cultures flourished elsewhere, particularly in the Americas. Furthermore, early metals weren't particularly good. Copper is soft; if hammered, it gets harder but more brittle. Bronze is better, but it still takes an edge far less sharp than flaked obsidian. I was much impressed watching an anthropologist eat a meal with an obsidian knife he'd made. He said (and I have no doubt) that he could shave with such a knife. Bronze, though, is less likely to crack when inexpertly used. That last point, in fact, may hold the key to the initial attractiveness of metals. Stone axes are hard, and flaked flints are sharp, but both are brittle, thus fragile, and their use calls for both skill and care.[13]

Finally, metals can be stamped, drawn, forged, cast, ground, sliced, and sawed. But living systems need to do none of these; they make materials into artifacts by internal growth and surface deposition. So our wonderfully diverse array of fabricational techniques may hold little appeal for nature. Once again, we're looking at two distinct but individually well-integrated technologies; an impressive aspect of one may have little relevance to the other.

HOW NONMETALLIC MATERIALS CAN AVOID CRACKS

Organisms may eschew metals, but human technology uses both metallic and nonmetallic materials. Through most of our history and prehistory we've been almost as nonmetallic as organisms; truly large scale use of metals is less than two centuries old. James Gordon[14] points out that even the main cylinder of Fulton's steamboat of 1807 (although not the boiler) was made of wood, and he claims that the tenfold drop in the cost of iron was the most important thing that happened during the long reign of Queen Victoria. The main materials of human technology have been stone, selected, shaped, flaked, or polished; ceramics, including brick, pottery, and glass; wood, from tied bamboo poles to seasoned timber and glued plywood; and a diverse assortment of natural fibers, mainly keratin from animal hair, and cellulose from plants. During the past century or so we've added a few things: rubber, starting with Goodyear's vulcanizing process; plastics, from Bakelite, the first, to ever more diverse polymers; artificial fibers, such as nylon and polyester; and more complex materials, such as chipboard and fiberglass. Our use of metals, relative to other materials, has probably passed its peak. During the fifty-odd years that I've observed such things, one metallic object after another has been re-created in plastic: file boxes, garden carts, even the side panels of our new car.

In this respect our technology is moving toward nature's. But I don't mean back to nature. We rarely use nature as explicit model, much legend to the contrary; as we'll explore in Chapter 12, we didn't learn spinning from spiders. We also subject naturally synthesized materials to increasingly complex processing before reusing them; for instance, we make both paper and rayon of the cellulose from wood, but neither much resembles nature's original product.

Of greater interest is how we're adopting nature's approach to dealing with the trade-off between stiffness and toughness in nonmetals. As noted, metals have decent values of both properties. (Nonetheless, improvement in one usually entails deterioration of the other. High-tensile steel stretches less than does mild steel, but it also cracks more easily.) Nonmetallic systems are worse, at least if made of ordinary, single-component materials. Glass is wonderfully stiff but cracks so easily that it probably wouldn't be approved as a new material today. Plexiglas[15] is less stiff, but it still cracks without much provocation. The softer plastics, such as the polyvinyl chloride of pipes, are nicely tough, but "softer" is just a gentle way of saying "less stiff."

Something odd, though, is at work. Our stiff, nonmetallic, single-component materials, such as glass or Plexiglas, bricks or ceramic tiles, are much more brittle than nature's wood, horn, or bone. Even seemingly brittle natural materials are often tougher than they look. Mollusk shell may look like ceramic tile, but it resists shattering much better when hit or drilled. I once made hanging ornaments from a bunch of scallop shells left over from lab work. Drilling a small hole near the thin edge of each took no special tools or care, and no shell shattered in the process. A number of primitive cultures fashioned fully functional fishhooks out of shell.[16] Somehow organisms deal with the awkward tendency of stiff nonmetals to crack.

From measurements of the strength of a material such as steel, one can calculate the strength of beams made of the material. But by the early 1900s sad experience showed that such calculations provided poor guidance for how beams behaved in use. Ships broke in half at stresses that their steel should have easily withstood. The culprit turned out to be force concentrations—small places where the stress was unusually high—and cracks that spread from such places. At least steel suffered less from crack propagation than many other materials, particularly stiff, nonmetallic ones. Sharp corners and preexisting cracks were bad things, but the exact nature of the problem remained mysterious.

Around 1920 a British engineer named A. A. Griffith, working at something other than what he was supposed to be doing, found that glass fibers could be pulled harder than could glass rods—that is, fibers could withstand great stress. The thinner the fibers, the more stress they could take, down to the finest that could then be made, with diameters around one ten-thousandths of an inch.[17] Put another way, a bundle of fine fibers would support a lot more load than a single rod of the same thickness, even where no bending was involved. Still, the strength of the finest fibers didn't quite reach the strength of their chemical bonds. So the real issue wasn't the strength of fibers but the weakness of rods. What made the difference were microscopic cracks.

Griffith went on to provide the basis for our present understanding of what makes cracks extend. The tip of a crack is a place where force is concentrated, as mentioned earlier and crudely illustrated in Figure 6.4. If you cut halfway through a sample of a stiff material and then pull on it, it usually breaks at far less than half the force needed to break an uncut sample. The concentration of force makes the crack extend, the sample is fractured further, more force is concentrated at the crack tip, and in an instant the unbroken sample is history. Whether a crack will propagate depends on the initial depth of the crack and the sharpness of its tip; the deeper and sharper the initial crack, the more likely it is to extend farther.

Making additional surface on either a solid object or a body of liquid absorbs energy—whether slicing bread, cracking eggs, or even as a water strider's leg makes a downward dimple on a pond. Since cracking makes surface, it absorbs energy. But cracking also releases energy as it relieves the force concentration at its tip. As a crack gets deeper, more force is concentrated at its tip. Thus, as a crack progresses, it relieves ever more force and releases more and more energy. Eventually the energy released

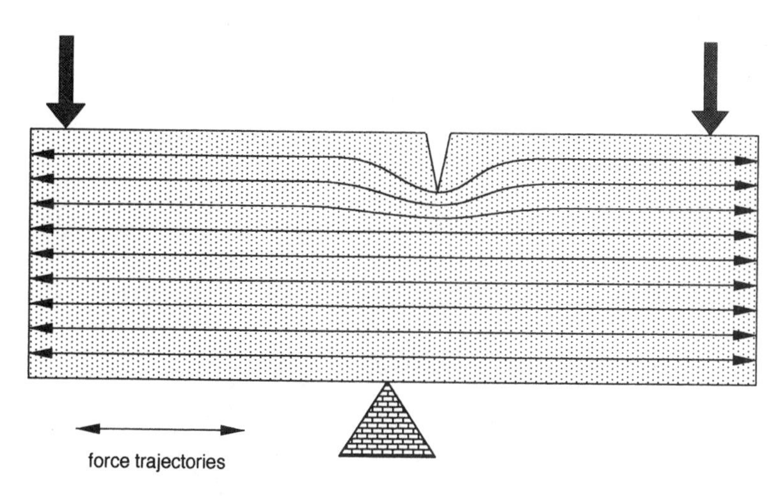

FIGURE 6.4. *Force trajectories bunch just beneath a crack, so force is more concentrated (stress is higher) there.*

by relieving the force concentration at the crack's tip surpasses the energy needed to make new surface as the crack propagates. Then stand back, for the system is out of control, and the crack will spontaneously extend at a fair fraction of the speed of sound. You deliberately crack a brittle material such as glass or plasterboard by making (scoring) an initial groove—a potential force concentration. You then apply a slight force, and the material breaks in two where you've started the crack. Breaking a piece of glass incompletely isn't at all easy.

All real objects have cracks. What matter are their depths, the sharpness of their tips, and the loads on the objects. Griffith's critical crack length is reached when the energy released becomes enough to keep a crack advancing. Thinner objects normally have shallower cracks, which make them stronger, able to withstand greater stresses. That's why a bundle of skinny glass fibers is stronger than a thick glass rod. Using a bundle of skinny fibers instead of a single rod gains something else as well. If the stress is high enough so a fiber does crack in two, that's usually the end of the matter. What chance that the next fiber will have a preexisting crack just where necessary for breakage to continue? When you break a bundle of spaghetti, the individual strands don't all break at the same place.

Stress a piece of metal, accommodatingly ductile, and its plastic stretch blunts the tips of its cracks. As a result, critical crack lengths in metals are between ten thousand and a million times longer than those in stiff nonmetals like glass. Cracking still matters—ships do break up, so portholes and hatchways still need to be rounded—but a steel rod is far less fragile than one of glass.

How might this crack-prone fragility of stiff materials, especially nonmetallic ones, be avoided? The obvious fix is to use very skinny fibers and run lots of them in parallel, as a kind of ropework. It works as long as the individual fibers can slide fairly freely against each other. A rope is weakened if the individual strands are glued together—perhaps a problem when ropes get iced up. But ropes take pulls and not bends. A second and better (or more general) fix uses an interface as a crack stopper. An interface is nothing special; in fact, it's no thing at all, just the place where one material stops and another begins—where the rubber meets the road, as a recent ad put it. But an interface can act in a curious way.

Whether a crack extends or not depends, again, on the sharpness of its tip. You can often stop a crack by making a round hole at its tip; an elderly machinist introduced me to the trick when I was a graduate student, and I suggested how you might demonstrate it with aluminum foil in Chapter 4. How, then, to blunt the tips of cracks? Think about that ubiquitous material Styrofoam. For its weight, it's strong stuff. What keeps strength up and weight down are the microscopic holes that make it a foam. Plastic and air come together at the edges of those holes; they're interfaces. A crack moving across the stuff can't avoid a hole, which blunts its tip. So the crack can't build up a sufficiently concentrated force on the other side of the hole to keep itself going, as put diagrammatically in Figure 6.5. Thus a simple way to build in crack stoppers is to make a foam of the material. The small, hard elements that stiffen and support such echinoderms as starfish are formed of such a foam. They're made of calcium carbonate (as crystals of calcite), which is pretty brittle stuff. But the interconnected holes, between one two-thousandths and one fiftieth of an inch across, have rounded, smooth surfaces, as in Figure 6.6.[18]

One crack may even stop another; a crack is just an interface between a solid and air, and interfaces can have the same blunting effect as round holes. Imagine a material with a lot of lengthwise cracks, perhaps ones so small and subtle that you'd not normally notice them. Cracks running lengthwise are innocent bystanders as far as load bearing is concerned, with about as much effect as the medians of divided highways have on traffic capacity. But when a crosswise crack, the kind that breaks an object, runs into a lengthwise crack, it's stopped—like a car running into one of those medians.

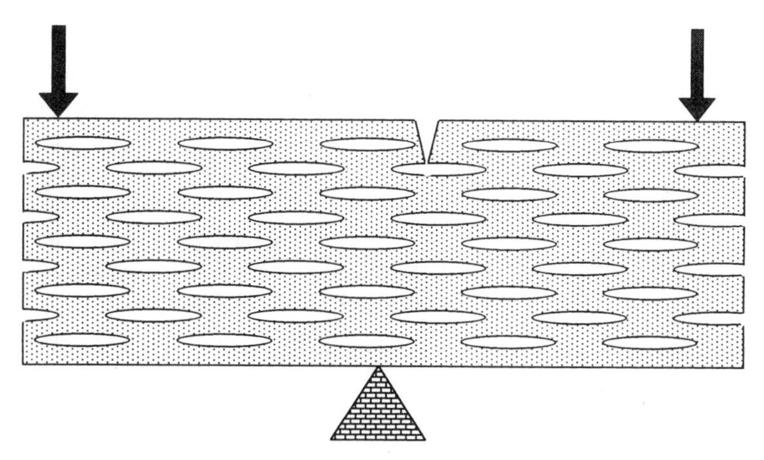

FIGURE 6.5. *Voids, if oriented lengthwise or if rounded, can reduce the ease with which a crack propagates across a loaded piece of material.*

FIGURE 6.6. *The hard elements (ossicles) of echinoderms are really foams of calcite and round-surfaced voids.*

How might lengthwise cracks be built into a material? Bundled fibers might make good ropes, but they won't make good beams and columns. For the latter, less dramatic interfaces work better, interfaces between two solids of very different mechanical properties rather than between a solid and air. Fiberglass uses just this trick. Glass still bears the stress, but the strength of a fiberglass pole dramatically exceeds that of a glass rod. In fiberglass, as shown in Figure 6.7, hard, stress-bearing glass fibers are joined with a relatively soft glue. After passing through the hard stuff, a crack hits the soft stuff, which stretches (gives) rather than transfers the crack across to the next hard element.

Materials that use this scheme are called composites, the name alluding to having more than one component. If you want to experi-

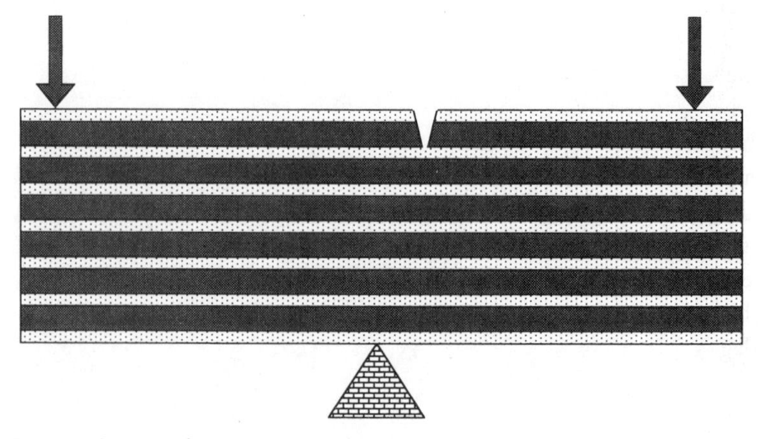

FIGURE 6.7. *Tough composite materials are made of fibers or layers of stiff material separated by material of much lower stiffness.*

ence a composite, mix wheat bran, for fiber, and egg white, for glue, and bake the product until hard; shaping can be done either before or after solidification. Adjusting the mixture, baking time, and temperature changes the character of the product, while adding some sweetening and flavoring gives you a satisfyingly tough, munchable material. (Don't do as several students did when I challenged them to produce an edible composite. Mixing marshmallow and Rice Krispies makes a material midway between solid and fluid, so shapes slowly sag into amorphousness.)

Fiberglass is neither the most common nor the best composite material that human technology has produced. Particleboard combines wood chips and binder; it's now the stuff beneath the veneer of our furniture. Soft steel in the reinforced concrete of our highways and public buildings offsets the brittleness of pure concrete. Composites compose a versatile technology. Individual fibers need not run the entire length of a structure; short ones (whiskers) are effective as long as the glue binds well. Nor must all fibers run in the same direction, as in a fiberglass fishing pole; the direction is random (in two dimensions) in fiberglass sheet. Nor are fibers the only useful geometry for the hard material; thin, flat plates or sheets work well.

For human technology, the main drawback of composites is cost, typically greater than that of ordinary metals or single-component plastics. In addition, material and structure usually have to be made together, something a little complicated and unfamiliar. Sheets or rolls of metal can be cut, bent, and riveted together to make an aluminum canoe. By contrast, a felt of glass fiber is combined with epoxy glue over a mold to make fiberglass and canoe simultaneously.

Nature uses composites for all her hard materials, usually quite complexly organized composites; Figure 6.8 illustrates just two. Wood is a composite of cellulose, a hard, fibrous material, and lignin, a glue. The cuticle of arthropods is a composite of chitin fibers in a proteinaceous matrix, with some calcium carbonate salts used to stiffen the larger crustaceans. The shells of mollusks are made of layers of hard mineral separated by a critical few percent of protein. Bone is a composite of the protein collagen, some other protein, and calcium phosphate salt. Even your teeth are composites of mineral and protein; when teeth are drilled, burning protein makes the sulfurous smell. In each case the structure is highly tuned for its specific applications. Not only does wood vary from one kind of tree to another, but the wood of a tree's roots has very different

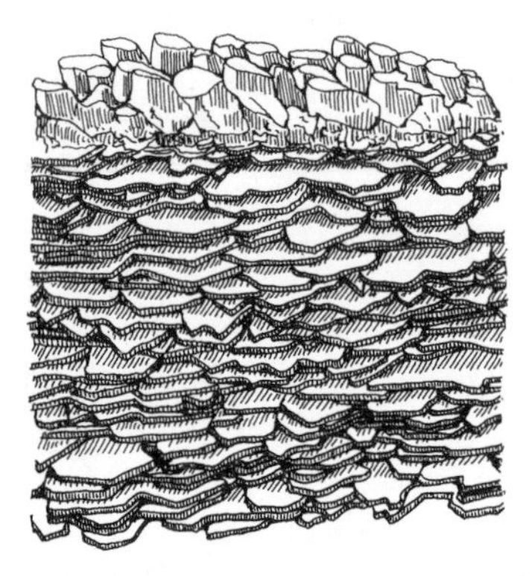

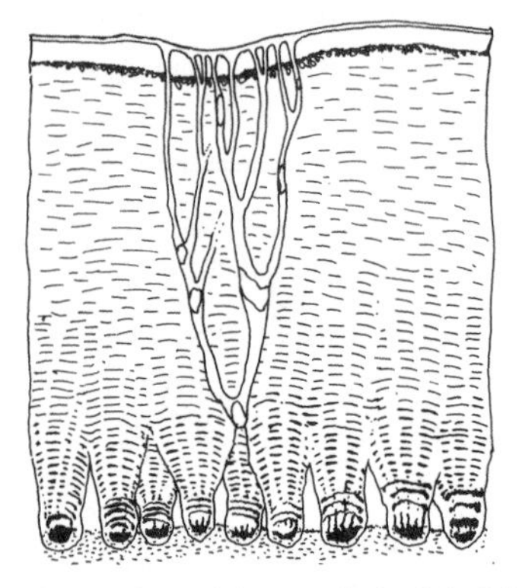

FIGURE 6.8. *Natural composites: the shell of a mollusk (left) and the egg of a bird (right). Both are made largely of brittle calcium salts with small but critical amounts of softer material to toughen them and prevent crack propagation.*

properties from the wood of its trunk. Bone isn't just bone, even in an individual; its composition and properties depend very much on what it's called on to do.

A NONMECHANICAL INTERLUDE

Let's briefly leave the world of materials and structures. Important nonmechanical differences between our two technologies arise from the absence of metallic materials in organisms. In particular, metals and nonmetals differ greatly in thermal and electrical conductivities. Both kinds of conductivity are hundreds to thousands of times greater for metals than for nonmetals, and both properties determine how we do many things.

To give a few examples, the thermal conductivity of copper is more than 3,000 times the conductivity of wood, 500 times that of glass, 660 times that of water, and 1,000 times that of fresh leaves. The consequences aren't minor. The higher the conductivity, the faster heat moves from the hotter parts to the colder parts of an object, so the object will more rapidly approach a uniform temperature. If you cook on an electric stovetop, pots and pans of high conductance[19] give a more even heat distribution and take much less stirring and scraping of their contents. Relative to its volume, aluminum has about two and a half times the conductance of iron or steel and thus works better as cookware. Pans of pure stainless steel are a nuisance, and cast iron is satisfactory only when thick and heavy; this slows its response to your adjustments of the stove and trades one disadvantage for another. Ceramic vessels, even if heatproof, heat up still less uniformly and mainly find use (their manufacturers' claims notwithstanding) as ovenware.

When the food reaches the table, though, high thermal conductivity becomes a nuisance. The coldness of metal and the warmth of wood—celebrated in aphorisms and figures of speech ("cold steel," etc.)—is the perceptual signal of the difference in thermal conductivity. A piece of aluminum feels colder than a piece of pottery because the aluminum more rapidly conducts heat away from your hand. Similarly, a silver or aluminum dish conducts the heat away from the food, increasing the area over which the heat is shifted to the surrounding air and moving heat to the place where you're holding the dish. We ate cold food from metal mess kits on scout trips but blamed it entirely on the weather. Tea or coffee cools more rapidly in a metal pot, especially in aluminum, copper, or silver, which have the highest conductivities. According to at least one guide, the governor of colonial Williamsburg, in Virginia, ate off silver dishes—the worst of all possible materials. Since cooking was done in an outbuilding, one suspects that no governor ever enjoyed a nice hot meal at home. Conversely, metal handles often get uncomfortably warm. Stirring a cup of boiling liquid with a silver spoon shows the downside of high thermal conductivity. The upside is good value as radiators for cooling internal-combustion engines and for steam heating in homes.

The high thermal conductivity of metals would be useful to living things in a number of situations. When large and moderate-size animals do heavy physical activity, they generate waste heat that they have to transfer to their surroundings. High conductivity would help get the heat from muscles to skin, but human conductivity is low, about that of water. So we resort to convection and evaporation instead of conduction. In convection, heat is moved from one place to another by physically moving some heated material rather than just shifting the heat from one bit of material to the next. The usual materials moved are fluids—hot blood and hot breath—and moving them requires pumps and energy. In evaporation, heat is moved by making water evaporate and then moving or discarding the water vapor. Evaporation takes energy, which the vapor contains; when the vapor drifts away, the body is left cooler. We sweat; dogs pant; we both lose water thereby. But water must be obtained and carried around, and both activities can prove troublesome.

Leaves do likewise. When the wind drops and they're still exposed to the sun, they get considerably hotter than the air around them; they can warm as much as 20° C. Their centers, less exposed to what air movement remains, face the problem at its worst. Hot centers would be less hot if leaves were made of nicely conductive metal,[20] but no such luck. Instead plants do things that must limit other activities or interfere with other desirable features of their design. They make elaborately lobed leaves to get a lot of edge and perhaps absorb less light. They use a lot of water for evaporative cooling, which requires a decent water source. They make thickened leaves, which takes material, so they heat more slowly and can endure longer lulls in the wind. And when the sun is strong, the wind is low, and water is scarce, many leaves droop down to a less sunstruck and more airflow-exposed vertical position, which compromises photosynthetic activity.[21]

We also take advantage of the high electrical conductivity of metals. Every electric wire is metallic; the only nonmetallic components of most electronic devices are active ones, such as transistors, or components, such as resistors, whose function depends on low conductivity. What about nerves? We may think of them as wires since nervous systems and electronic devices do similar tasks. But the analogy between nerve and wire misleads. A nerve impulse moves along the axon of a nerve cell by a different physical scheme from that of a pulse of electricity traveling along a copper wire. Which is better? I'd go with the wire. For instance, nerve conduction is glacially slow compared with that along wires. One hundred twenty meters per second, a mere 270 miles per hour, is about as fast as any nerve conducts impulses; by contrast, wires transmit electrical pulses some *five million* times faster.[22] A thousand impulses each second is about the limit for a nerve; wires carry millions of pulses per second. All the fancy stuff we do with our brains requires the most massive parallel processing—using many nerves running alongside each other and great numbers of circuits working simultaneously.

When rigidity is the aim, we make great use of metals and a little of composites; nature makes elaborate use of composites but none at all of metals. I've suggested several reasons why nature doesn't use metals, but none was fully persuasive. Two further explanations, alluded to earlier, share the same uncertainty—admittedly an unsatisfying state of affairs.

Perhaps nature doesn't use metals simply because she has something every bit as good. If we ask whether natural composites are in the same class with metals, we have to say yes. If, for instance, stiffness relative to density is the criterion, wood and steel are about equal. If work of extension is what matters, and weight rather than volume is the reference, then yew wood (as used in English longbows), collagen (as used in Roman catapults or ballistae), bone (much used by Inuit), and horn (used in Chinese composite bows) all surpass spring steel. One can pick alternative comparisons in which our manufactured materials come out ahead. Overall, the rigid materials of the two technologies, though dramatically different, do about equally well. Our present aggressive development of composites is driven not primarily by the excellence of nature's but by the good performance of even the crude ones we already use. At least in small quantities, we've now made composites that, in the ways that matter to us, do better than anything in nature.

How curious a process is "design" in nature! For better or worse, composites are what a mindless, blundering, information-starved, and minimally coordinated system might be expected to make. Specifically, their properties are highly sensitive to tinkering with the amounts and arrangements of their constituents on a microscopic level. In short, they're what one ought to expect from microscopic improvisation, nature's way, as opposed to macroscopic deliberation, our human mode. And that brings us to the last possibility.

Maybe in noticing that nature uses composites rather than metals, we're simply looking at a result of the conservative, noninnovative, stick-with-the-tried-and-true process of evolution by natural selection. Once organisms of nonmetallic construction became established, what chance had initially crude metallic forms? Future benefit is something evolution knows nothing about; she has precious little venture capital. But even if evolutionary inertia is a reasonable explanation, defending it is difficult. How could one ever come within shouting distance of either proof or disproof? Perhaps our best bet is to regard evolutionary inertia as a hypothesis of last resort—distasteful but unavoidable, its mention deferred as long as possible.

Chapter 7

PULLING VERSUS PUSHING

Ropes resist having their ends pulled apart; bricks resist having their ends pushed together. To formalize the distinction, let's call structural elements that resist pulls *ties* and those that resist pushes *struts*. For ties we use cables, ropes, belts, and some glues as well as some metal, wooden, and plastic rods and bars. For struts we use walls, columns, rods, and bars of brick, stone, concrete, wood, and plastic. Nature's ties include muscles, tendons, ligaments, strands of silk, and the stems of fruits. Most bones, hard coral, and tree trunks and a lot of insect cuticle serve as struts. In this simple—perhaps oversimplified—distinction between ties and struts lie some interrelated contrasts between human and natural technologies. The contrasts aren't as tidy as those between metals and composites, but they're equally pervasive and no less significant. What human technology does about pulls and pushes is ancient in origin and cross-culturally consistent—so ordinary and familiar that only viewing an alternative reveals its peculiarity.

Two chapters back, the reader's toughness and resiliency were stressed and strained by a great batch of information about mechanical properties. In talking about all the properties, though, we considered only what happened when we pulled on things—tensile tests on samples of different materials. But when push comes to shove, more than tension deserves our attention. One can also press on a material, loading it in compression, and one can shear a material, distorting, say, a rectangular block into a so-called parallelepiped, as in Figure 7.1. So three stresses matter: tensile, compressive, and shearing.

In reality one doesn't test a material per se but only a particular piece with a particular shape. In a tensile test, shape matters very little. After you pull on a round rod or an I-beam or a tendon, you divide the force of the pull by the cross-sectional area of the piece to get stress. For tensile stresses, the shape of that cross section doesn't matter; a rope, a rod, and an I-beam give the same results. By contrast, in a compressive test, shape matters a lot. Push on something short and fat, and it squashes. Push on something long and thin, and it bows out to one side before it breaks. Push on a tube with a very thin wall—such as a drink can—and the walls

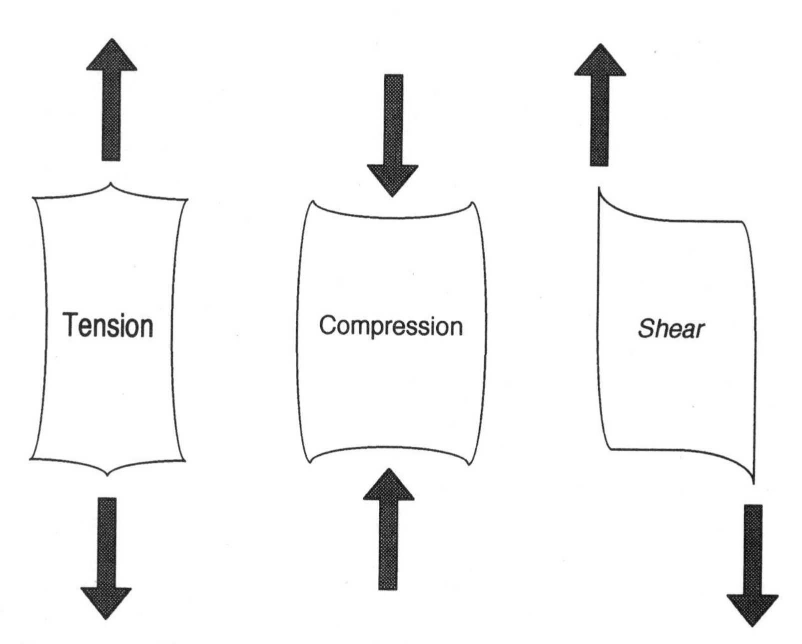

FIGURE 7.1. *Three ways to stress a sample of material.*

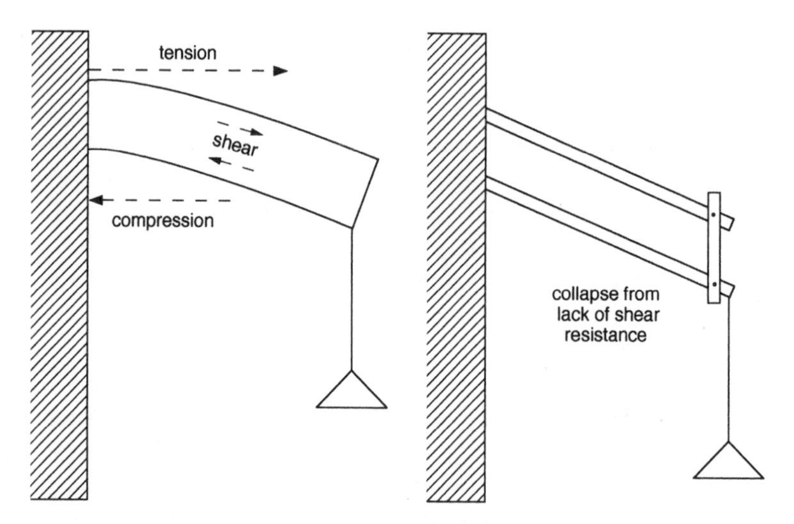

FIGURE 7.2. *The stresses on a protruding beam bent downward by a weight at its end and what happens if the beam lacks resistance to shear.*

crumple suddenly from a single point. Round, square, and flattened rods of the same cross-sectional areas collapse under different loads. Pulling is simple; pushing is anything but. When we worried only about differences between materials, we could (usually) get away with tensile testing alone. But materials serve as components of structures, and to understand structures, we have to worry about shape.

Besides tensile (pulling), compressive (pushing), and shearing loads, two more complicated loads matter for structures: bending and twisting. When a structure as a whole is either bent or twisted, parts of it experience each of the three simple loads. Think about what must happen when a weight on its end bends a long, protruding beam, as in Figure 7.2. The weight is eager to move toward the center of the earth, while the right-minded beam opposes any such tendency. In the process the top of the beam stretches, at least slightly. Stretching—that's what ties resist, so we can view the top as a tie. A little less obviously the bottom of the beam compresses, again at least slightly. Compression—that means the bottom works as a strut. But if we actually substitute tie and strut, disaster ensues: Instead of the tie taking tension and the strut taking compression, the whole thing collapses downward and inward. The beam, initially rectangular in side view, has become a parallelogram. Subtly but crucially the middle of the beam has been resisting shear. Thus, when a beam bends,

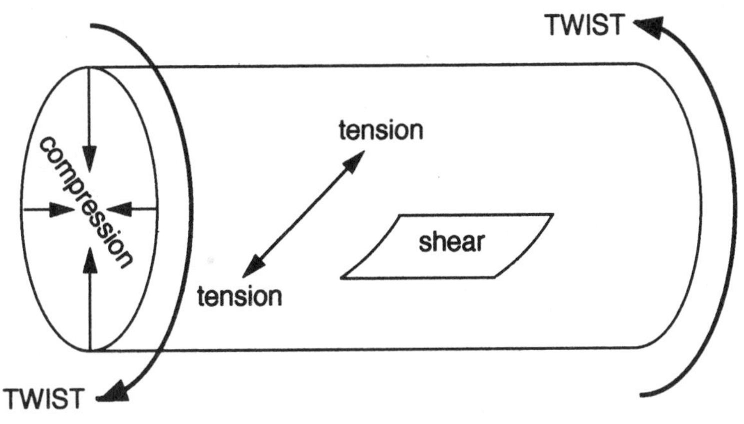

FIGURE 7.3. *The stresses—tensile, compressive, and shearing—on a cylinder as you twist one end one way and the other end the other.*

all three stresses—tensile, compressive, and shearing—load its material, and the amount of each stress varies from place to place in the beam.

Something analogous happens when a structure, such as the cylinder in Figure 7.3, gets twisted. Tension loads the outside, and compression loads the inside. You can see both by twisting a wet cloth: Tension on the outside is obvious, and compression of the middle is what squeezes out the water. In addition, shear loads the whole thing (except the exact center line). That's clear enough if (as in the figure) you draw a square on a long balloon and then twist it; the square becomes a parallelogram. Yet again, a single load on a structure induces three stresses in the material of which it's made.

We humans take great advantage of this combination of stresses in twisted cylinders. We lay alongside one another short natural fibers, such as linen, cotton, or wool, and twist (spin) them into long threads or ropes. Even though their fibers aren't joined end to end, these ropes are remarkably strong. The fibers are kept from pulling apart by their resistance to shearing (sliding) across one another when they're compressed. Pulling on a spun thread or a twisted rope has the same effect as shearing a structure; it presses together the fibers in the middle. If the fibers aren't too slick, they'll break before they slide apart. This tension-induced compression is especially obvious with fluffy yarn. Spinning, either for rope-making or prior to weaving, dominated many ancient cultures, consuming a large fraction of the working hours of women, who in almost every case performed the task.[1]

Nature never does this wonderful trick. She makes her long threads and ropes as continuous strands—spider silk, for instance. Or, as in vines, she uses a system of joinery based on something other than helical twisting and frictional resistance to shear between fibers. Not that multiply stranded helices are at all rare in nature's load-bearing structures—the double helix of microtubules (Figure 2.2) and the triple helix of collagen are far from obscure—but these twists don't gain tensile strength by resisting shear. Spinning is a human activity, despite our misuse of the word for thread extrusion by silkworms and spiders.

Our protruding beam bent simply and obviously when weighted at the end. Less obviously almost the same thing can happen when a long structure gets its ends pushed together. The classic case is an erect column, what the old Egyptians and Greeks liked to put around temples to keep their roofs aloft. The force of gravity pulls down on the column; since the column doesn't accelerate downward, forces must be balanced, and the ground must be pushing upward. Thus, overall, the column faces a compressive stress. But as soon as it starts to bend, a compressively loaded column develops tension in the side that's outermost, as in Figure 7.4. The side inside the bend experiences additional compression, and shear develops within the column just as it does in a beam. So except for very short, fat columns, where the significant hazard is simple crushing, columns, just like beams, get loaded in tension, compression, and shear. The main difference is that the column can buckle in any direction, whereas most beams know which side is up and which is down and therefore where tension and compression will occur. A proper I-beam should be higher than wide, while columns can't deviate far from circularity.

TENSION VERSUS COMPRESSION

Say you're designing a structure and you're free to use struts and ties in any combination. What factors bear on your choices? Primary among them are the properties of the materials at your disposal. Of these properties, what matters most is relative performance under tensile and compressive loading.

Stone, bricks, masonry, and other ceramic materials are terrific at resisting compression. If you put something on top of them, they stay put and stay intact. An ordinary brick can support several hundred thousand pounds. Conversely, these ceramics resist tension poorly, with at least ten times less strength.[2] In part that goes along with their stiffness; again, stiff

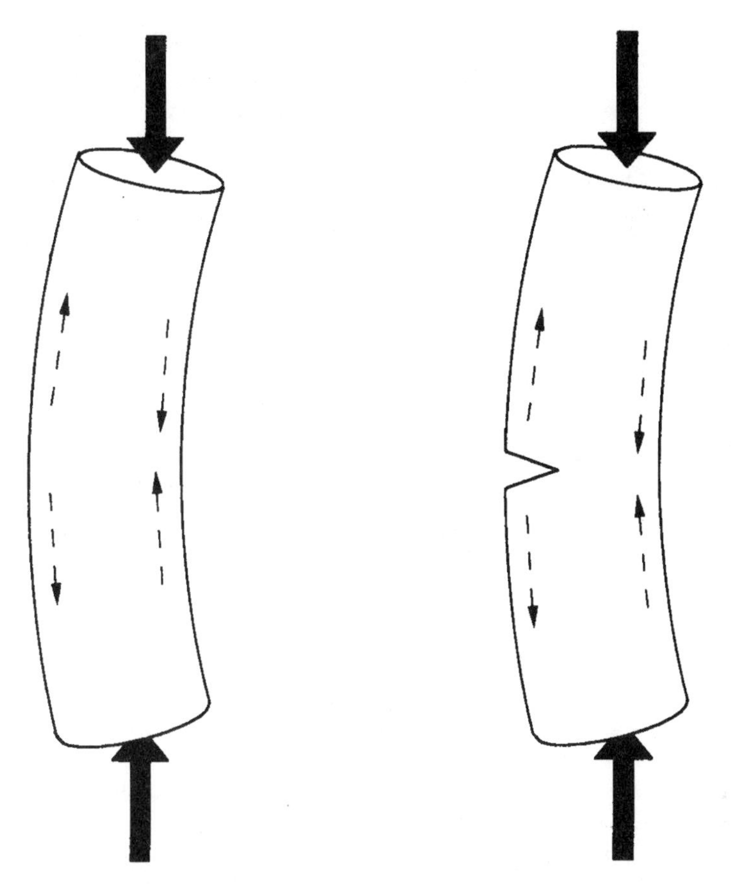

FIGURE 7.4. *Counterintuitively, local tension can crack a column that overall is loaded in compression.*

materials tend to be brittle—which is to say that they don't resist crack propagation. Cracks present far more of a problem in tensile loading, which tends to spread them apart, than in compressive loading, which has no such reprehensible habit.[3] In addition, the traditional cements and mortars used to join bricks and stones are only weakly adhesive as well as being brittle themselves. Since in our uncertain world, purely compressive loading can't be guaranteed, a little tension insurance pays, presumably why (as one learns in Exodus 5:7–5) sun-dried bricks do best with a little straw. It's also why plaster for cornices was made with horsehair.

However, if you can be assured that all loads will be compressive, and if weight economy isn't uppermost (as it is in structures that move), then ceramics are excellent materials. Their high stiffness makes it safe to ignore their weight in all but very large structures; bricks at the bottom of a wall aren't squashed much by the ones above. Gothic cathedrals, truly spectacular structures, must be the greatest things humans have ever built that tolerate only compressive loading.[4] Vaulted roofs press outward on the walls, and external buttresses press inward, with the two in very close balance. The less massive the walls, the better must be that balance, or the walls—just piles of stones—will topple inward or outward. Building a cathedral entailed a lot of mid-course adjustments of vaulting and buttressing to keep everything in proper compression.[5] Failures were common during construction, in part why building one sometimes took a century. St. Peter's in Rome took no less than 181 years to construct, but St. Pierre's of Beauvais, France, retains the record. Two spectacular collapses punctuated 350 years of intermittent building, and the cathedral remains unfinished to this day. But once correctly done, a cathedral proves durable and maintenance-free. The same can be said for stone arch bridges, aqueducts, and, of course, pyramids.

Ropes, cables, chains, and vines resist tension, but compression causes immediate flabby collapse. Hammocks are loaded in tension, and suspension bridges are tensile structures except for their towers and roadways. Tensile structures form subsystems that support elevators and the masts of cranes. Tension in the rubber walls of tires—as much as air pressure—keeps automobile bodies off the road, and tension in their outer fabric maintains the shape of blimps. But compared with compression-resisting structures, the items are fewer and smaller, and more of them are contemporary.

One of the best things about metals is that they take pulls and pushes about equally well. That suits them for use in many of our best-loved devices, which must withstand both stresses, sometimes in rapid alternation. Consider the framework that resists a car's weight and motion as the car speeds along a bumpy road. Or think of the hull of a ship in a wavy ocean. If the ship spans two wave peaks, its ends get pushed upward and its middle sags. If the ship straddles a single wave, its middle gets pushed upward and its ends sag. The fine ability of metals to deal with both compression and tension means that the casual observer often can't guess from structural details what kind of load a particular member was designed to resist.

Biological materials differ only slightly. Like cables, our muscles, tendons, and ligaments resist only tension. Muscles pull and are pulled on, and you suffer pulled muscles, not pushed ones. Like steel beams, our bones and cartilage resist both compression and tension. The differences, though, are significant. For one thing, purely compression-resisting elements, such as bricks, almost never occur. Even teeth must resist tension; the way we load our teeth, imagine how much more prone to breakage they'd be if weaker in tension. For another thing, biological materials don't behave as similarly in tension and compression as do metals. Wood, thoroughly fibrous stuff, is about twice as good at taking tension as compression: Bend a live twig or a thin dowel, and it buckles on the inside of the curve before cracking on the outside.[6] Bone, by contrast, is slightly stronger in compression than tension. Not surprisingly, mollusk shells are stronger in compression. What's interesting about shells and other biological ceramics is that most are only about three times weaker in tension, far less than the tenfold of conventional ceramics of human manufacture. That's of course the advantage of being composites or foams.

TIE VERSUS STRUT

Not surprisingly, both human and natural technologies use both struts and ties. But functional requirements don't fix the exact mix of the two. The designer of something complex, like a large bridge, has considerable flexibility in deciding how much to rely on struts and how much on ties. I think each technology has a perceptible bias in its designs. Historically, humans have done better at handling compressive loading; in struts we trust, not in the ties that bind. Nature prefers to pull and is fonder of ties.

Such a bias will have a dramatic impact on the structures you'll make. If you stand with struts, you might build an arch bridge (Figure 7.5), in which almost everything is compressively loaded, and the ends push (unsuccessfully) the banks of the river apart. That's why arches figure so prominently in stone-based classical architecture. If you're hung up on ties, you'll be more inclined toward a suspension bridge, where the ends try to narrow the river. Besides cost, properties, and availability of materials, several other factors might bear on the choice between struts and ties.[7]

- As we shift attention from materials to structures, size takes on renewed importance. But it does so in counterintuitive ways—the general point of Chapter 3. Fishing line labeled "Ten pound test" can take a ten-pound pull no matter how long a piece you unreel.

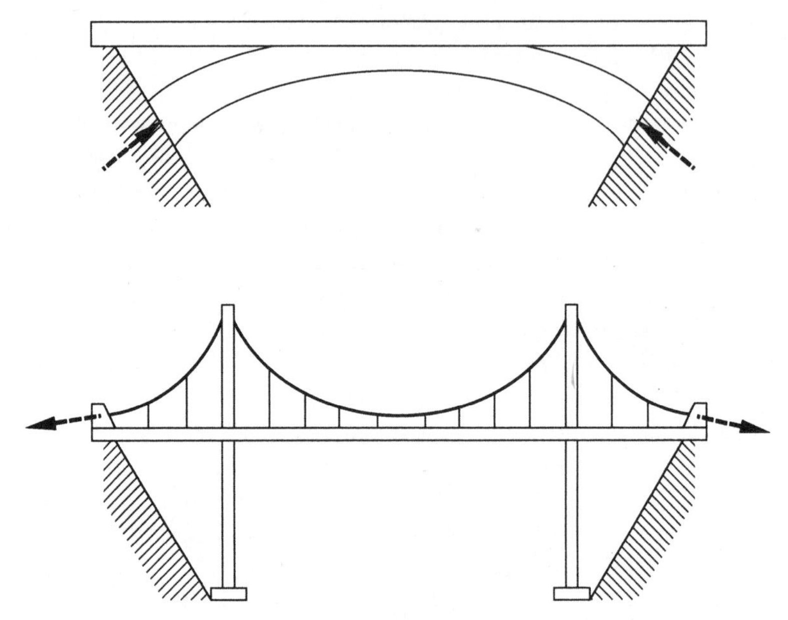

FIGURE 7.5. *The arch of an arch bridge is loaded in compression, while the main cables and suspenders of a suspension bridge are loaded in tension.*

To support a given load, a long tie need be no thicker than a short one, at least until the weight of the tie itself—self-loading—becomes significant. (Digression: In theory an earth satellite can dangle a rope down to the earth's surface yet stay in orbit. If the satellite's center of mass is high enough so the orbit is geosynchronous, the rope won't even shift around. Terrific. For something to get into orbit thereafter, it simply climbs up the rope, pushing upward just enough so the center of mass of the whole system doesn't drop. So what's wrong? Physics isn't offended, but material science fails. No known material has sufficient tensile strength for a rope that long not to break of its own weight).[8]

For compressive loading, though, making something longer makes it weaker. A short stick is harder than a long one to break by pushing the ends together. Try it with any long, thin, stiff thing such as a dowel, a soda straw, or a strand of spaghetti. The guilty party is sideways buckling, to which the long one is more vulnerable. Proportionately increasing the thickness of a strut as its length increases doesn't solve the problem. As noted in Chapter 3, doubling the size of a strut-supported system without changing its shape doubles the stress on the struts. To keep the stress constant, their thickness must increase disproportionately.

Nature appreciates the problem of scaling struts. Among trees an increase in height comes with a disproportionately great increase in trunk diameter.[9] The bones of large terrestrial mammals make up a larger fraction of their body masses than do the bones of small mammals. The skeleton of a 10-pound cat makes up about 7 percent of its weight; a 130-pound person is 8.5 percent skeleton; a 1,300-pound horse is 10 percent; the skeleton of a 15,000-pound elephant is all of 13 percent of its weight. In fact, even these increases in skeleton don't keep all mammals equivalently sturdy, and the larger ones are more fragile than they appear.[10] As D'Arcy Thompson poetically puts it, "Elephant and hippopotamus have grown clumsy as well as big, and the elk is of necessity less graceful than the gazelle."

Thus when scaled up in size, the strut gets into trouble that the tie knows nothing about. Gibbons provide an instructive case, since they both walk on two feet and swing branch to branch (brachiate) from two arms. Their arms are loaded almost entirely in tension as they swing.[11] The weight to be managed may be the same, but the bones of their arms are longer and thinner than the bones of their legs. Ties can be long, but struts should be kept as short as possible.

- The stability of struts and ties differs in a curious but entirely reasonable way. Once a strut starts buckling, it does so more and more, and it quickly collapses. The failures of the cathedral at Beauvais must have been scary. Problem: As soon as a strut begins to bend, it loses strength, and the more it bends, the weaker it gets. In part that instability comes from the increasing leverage of the bent strut as it moves outward. In part it comes from the way crosswise cracks extend ever faster as buckling proceeds. As an engineer, Michael French, pointed out, once Samson pushed enough to bow the columns of the temple by some critical amount, he could ease up and let the roof do the rest.

  In this respect a tie is intrinsically stable. Stretch a tie between two posts, and hang a weight in the middle; the tie will deflect downward. More weight means more deflection, as either the posts bend or the tie stretches. But at no point does the process become self-sustaining; clotheslines pose few hazards. For stability as well as for scaling up to larger size, ties do better than struts. Once again, nature responds rationally. Larger mammals may be bonier, but they're no more muscular. Forty percent muscle does for us all.

- But building with ties is trickier than building with struts. Struts go together easily; they stack, with gravity providing the glue. Ties are hard to connect to each other or to struts, as we've noted several times already. The technologies of fasteners and adhesives have challenged every culture that has built things by any scheme other than piling rocks or blocks one upon another. In a dome, for instance, tension loads the lower periphery. Great tensile chains of iron, hidden within the masonry, encircle Brunelleschi's great dome of the Cathedral of Santa Maria del Fiore in Florence, Italy.[12] By making rings in which the chain connects only to itself, Brunelleschi circumvented the problem of tensile fastening. Another problem for ties is gravitational loading. Most of our structures extend upward from the earth's surface, something that can be done with struts alone but not solely with ties. Wooden ships provide an especial challenge to effective joinery, and by long tradition shipwrights enjoy more prestige than carpenters; according to an old saying, even a poor shipwright makes a good carpenter.

Not that ties aren't used in ordinary construction. Conventional frame houses with peaked roofs have horizontal members (stringers) that run at right angles to the ridge of the roof; they often form the floor of an attic. Two roof supports (rafters) and a stringer form a triangle (Figure 7.6). Gravitationally loaded, the roof will tend to force the walls apart; the stringer is the tie that binds the walls. Alternatives are available, so a peaked roof needn't be tied by stringers. One, again, is external buttressing, providing a counterforce pushing inward at the tops of the walls, as in Gothic cathedrals and some A-frame houses. Another, which gives my house the pleasant openness of so-called cathedral ceilings, is a large, stiff beam just under the ridge that's held up by inconspicuous vertical columns. That lengthwise, horizontal ridge piece then keeps the rafters from spreading.

For some reason (or, more likely, reasons), tensile joinery doesn't faze nature. Once in a while a tendon pulls loose from a bone, but

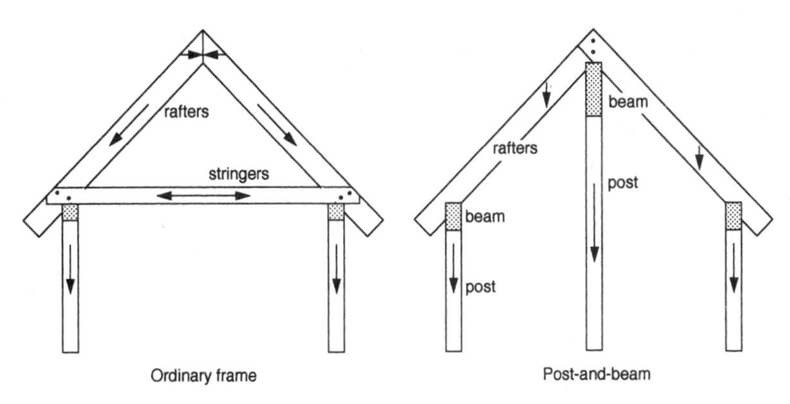

FIGURE 7.6. *A conventionally framed house in which stringers act as tension-resisting elements and a post and beam house that manages without stringers.*

the commonest examples of tensile disconnection—leaves, seeds, and fruits from their parent plants—can't be dismissed as cases of failure.

- Getting the stiffness that we humans like presents difficulties for a structure that makes elaborate use of ties. A chain or steel cable may be stiff in our original sense of material stiffness—that is, in resisting stretch when pulled lengthwise. But it isn't stiff in a structural sense; it bends with little provocation. That's the downside of the length and slenderness tolerable in ties. Since nature tolerates low structural stiffness, she's better poised to take advantages of ties.

Nowhere have we used tensile ties on a larger scale than in suspension bridges. Aside from primitive suspension bridges built of vines, the scheme goes back only as far as the availability of reasonably cheap metal—first wrought iron and then steel—that could resist tension. (Most suspension bridges use steel wire for both the main cables, in wrapped bundles, and the road hangers. But a lovely wrought-iron one spans the gorge of the Avon at Bristol, England. It has no wires. Bone-shaped links (eyebars) make up the main cables, with transverse pins connecting them, and the hangers are very long rods that descend from those connecting pins. The great engineer Isambard Brunel designed it in 1831; when it was built in 1864, steel had become cheaper, and the bridge was anachronistic.) Suspension bridges are flexible things. More than a century ago railroads found them unsuitable. A steam locomotive is a heavy, moving load—a bad combination for the continued integrity of both bridge and railbound train. On at least one occasion the flexibility of a suspension bridge proved its undoing. The best-known American bridge disaster was the plunge of the Tacoma Narrows suspension bridge into Puget Sound in 1940. The immediate cause wasn't just wind but the violent oscillations that resulted from the interaction of wind from a particular direction with the deck of this narrow, graceful, but flexible bridge.[13] Collapse took an hour or so as spectators were awed, and a movie—shown to almost every engineering student since—recorded the event.

TENSION VERSUS COMPRESSION, AGAIN

Recall that a load stretches the top of a protruding beam, compresses the bottom of the beam, and shears the middle. How might we design an efficient beam? One popular solution is the I-beam, so called because its cross section looks like the letter *I*. Its excellence depends in part on the similarity in performance of metals under tension and compression; the identical upper and lower flanges (the serifs of the *I*) represent lots of material well above and below the center line so they can better oppose the beam's bending. Vertical "webbing," which takes the shear, separates the flanges.

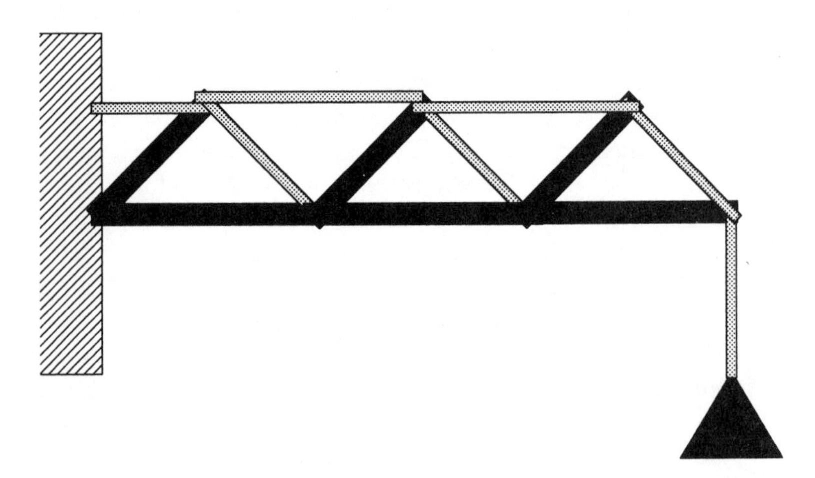

FIGURE 7.7. *A protruding truss. The compressively loaded elements are the darker and fatter ones.*

The word "webbing" hints at how one can do a bit better. We can and do replace it with a latticework of diagonal elements—an actual web. To withstand shear, the elements of the lattice have to run diagonally, since a set of verticals would fold very nicely as the protruding beam sagged into a parallelogram; recall Figure 7.2. When the beam is equipped with this latticework, as in Figure 7.7, we call it a truss, but little has really changed. Consider that lattice more closely: Half its diagonals are loaded in tension and the other half in compression. As with the flanges, if we're using metal, the diagonals can be identical. But weight might be saved if we use cables for the tensile diagonals. Of course we'll lose a little overall stiffness, and we're limited to downward loads lest the cables come slack. In fact, we can push the logic a step farther. If the load is consistently downward, then the top flange always feels tension and can also be replaced by a cable—again if we accept the loss of stiffness. Struts and ties thus emerge as identifiable items, solid bars and cables.

Nature uses such specialized beams to support the heads of many mammals, as D'Arcy Thompson pointed out long ago. As in Figure 7.8, the lower, compression-resisting flange is provided by the bony centers of the vertebrae of the neck and thorax and the cartilaginous intervertebral disks between them. The tension-resisting upper flange is provided by muscle and by a ligament, the one mentioned in Chapter 2 as having a lot of stretchy elastin. The compression-loaded diagonals are bony exten-

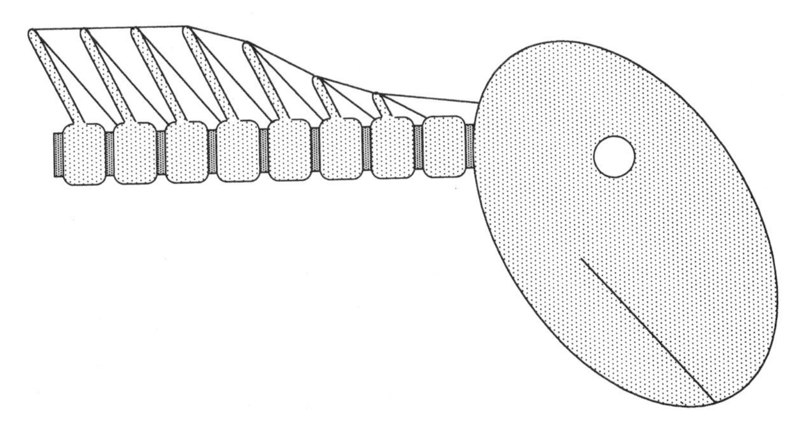

FIGURE 7.8. *Another protruding truss, this one (thoroughly idealized) the vertebrae, intervertebral disks, muscles, tendons, and ligaments supporting the head of a large mammal. Tension-resisting elements are shown as lines.*

sions of the vertebrae, and the tension-loaded diagonals, running oppositely, are made up of more muscle, tendon, and ligament.[14] The whole thing may be flexible, but being stiff-necked isn't the object. The trusses used in human technology, by contrast, rarely make any distinction between their struts and their ties. In a sense, we construct only struts and use half of them as ties, accepting a slight loss in weight economy for simpler construction and greater stiffness.

A horse carries most of its weight on its forelegs. These don't attach to the backbone and rib cage any more directly than the attachment between the wheels and body of a car. Even the best of roads have bumps, and so teeth don't get rattled or torsos tossed, proper vehicles need suspensions—springs, at least, and perhaps dampers (as mentioned in connection with spider silk) to keep the springs from repeated rebounding. When road and body move closer, the springs are compressed; when they move apart, the springs are stretched. But both horse and car have weight, so the springs bear loads even when the road behaves itself—what's called preloading. A suspension can be designed so either preloading and active loading can be either tensile or compressive. A weight (such as a rider) or

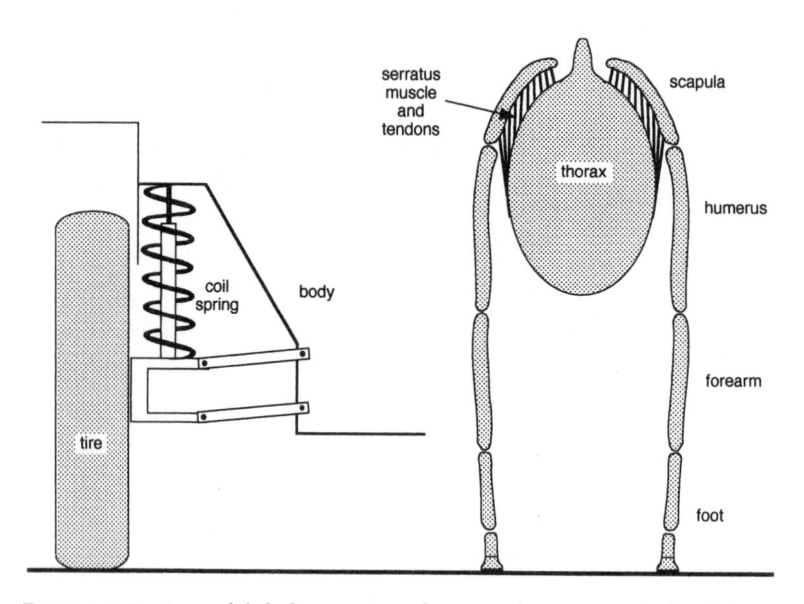

FIGURE 7.9. *Automobile bodies are commonly supported on compressed coil springs near each wheel, while horses and other quadrupedal mammals have tensile suspensions; their torsos effectively hang from their pectoral girdles and forelegs.*

an antiweight (such as a pothole) can be made to work either way, depending on how the suspension's components are arranged.

Horse and car happen to use opposite loading and preloading. As Figure 7.9 shows, a horse's body is hung from its pectoral girdle. Sturdy muscles, tendons, and ligaments descend from the scapulae (shoulder blades, in humans) to the backbone and rib cage. An increase in weight on the horse raises the tension on these components above their tensile preloading; compression of the whole system stretches its springs! Most cars have springs coiled underneath, arranged so an increase in the car's weight compresses them. Metal springs work well in either tension or compression, and cars have been built that (like horses) are really suspended from their suspensions. Such cars have a mildly disconcerting way of leaning inward rather than outward on turns, since their masses are centered below the level of attachment of their suspensions.[15] Putting compressed springs underneath is simply more convenient, given the other imperatives and constraints on the designer. We make wide use not only of tensile and compressive springs but of bending, torsional, and even occasional shearing ones as well—springs that resist every load that's been mentioned.

Nature's springs are much less diverse. Only a few, such as the hinge pads of scallops and other bivalve mollusks, absorb energy by compressing rather than stretching. Even less often, as in the wishbones of birds during flight, do bones or exoskeletal components act as bending springs like the leaf springs on some cars.[16] The vast preponderance of natural springs works by stretching.

Deliberate preloading, though, is more than a minor convenience for vehicle suspensions. Both technologies make wide use of the device to ensure that materials will be loaded in the ways they withstand best. We put preloading to particularly good use in what we appropriately call prestressed concrete. We cast a large piece with steel rods running through it and then tighten the rods, or else we stretch and hold rods in tension while concrete hardens around them. Either way, the steel is preloaded in tension, and the concrete in compression. Stretching the concrete then just relieves preexisting compression and doesn't tickle its dangerous weakness in tension. We sometimes get even fancier. Using curved rods, we can prestress a long beam with a bend that will then be offset by its weight and live load. Prestressing doesn't simply provide reinforcing and crack stopping, as when concrete is cast around rods of steel; it's a finer thing altogether.

Just as a material that behaves badly in tension can be precompressed, so a material that does worse in compression can be pretensioned. Wood is such a material, and trees do exactly that. The inner portion of a tree trunk, which bending affects little, is normally compressed by more than just the weight of wood above it. Additional compression comes from tensile squeezing by the wood surrounding it. When the trunk bends, the outside of the bend is stretched, a load wood withstands well. The inside of the bend is compressed, a load wood takes poorly. But that compression mainly relieves preexisting tension, thus making better use of wood as a structural material.

THE FORMS OF TIES AND STRUTS

Cables and bars are unmistakably different, even when both are made of the same material. The best shape for a tie will rarely coincide with the best shape for a strut—surely no surprise at this point. At the least, compression elements should be relatively short, whereas the length of tensile elements doesn't much matter. Both technologies recognize the principle, but they apply and compromise it in different ways.

Suspension bridges, again, give a good view of our best efforts: They're big, their builders care about minimizing cost and self-loading, they use both tensile and compressive components, and most of their components are visible at a glance. The tensile elements—the main cables and the road-suspending descenders—are long and thin. The compressive structures, the towers, have a sturdier look. If made of masonry, as in nineteenth-century designs, the towers are imposing piles, tall but far from thin. Even if made of steel, they're much fatter than even the main cables. Steel towers may be made of latticework grids or their equivalent in plated tubes. But they work much like solid columns in that individual struts (or their equivalent in the walls of the plates) are fairly short. Buckling must be prevented, something that just isn't a problem with tensile loading.

Other structures also follow the general rule of keeping individual compressive elements short. The two-by-fours that form the vertical studs in the walls of a house may be eight or ten feet long, and individually they're prone to side-to-side buckling. But we nail siding, Sheetrock, and various kinds of paneling to the studs and thus make them into an interactively self-bracing array. Even flimsy siding can keep the studs from buckling; preventing the initial lateral movement is easy even if the process gets more forceful further on.

In nature the picture is less clear, even with a great diversity of tensile elements to look at. Purely tensile structures may be quite thin relative to their lengths; look at the threads of a spider web or at a hundred-foot-long seaweed. But compressive structures aren't always fat—consider the trunks of tropical trees and the leg bones of gazelles and storks—although compressive structures are thicker relative to their lengths than are tensile ones. Tall, woody terrestrial plants have evolved on several occasions. No lineage has made extensive use of external tensile braces—guy ropes—despite the fine strength of wood loaded in tension.[17] Still more peculiar, to me at least, is how arthropods so often use very skinny compression elements. At least some members of each major group of arthropods have what seem to be counterproductively thin appendages: Long, skinny legs characterize crane flies (insects), harvestmen (arachnids), spider crabs (crustaceans), some centipedes (myriapods), and sea spiders (pycnogonids). All are jointed and have muscles inside, so they must be compressively loaded. Figure 7.10 shows a few of these mechanically bizarre creatures. Our book of biomechanics still lacks whole chapters!

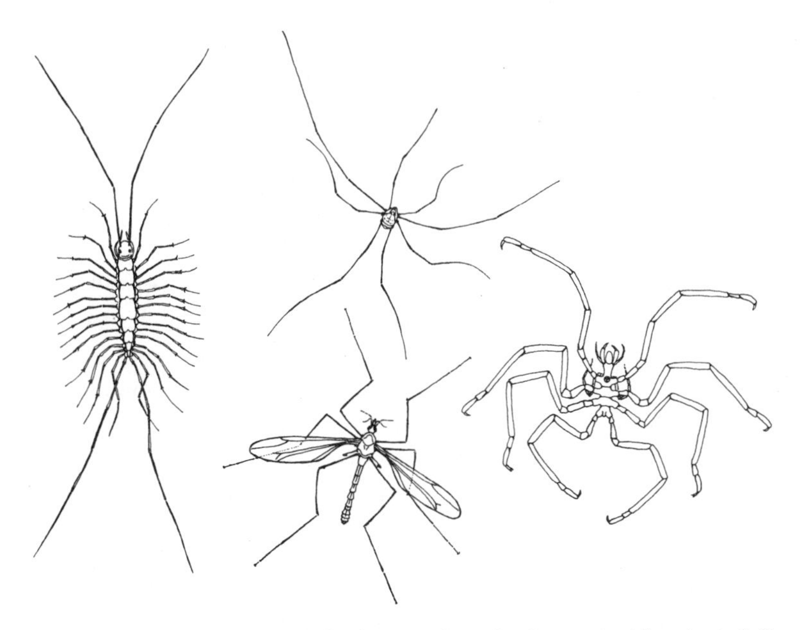

FIGURE 7.10. *Some arthropods that have very long, thin legs: scutigerid centipede (left), harvestman or daddy longlegs (center top), crane fly (center bottom), and sea spider or pycnogonid (right, considerably enlarged).*

One group of animals deserves special mention. Most sponges, as described in Chapter 4, are supported by tiny hard spicules of calcium salts or silica (essentially glass) laced together by flexible pads of a protein much like that of our tendons. The entire group of animals lacks substantial muscles, so they don't need long, stiff struts for their ties to pull against. With that major constraint relaxed, the general rule about short compressive elements becomes decisive. Their hard parts, the compression-resisting spicules, are less than a millimeter long.[18] Sponges aren't all small or weak—individuals three feet high survive hurricanes and typhoons—but sponges built this way are fairly flexible.

A small jump from sponges leads to something we haven't yet encountered: an attractive-looking scheme that's little used by either technology. Where struts and ties are distinct elements, we expect the struts to form a continuous, interconnected network with ties inserted here and there. How else to resist anything but purely tensile loads? In fact, the continuously interconnected network of struts isn't crucial. Structures that resist bending, twisting, and compression can be made in which the continuous array—one hesitates to call it a framework—consists only of ties. The struts need make no contact with one another. Whether such a structural scheme was invented by a specific person isn't clear, but R. Buckminster Fuller took out the key patents and named the scheme tensegrity.[19] In rich language for a patent, he spoke of "islands of compression in a sea of tension."

A picture such as Figure 7.11 serves better than words for explaining how a mast or tower, for instance, can stand up without interconnected struts. The key is putting ties where pushing, bending, or twisting the structure as a whole increases their tensile loading. Ties, after all, resist only tensile loads and collapse helplessly under any other kind. Masts and towers aren't the only possible tensegrities; domes, for instance, aren't hard to design. One can, I'm sure, make a fabric tent with noninterconnecting pockets for stiff battens; perhaps it has already been done. The main disadvantages of structures based on tensegrity are those of tensile construction in general: the lack of stiffness and the dependence on fasteners. So tensegrities remain mainly an art form.

Sponges may make some use of the concept; at least their spicules most often don't interconnect directly. But the arrays of spicules bear only limited resemblance to Fuller's designs, probably because even tensegrities are stiffer than optimal for sponges—animals that accept rather than minimize flexibility. Spicular skeletons—that is, tension-resisting tissue with small pieces

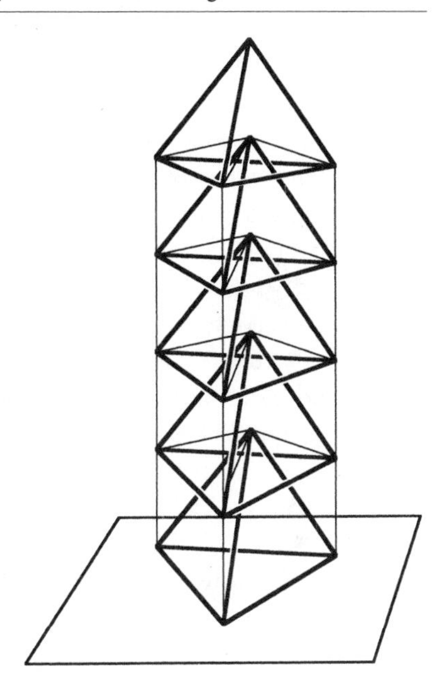

FIGURE 7.11. *A tensegrity tower or mast. The strutted tetrahedrons don't touch except through sets of ties, yet the thing will stand erect. Imagine pulling on a rope and raising the flagpole as well as the flag; a tensegrity tower permits such intuitive preposterousness.*

of hard material embedded in them—aren't limited to sponges. They occur in some soft corals, sea cucumbers, stalked barnacles, and elsewhere. These look even less Fulleresque and more like steps in a continuum between deviant tensegrity structures and composite materials.[20]

My best guess about why tensegrity hasn't been elaborated by nature is as much of an evasion as those guesses I gave for the absence of metals. Where stiffness is important, tensegrities are too flexible. Where flexibility is tolerable, something better is available. We turn, then, to the ultimate expression of tensile support, hydrostatic systems.

FLUID STRUTS AND HELICAL TIES

Something close to home for half of us. Consider the penis of yourself or your nearest and dearest; in an immediately evolutionary sense fitness depends on stiffness. Discontinuous struts were odd enough, but here's a stiff structure with no solid strut at all! Does a strut really have to be solid to resist compressive stress? In fact, it doesn't. Consider how the three ordinary states of matter behave. Solids resist tension, compression, and shear; liquids resist tension and compression; gases resist only compression.[21] Thus all three states of matter fight back—they resist compression—when pushed into a corner. So a fluid, whether liquid or gas, might serve as a strut. That sounds peculiar, but maybe we just haven't taken a sufficiently broad view of possible struts. Fluid struts would certainly be economic; water and air are boundless resources. How, then, can we make them?

Making a fluid strut is simple. A tension-resisting sheath need only be wrapped around a body of compression-resisting fluid to get a structure that has a discrete shape as well as appreciable stiffness, strength, and so forth. We're talking about nothing more than a balloon of air or a canvas fire hose of water.[22] Not to prolong any suspense, we're also talking about blimps and buildings supported by air pressure as well as about lots of worms and a host of cells—plus plenty of penises. Of course the character of the structure depends a lot on whether it's filled with gas or liquid. Water is eight hundred times denser than air and much more resistant to compression. A water-filled balloon has an almost constant volume whatever the load, but its shape reflects gravity's downward pull. The shape of an air-filled balloon cares little about gravity, but its volume varies more. We lack a single word for such inflated, pressurized, balloonlike structures: If water-filled, they're hydrostatic; if air-filled, they're aerostatic.

As practical pressurized bodies, rubber balloons have serious limitations. For one thing, rubber isn't too stiff, so the whole structure ends up a bit flabby. For another, rubber's stress-strain plot is nearly straight rather than concave upward. As a result (and as brought up in Chapter 5), balloons are prone to aneurysms; a cylindrical one doesn't inflate uniformly. Furthermore, inflation stores up a potentially dangerous amount of elastic energy; a balloon fails violently. Using a less stretchy wall material improves the situation. But practical and versatile aerostats and hydrostats depend on complex, multicomponent walls containing tension-resisting fibers. These most often run not lengthwise or in circular rings around the cylinder, but in helices, as in Figure 7.12

The two technologies use this general scheme in different ways and to different extents. Humans make relatively limited use of it, and a gas usually serves as the fluid of choice, as in air-supported buildings. Sometimes, though, we fill the inside with liquid, as in collapsible fire hoses and fiber-reinforced storage tanks. Nature uses the device extensively, and her fluid of choice is either water or some muscle or mucus. Here and there she does use gas, as in various floats, such as that of the notorious Portuguese man-of-war.

While the classic zeppelins had rigid frames and internal gas bags, modern blimps are proper aerostats, with the helium inside doing struc-

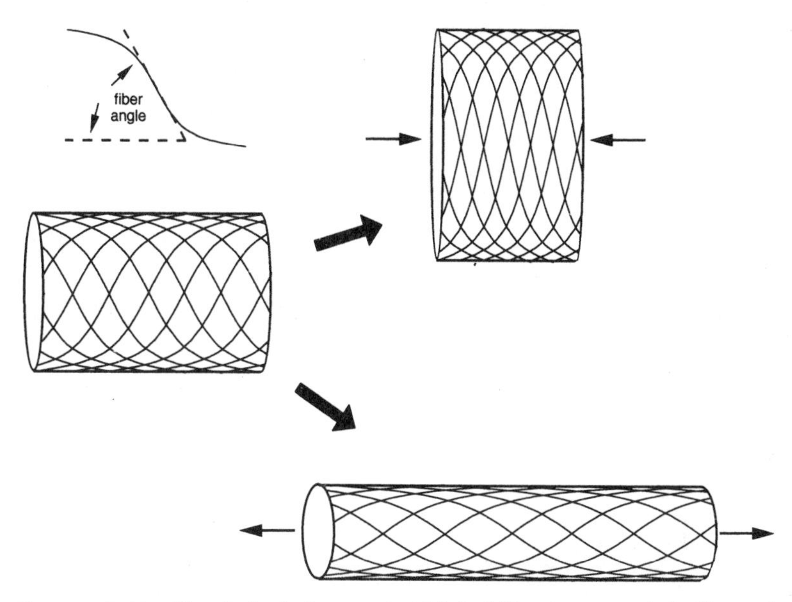

FIGURE 7.12. *If a cylinder that's wrapped with helical fibers is compressed, the fiber angle increases; if it is stretched, the fiber angle decreases.*

tural service as the compressive counterpart to the tensile skin. Walking into an inflatable building through the antechamber—a real breezeway—may feel a little strange, but you can't feel any increased internal pressure once you are inside. All the inflatable rubber tires of our vehicles do the same thing on a smaller scale but with higher pressures. But none of these applications is both common and large. Hydrostats are scarcer still. Besides the hoses and tanks already mentioned, a few barges for carrying liquid cargo have been built as tensile bags, as have the skins of large rockets with internal fluid propellants.[23]

Nature may rarely inflate with air, but her water-filled hydrostats are wonderfully diverse in appearance and function. In addition to cases already mentioned, they provide the main stiffening for the tiny tube feet of starfish and sea urchins. They add stiffening to that provided by the skeleton in swimming sharks. And they contain the pressurized water that a squid squirts in its jet propulsion system. To understand how these schemes work, we need to look at how the volume of a cylindrical hydrostat changes with alterations in its length, as in Figure 7.13. At one extreme the cylinder is squashed into a volumeless disk with the fibers

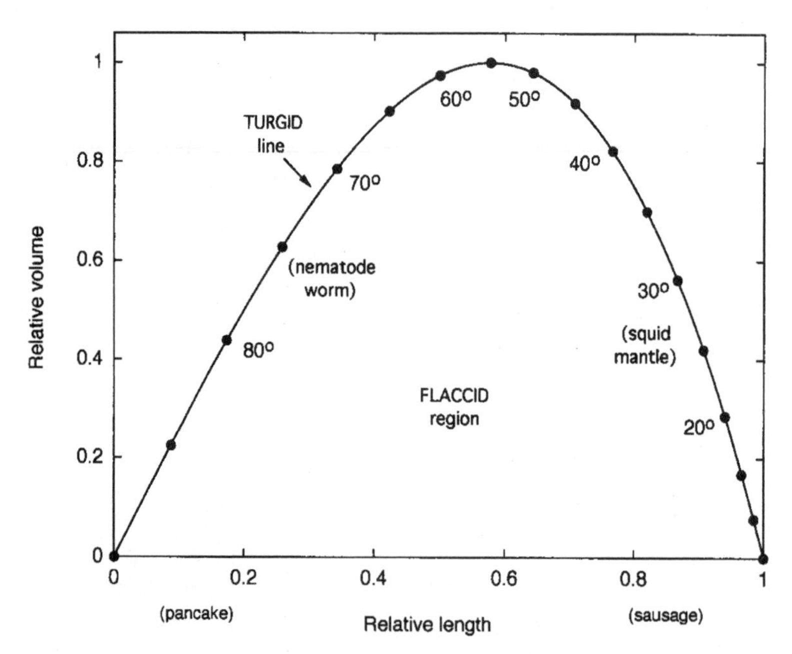

FIGURE 7.13. *The relationship between volume and length as fiber angle (noted on the curve) is changed. The helical fibers of the hydrostat are assumed to be inextensible. Beneath the curve the hydrostat is limp; above the curve it has exploded.*

running circumferentially, while at the other end it's a volumeless line with the fibers lengthwise. The cylinder has its greatest volume almost midway between, where reinforcing fibers run helically at a fifty-five-degree angle—a little more circumferentially than lengthwise. At that angle an increase in internal pressure has the least chance of increasing the volume of the cylinder and an equal tendency to make the cylinder get longer or fatter. That specific angle represents a divide between two ways that nature uses hydrostats.

Consider how a squid squirts a jet of water. It has an outer mantle with muscles running almost circumferentially around its girth and helical fibers running more lengthwise than circumferentially. Contracting these muscles might make the mantle get longer (as when you squeeze a cylinder of clay or dough), but that would make the fibers run even more nearly lengthwise, moving the system to the right on the graph and reducing its volume. Water doesn't compress; instead it squirts out as a jet, and the squid is suddenly elsewhere.

Alternatively consider a limp worm that wants to burrow. If it's a fairly simple, unsegmented worm, fibers in its outer cuticle run almost circumferentially. Contracting lengthwise muscles might make the worm get fatter, but that would make the fibers run even more hoop-wise, moving it to the left on the graph and reducing the worm's volume. Since the worm is closed, what happens is that the muscle contraction greatly increases its internal pressure, stiffens it, and facilitates burrowing. Contracting muscle on one side of the body curves as well as pressurizes it, so the worm can move by manipulating its hydrostat.[24] That's good for the worm, although perhaps not so good for us; we're explaining (among other things) how some parasitic worms can make their way through our flesh.

Another application of this mechanism doesn't involve a passive core of compression-resisting water, blood, and guts but instead uses active muscle throughout. Muscle of course can only shorten, so all muscular devices need some way to extend—either initially or when getting reset for their next pull. If a cylinder has a constant volume, then decreasing its diameter must make it lengthen. That's how our tongues, the arms and tentacles of squid and octopus, and even the trunks of elephants manage to extend.[25] These "muscular hydrostats" are versatile. Muscles can be arranged so the overall cylinder extends relatively slowly and forcefully for a short distance (as does an elephant's trunk) or relatively rapidly and less forcefully for a longer distance (as does a squid's tentacle). We'll defer more specific explanation until we consider the general topic of leverage. The general theme is what matters here; resisting compression, what struts do, doesn't require solid materials.

Two final questions about aerostats and hydrostats. First, why does nature make elaborate and creative use of hydrostatic support in aquatic systems but rarely use them in terrestrial ones? While hydrostats can't be made without water, even terrestrial creatures are full of the stuff. Our tongues and penises, some plant cells and small stems, and a few other cases don't come close to the several phyla of worms, to snail and starfish feet, to all cephalopod arms, and to sharks. I suspect, and the suspicion may be relevant to human use of the mechanism, that gravity poses a heavy problem. Hydrostats can't achieve great stiffness, penises notwithstanding, a serious disability where gravity dominates and organisms must extend above the ground. Tree trunks and long bones, for instance, must not be prone to bending if trees and mammals are to remain erect.

The second question concerns one of those odd omissions in nature. She makes no blimps—no lighter-than-air craft. Flight by heavier-than-air craft doesn't come cheaply, dispersal matters in nature's scheme of things, and a lighter-than-air flier should be splendidly well suited for completely passive dispersal. Is making hydrogen beyond nature's ability? Unlikely—photosynthesis in every green plant starts by splitting water into oxygen and hydrogen. Every one of our cells continuously removes hydrogen from fat or carbohydrate as it rechannels stored chemical energy. My best guess, but one lacking great conviction, invokes scaling. Nature starts small and perhaps finds a tiny blimpish seed or fruit hard to make or not too useful. To keep the thing buoyant, wall thickness must not exceed some fixed fraction of diameter, so a very small blimp must have a very thin wall, raising difficulties of gas penetration and mechanical integrity.

This book uses natural technology to provide perspective on our own. Going beyond a scientist's analysis to an engineer's synthesis can aid in gaining that perspective. As an exercise in creative synthesis, you might try devising (on paper) an alternative technology based on ropes, hydrostats, and other tensile schemes. Imagine a culture whose solid materials resist only tension and not compression, a culture in a world in which compression can be resisted only by fluids, either gaseous or liquid. How might buildings, vehicles, furniture, and other everyday structures and machines look and work?

Chapter 8

ENGINES FOR THE MECHANICAL WORLDS

Up to now we've mainly looked at static systems—at the geometric, material, and structural aspects of buildings, bridges, and their living counterparts. We turn now to dynamic systems, ones that move themselves or their parts. Whether they are living or not, the same basic principles must apply, but once again the practicalities will prove remarkably different.

Making moving machinery involves two or three kinds of component. First, energy must be fed into the system. Neither natural nor human technology really *makes* energy in any normal sense, but instead each manages to get it from external sources. Bending the ordinary definition a little, we'll label as "engines" all devices that feed nonmechanical energy into mechanical systems. These engines use forms of energy, such as heat and electricity, to push, pull, swell, shrink, bend, turn, or slide. Second, the mechanical energy has to be applied to a specific task, such as galloping, grinding grist, or gathering gold. Stretching another definition, we'll call all the requisite coupling devices, from simple cables and shafts to complex pulleys and gears, transmissions. Third—at least sometimes—mechanical energy is stored for a while between being generated by an engine and being used to move something around. The energy may be supplied in pulses (as with the separate firings of an engine's cylinders), or its task may be intermittent (as with any back-and-forth saw) or it may be needed in more concentrated packets (as when the slowly developed momentum of the hammer suddenly assails the nail). We'll call all such energy accumulators batteries. The focus of this chapter will be engines; of the next, transmissions and batteries.

Both technologies use energy for purposes other than moving matter around: for such things as biological growth and the synthesis of useful chemicals, for heating bodies and buildings, and for communications with nerves or with wires. The big bang, where it all began, left aside, the ultimate source of almost all this energy is the sun. Living systems capture it through the photosynthetic activity of plants and bacteria. Human technology gets most of its energy from the same process, preferring, though, the well-aged product, fossil fuels. Both technologies make a little use of aero- and hydropower, ultimately just different pathways for obtaining solar energy. We get some additional energy from nuclear reactions and a very tiny bit from geothermal sources, neither of which is a significant source for nature. Finally, we co-opt nature's engines themselves—our own muscles and those of our domestic animals. Figure 8.1 puts all this energy transfer in diagrammatic form. It should point up two things: the central position of photosynthesis for both technologies and the greater diversity of processes used by humans.

Another digression: Talking sensibly about motility requires a few additional terms in their scientific senses. So then . . .

Work. It gets done when something that resists being moved is nonetheless moved some distance. Quantitatively it's the force that's used times the distance the thing moves. If you hoist a ten-pound something or other a foot upward, you've done ten foot-pounds of work. The resistance you're working against is the gravitational attraction between earth and weight. Lifting five pounds two feet does the same amount of work, irrespective of how the lifting is done. The same work is done on the object if, with a lever, pulley, or windlass, you exert only one pound while moving your end of the device ten feet. Only force times distance matters, and both must be in the same direction. Lever, pulley, and windlass are just transmissions.

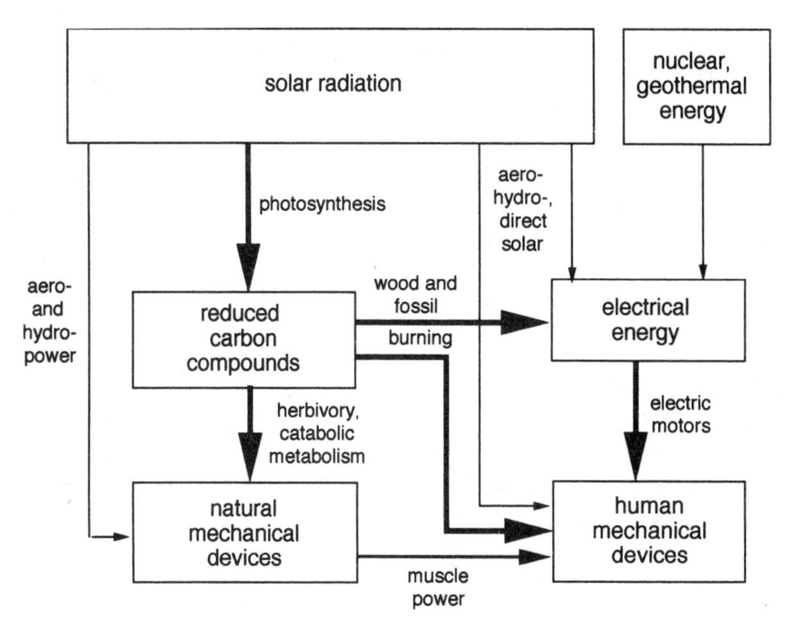

FIGURE 8.1. *How energy gets into the mechanical domains of the two technologies. The thicknesses of the connecting lines give a crude indication of the relative magnitude of the pathways.*

Energy. This is mysterious stuff. But the standard evasion, "the capacity for doing work," will do for our mundane purposes. What matter here are two points. First, energy and work are nearly the same thing, and we measure both on a yardstick marked with the same units: foot-pounds, or calories, or (in official scientific usage) joules. Second, the notion of energy permits statement of some general rules about our physical universe. The first is what's called a conservation law: In any real process energy is neither created nor destroyed but is at most shifted from one form to another. When we say that energy is "used," we really mean that it's transformed from a potentially useful form (the electricity in a battery or the high water behind a dam) into some less useful form (heat, in particular). Conservation laws serve us so well that a quantity such as energy is worth inventing just so we can state one.

Power. It's simply how fast a process uses energy or does work—energy or work divided by time. We measure it in horsepower, or foot-pounds per second, or (officially) watts. Quite straightforward, an engine's output power is how fast it does mechanical work, and a more powerful engine takes less time to do the same work. The elevator with the bigger motor gets you up in shorter order. Power can be viewed another way too: Force times distance divided by time is the same as force times speed. So, just as a certain amount of work can be done with any combination of force and distance, so a certain amount of power can be expended in any combination of force and speed. Things that swim and fly care about this last trade-off. One can invest an airplane's fuel in big, slowly turning propellers or small, rapidly spinning jets.

Efficiency. Even though the notion isn't complicated, we need to be careful to use it precisely. Efficiency is what you get out of some device divided by what you put in. It's the weight of the elevator times the height it ascends divided by the energy consumed by its motor. If work or energy is the currency (energetic efficiency), then that conservation law just mentioned limits the quotient to one. You can't get an efficiency over one, or 100 percent. In short, you can't get something for nothing, or shorter, there's no free lunch. In our real world all mechanical devices have energetic efficiencies below 100 percent, which means that some energy ends up where it does you no good. If you run, you get hot, and the engine of your car needs a radiator to dispose of heat.

The main measures of an engine are power output (or power relative to weight) and efficiency. Relevant also, if a little less so, is how the power output of an engine changes as its operating speed changes. For instance, many of our electric motors deliver proper output with decent efficiency only within a very narrow range of speeds. The operating speeds of automobile engines, while less critical, are still so limited that the engines need complicated and costly transmissions. Old-fashioned steam engines were much more tolerant; steam-driven cars and locomotives connected their engines directly to their wheels.

THE VARIETY OF ENGINES

Nature and humans have engines of great importance and engines with minor roles. For contemporary humans the main players—the prime movers, so to speak—are heat engines. These include all the external- and internal-combustion engines that burn fuel to make steam or that move pistons or turbines directly—almost all the engines we use to move our cars, boats, and airplanes. Nuclear power must be included here since it gets fed into our systems by transferring heat. A nuclear power plant uses heat from its reactor to generate steam, which then does the same job as the steam in a coal-fired plant. Our second team consists of electric motors, although we don't take electricity as such from the environment, so these aren't quite analogous to heat engines. Instead they're paired with generators as systems to transfer power from generating plant to kitchen. Electric motors serve the same role as the array of overhead shafts and descending belts that, in a nineteenth-century stream-side factory, drove all the machines from a common waterwheel. Beyond heat engines and electric motors, the list of engines grows long, but the overall contributions are slight: hydroelectric plants, windmills, tidal and wave-energy plants, and so forth.

For most of human history, the principal engine has been muscle—first human and then, increasingly, animal. To be useful, animals must be domesticated, trained, and their power harnessed toward plowing, hauling, or whatever else we want them to do. Harnesses—transmissions, essentially—were rudimentary before the Middle Ages. Waterpower—as waterwheels—played some role in the classical world, but wide use of wind power, except in the sails of boats, is more recent. By the Middle Ages both windmills and waterwheels were common.[1] Hero (or Heron) of Alexandria invented a steam engine in the first century C.E. but the design wasn't auspicious for practical applications, as will be clear when we look at jet engines specifically. Thus our major engines are recent. Practical heat engines appeared in the eighteenth century, and electric motors came into wide use barely more than a century ago.

Nature's engines are of no greater fundamental diversity. The most familiar is muscle, as noted earlier, about 40 percent of the weight of a typical mammal.[2] We also use motile cilia, microscopic cylinders that can actively bend, for such things as propelling sperm and moving mucus in our respiratory passages. Such cilia are widely used by animals both big and small for locomotion, pumping, and other purposes. (Curiously, motile cilia are absent in the entire phylum Arthropoda—insects, crustaceans, arachnids, centipedes, and such. Recall the comments in Chapter 2 on the limitations imposed by one's lineage.) Beyond these relatively rapid major engines loom some slow but not insignificant ones, at least if we relax our zoocentric sense of speed enough to give plants proper attention; over its lifetime a corn plant or a tree lifts many times its own weight of water from soil to leaves. In addition, a few engines extract

ENGINE	OUTPUT, WATTS/KG	OUTPUT, HP/LB
18th-century steam pump	10	0.005
Cilia	30	0.02
Skeletal muscle	200	0.12
Electric motor	200	0.12
Automobile engine	400	0.24
Motorcycle engine	1,000	0.6
Aircraft engine, piston	1,500	0.9
Aircraft engine, turbine	6,000	3.6

TABLE 8.1. *The power output relative to mass of a variety of engines.*[3]

power directly from moving air and water, doing what windmills and waterwheels do, if in different ways.

If our criterion of excellence is power relative to mass, nature's best engines match only low-level examples of human technology. By this measure, modern heat engines are truly superb. Table 8.1 compares the power output (watts or horsepower) per unit mass (kilograms or pounds) for some living and nonliving engines—devices we rarely consider at the same time.

COMBUSTION ENGINES

At some risk of oversimplification, heat—that is, combustion—engines can be divided two ways. Combustion may be external, with some working fluid carrying energy from the heater into the engine proper, as in all steam engines, whether piston-driven locomotives or modern steam turbines. Or it may be internal, as in our automobiles, where the burning of fuel in the cylinders generates the pressure that moves the pistons. The motion may be intermittent and reciprocating, as when pistons go to and fro. Or it may be continuous and rotational, as in all turbines, whether the steam-injection ones of power plants or internal-combustion jet engines.

Heat engines are unique to human technology, and the obstacles to their development were formidable. No natural analogs provided hints for their design, understanding of the underlying science came slowly, and metallurgical and fabricational limits long precluded steady operation at high temperatures.

The earliest practical combustion engines were about what one might guess. Thomas Newcomen's mine pump (Figure 8.2), popular during much of the eighteenth century, was an external-combustion piston engine using steam at atmospheric pressure as the working fluid. Motion of the piston was generated not by expansion of steam in a cylinder but instead by the press of the atmosphere on the other side of the piston when a spray of water lowered the temperature and made the steam condense. Thus especially high temperatures, good pressure-resistant fittings, and carefully machined crankshafts were not needed; the engine made modest technological demands. Conversely, the low pressure difference (one atmosphere at most) demanded huge pistons to produce much power, and wasteful heating and cooling of the cylinder's walls guaranteed very low efficiency. Still, these huge, slow machines gave fine service, especially for pumping water out of coal mines, where fuel supply was not an issue.[4]

Nineteenth-century steam engines were much better machines, efficient enough in weight and fuel consumption to move themselves and

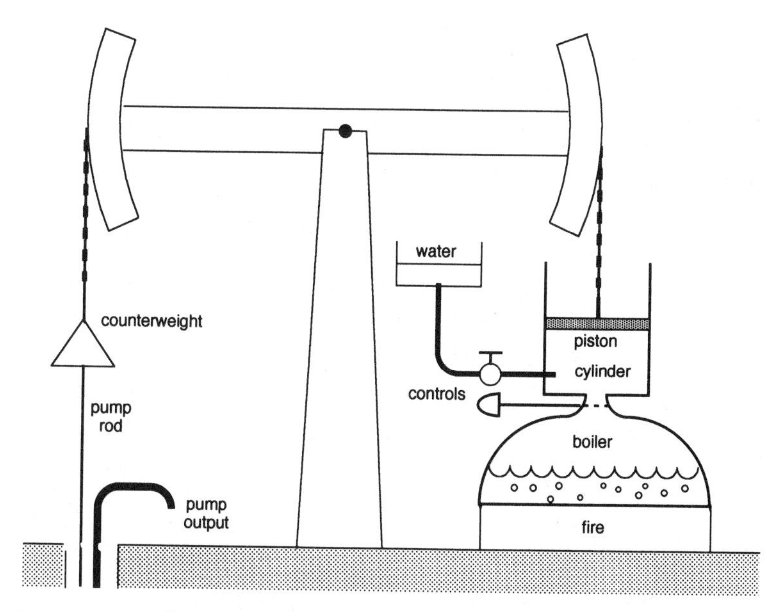

FIGURE 8.2. *The main components of Newcomen's steam engines of 1712 and thereafter. A huge rocker arm connected the piston's chain and the pump's chain. The power stroke of the piston was the downward one as steam was condensed by water sprayed into the cylinder. The counterweight then pulled the piston back up while more steam was let into the cylinder. In the early versions a person working the controls alternately sprayed in water and admitted steam.*

some payload around—at least on hard rails and with gentle grades. James Watt, late in the eighteenth century, took advantage of improved metals and better machining to make steam push pistons instead of condensing the steam in the cylinder and getting only the push of the atmosphere outside. His engines were better in other ways as well. Paired steam inputs pushed the pistons in both directions, incoming water was preheated with waste heat, the cylinders were kept hot by condensing steam outside, and piston rods turned wheels rather than just rocked arms. Watt and his contemporaries conceived of the steam turbine—after all, the principle is simply that of windmill or waterwheel—but both the precision needed to construct such an engine and the availability of transmissions adequate for its high speeds delayed its practical realization until the present century.

The story of internal-combustion engines is a similar tale, needing only a time shift of fifty or a hundred years. Piston engines came into wide use toward the end of the nineteenth century, and turbines around the middle of the twentieth. What must be emphasized, again, is the novelty of this whole technology next to anything in nature. Only the fuels are similar: hydrocarbons resulting from the biosynthetic activities of organisms. Metals, high temperatures, gases at high pressures, the large size of early engines, the high speeds of modern ones, pistons, crankshafts, flywheels—none has a close natural analog. Even the basic mode of operation of heat engines is unnatural; these engines work by either pushing (stretching the point just a little for Newcomen's atmospheric engine) or rotating.

Why doesn't nature use anything like our heat engines? Because the fundamental rules that govern heat engines are alien to the living systems on our planet. And why is that? A heat engine needs not only a high temperature but also a difference in temperature—both a hot source and a cooler sink. Down this temperature gradient energy will flow, as when hot coffee cools in air or warm water melts ice. The temperature difference sets an absolute limit on how good a heat engine can be. The basic rule is simple, but it does require that we use a temperature scale with a real zero point—the coldest possible cold. To convert a Celsius (centigrade) scale to one with a real zero, just add 273°; to do the same with a Fahrenheit scale, add 459°. Using such a scale, the maximum efficiency equals the sink temperature divided by the source temperature with the result subtracted from one. (Multiply by 100 to get percent efficiency.) Thus one could get perfect efficiency (1.0 or 100 percent) only if the sink were at an unrealizable -273° C or -459° F or the source at an unthinkable infinite temperature. For a steam engine, the hotter the steam put in and the colder the steam finally released, the better this underlying thermodynamic efficiency.[5]

Compare what our technology does with what nature might do. We can easily use a source at 1,000° C (1,800° F) and a sink at the boiling point of water, 100° C (212° F); thermodynamic efficiency is then 71 percent[6]—not at all bad. Nature might use a source at 40° C (104° F) and a sink of 0° C (273° K), about the maximum range of temperature tolerated by active animals. If so, the efficiency of her engine will be less than 13 percent. For a more easily realized ten-degree difference, ideal efficiency drops to a disastrous 3 percent. And these maximal thermodynamic efficiencies always exceed those of real devices.

Incidentally, nature's nonuse of heat engines carries a message that's familiar to almost every engineer but to few of the rest of us. Might we use the temperature difference between surface water and deep water in lakes or oceans to tap an unlimited and benign energy source?[7] Yes, the energy is there and available; after all, a temperature difference indisputably exits. From the rule about energy conservation one can easily calculate that monumental quantities of energy are available. But the result is monumentally misleading. What must be factored in is the thermodynamic tax imposed by the small temperature difference; the energy used to run such things as pumps would most likely be greater than the energy obtained from the underlying resource.

ELECTRIC ENGINES AND GENERATORS

The other major engine of contemporary human technology is the electric motor. Practical versions range from microdevices a fraction of an inch in length to ones weighing many tons that drive the propellers of large ships. They perform a wide range of tasks with remarkable effectiveness and reliability. My not especially high-tech household has more than sixty small electric motors. Try such a count yourself, including such easily forgotten ones as the fans, compressor, and defrost timer of the refrigerator; you'll be surprised at how many you happily harbor. Electric motors are more efficient than combustion engines and don't get too hot when running. Still, their high efficiency can mislead us since (as noted earlier) most electric motors just provide output devices for large heat engines, much like multiple dumb terminals for a mainframe computer. Their heat engines are of course the fossil fuel and nuclear power converters in generating plants. (Hydro- and wind-powered generators make only a small contribution to our production of electricity.) Thus a full accounting should consider overall efficiency, from fuel to mechanical power output, including substantial losses in the electrical generating plant and transmission lines. We'll come back to that shortly.

Once again we're looking at a distinctly human class of engines. Electricity is common among organisms; every cell has a charge across its outer membrane, for instance, with potentials approaching a tenth of a volt. Not too many cells in a series would give quite a good voltage, nor would all that many, working in parallel, deliver an eventful current. Several kinds of electric fish modify muscles into arrays with series and parallel connections that deliver stunning performances, reportedly up to 650 volts.[8] While that's an eclectic habit, it shows that producing electric power in quantity takes only minimal alteration of normal tissue and ordinary metabolic chemistry. Our muscles, even those of our hearts, may be electrically controlled, but they're not electrically powered.

Why doesn't nature build electric motors? Must these motors use unnatural wheels and axles and engage in unnatural rotational motion? Probably not. While all familiar electric motors rotate, rotation simply serves a technology that finds the wheel and axle easy to make and likes the versatility of belts and gears. Every kind of rotary electric motor has its linear or reciprocating counterpart. Linear engines have been developed for driving trains; the track forms half the motor. Short-stroke repetitive pullers (solenoids) do chores such as opening and shutting the water lines and drains of our washing machines; turn on the electricity, and a metal core moves forcefully through a coil. The earliest electric motors of Joseph Henry, in the United States, and Charles Wheatstone, in England, in the first half of the nineteenth century were reciprocating devices. One early reciprocating electric motor drove a pump the same way a Newcomen steam engine did.[9] Closer to home, motors in which a piece of metal vibrates rapidly back and forth sometimes power electric razors, massagers, and reciprocating sanders.

A more likely obstacle to living electric motors is that dullest of components' wire. Good conductivity comes hard without metals, as noted two chapters ago. The salt solutions of cells don't come close. For instance, a strong solution of potassium chloride (seventy-one grams per liter, a so-called one molar solution), an especially conductive brew, is still nine million times less conductive than copper.[10] Equal performance would require that a copper wire a mere tenth of a millimeter across be replaced by a pipe of potassium chloride at least a foot in diameter. Nothing remotely appropriate to get power from generator to rotor exists in nature's armamentarium; electric motors are mainly practical in a metallic technology. Neuromuscular systems use electricity in a way peculiarly adapted to or limited by (take your choice) their low conductivity. In a nerve, electrical currents flow only for short distances. Long-distance transfer of information takes place by an odd scheme in which a local electrical event incites a similar event adjacent to it—the way a wave travels across the ocean without moving any bit of water very far.

Electric power's popularity comes from its handiness as a transportable intermediate form of energy. We generate electricity in one place and then transmit it for long distances at (to reduce power losses) several hundred thousand volts. For safety and convenience, we then reconvert the high-voltage electric power to lower voltage (240 and 120 volts, mainly) near the sites of use.[11] We vertebrates do much the same thing to power our muscles, although we do it chemically rather than electrically. We move the sugar that results from carbohydrate digestion through a special set of veins to our livers, which store it (as a polymer, glycogen) and meter it out to the rest of the circulation.[12] In muscle cells the sugar charges up a more immediately usable energy transfer system: Breakdown of sugar to either lactic acid (short term) or carbon dioxide and water (in sustained activity) converts adenosine diphosphate (ADP) to adenosine triphosphate (ATP). This last then feeds power into the actual muscular motor. The coupling of sugar breakdown to ATP synthesis in muscle has about the same function as the conversion of high- to low-voltage electricity in an electrical transformer.

WIND AND WATERPOWER

First, we should remind ourselves that getting power takes more than just air or water in motion. A windmill on a freely floating balloon won't turn; the balloon moves with the wind, and no apparent motion of the earth beneath helps at all. Having its feet on the ground matters as much for a windmill as having its rotor in the air. Extracting power requires a device that experiences a difference in speed: between air and ground for a windmill, between water and ground for a waterwheel, or between air and water for a balloon towing a submerged turbine. The requirement has the same basis as that for two poles on a functional battery, for different water levels on the two sides of a power-producing dam, and for a hot source

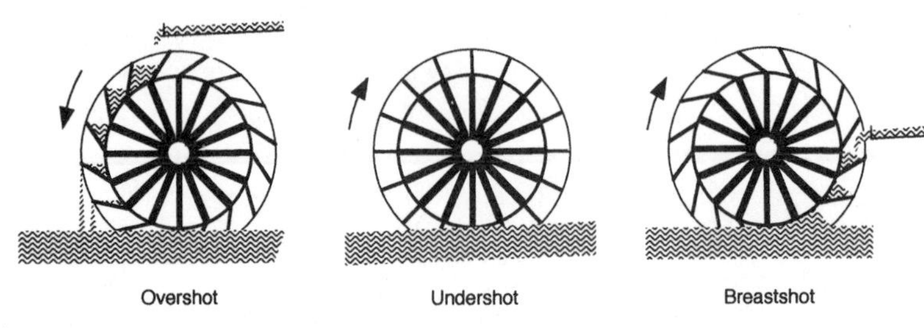

FIGURE 8.3. *Three kinds of horizontal-shaft waterwheels.*

and cold sink in a heat engine. It's incumbent on both our law-abiding technologies.

The classic water-powered engine is the waterwheel, of which Figure 8.3 shows several useful versions. Water runs over the top in overshot wheels, where the weight of the water (gravitational energy) pulls one side down. The water's motion (kinetic energy) pushes paddles on the bottom in undershot wheels. Water enters in the middle, tapping both resources, in breastshot wheels. Less commonly, a stream of water may fall against inclined paddles around a wheel with a vertical shaft.[13] None of these arrangements achieves great efficiency, and for at least a century they've been superseded by various kinds of turbines—rapidly turning rotors concealed within the ductwork of large hydroelectric plants. Small-scale hydropower as used by old grain and sawmills has been largely abandoned, as a result both of the widespread availability of electricity and good electric motors and of the change to the large-scale production of flour and lumber.

The classic wind-powered engine is the windmill, of which Figure 8.4 gives representative designs. Windmills are more recent devices than waterwheels for the same reason that internal-combustion engines followed external-combustion ones: The technological problems are more challenging. To develop much useful power, the rotors of windmills must be large, but if large, they become vulnerable to storms. The speed of atmospheric winds varies a lot more than does the pressure of water stored behind a dam. The basic setup is also less forgiving for primitive devices; a waterwheel needn't be completely immersed in water, but a windmill must work entirely in air.

Almost all windmills use one or the other of two physical mechanisms.[14] The oldest ones had vertical shafts and turned in a horizontal

FIGURE 8.4. *Windmills of various designs. Above are three with horizontal shafts: classic Dutch, American farm, and a modern high-efficiency design. Below are two less conventional ones with vertical shafts: Darrieus turbine and Savonius rotor.*

plane. Like whirling cup anemometers, they turned in winds coming from any direction. But turning depended on blades that had higher drag when facing into the wind than when facing away from it, a sure recipe for inefficiency. A recent version is the Savonius rotor, which enjoyed a brief countercultural fashion since it could be made from a metal drum sliced lengthwise.[15] By contrast, most windmills of the present (second, not third) millennium have horizontal shafts, with a set of propellerlike blades that rotate in a vertical plane. While the blades are useful during their entire circuit, the whole structure must turn to face the wind. Aerodynamically they work by producing lift, a force at a right angle to the wind direction, rather than depending on the difference in the drag of bodies facing into and away from the wind. Generating lift is a subtle business whose physical basis wasn't understood until the early part of this century. (The airplane's demand for really good wings and propellers of course provided the impetus.) Not surprisingly, old windmills were aerodynamically poor, and the drag on the blades and their supports was unnecessarily high relative to the power that could be extracted.[16]

All these waterwheels, turbines, and windmills use wheels and axles. So none of this sounds like a technology that has natural analogs, and in a strictly mechanical sense it doesn't. Nevertheless, it does have energetic analogs. Sometimes nature draws power from speed differences between the ground and flows of air or water. We know of a number of cases and several different arrangements.

Consider, as in Figure 8.5, a wind across the two openings of some U-shaped pipe beneath the ground, a pipe whose openings differ in height. The higher opening will ordinarily be exposed to faster flow, and, by Bernoulli's principle, the faster the flow, the lower the pressure. Since any fluid—gas or liquid—left to itself flows from high pressure to low pressure, then flow within the U-shaped pipe will go from low opening to high opening—irrespective of the direction of the external flow driving it.[17] The arrangement is used by the prairie dogs of the North American Great Plains to ventilate their long, deep, multiapertured burrows. I was involved in investigating the phenomenon in the early 1970s. At the time I thought that the ventilation supplied the animals with oxygen, but now I'm less sure; it's far better than necessary for that purpose. More likely, airflow through the burrow gives the animals the scents of what's above; the scheme is an olfactory sensation. With water as the moving fluid, various worms and crustaceans may use an analogous arrangement to irrigate their burrows in sandy substrata beneath shallow bays.

If air or water moves across an elevation, normally the air or water

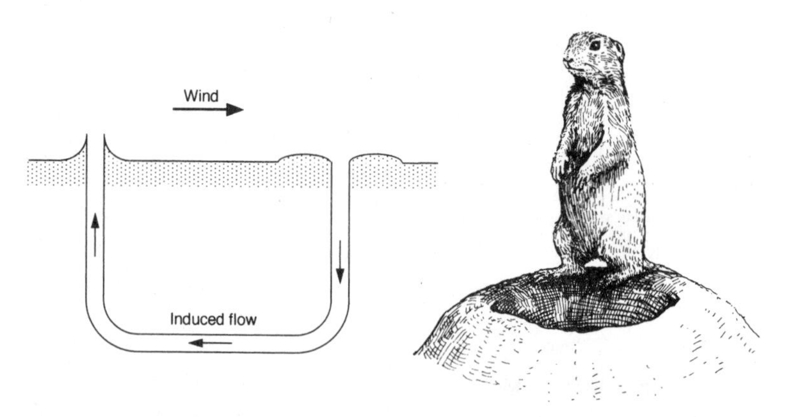

FIGURE 8.5. *A scheme for using ambient flows of air or water to induce a secondary flow in a burrow or other passage through the substratum, and a prairie dog viewing the world from the crater-shaped opening on one end of a burrow that uses the scheme for ventilation.*

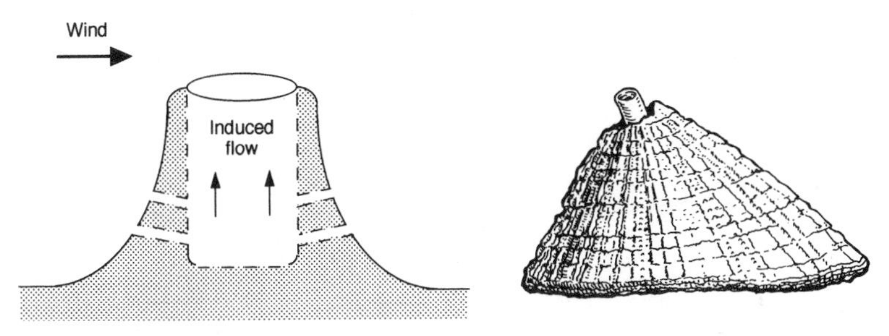

FIGURE 8.6. *Another scheme for using ambient flow to induce a secondary flow, this one through some elevated structure; and a keyhole limpet, which uses the scheme to draw water in beneath the edge of its shell, over its gills, and out the apical hole.*

will flow fastest at the peak; ridges are more windy than valleys. If pipes or a porous medium (such as sand) connect a valley with a peak, as in Figure 8.6, then air or water inside will move from one to the other. Again, the direction of the external flow makes no difference, and the physical basis remains Bernoulli's principle (plus some secondary effects). This second arrangement makes wind draw air and thus oxygen through giant termite mounds in the open country of East Africa. Similarly, sponges, which live by filtering microscopic organisms from seawater, use the flow of water around themselves to reduce their cost of filtration. That improves the (net) profit that determines if they're viable in a particular place. And slotted sand dollars form slight elevations on sandy bay bottoms, so water flowing across them draws water and edible particles up though their slots from interstices in the sand.

In a still simpler arrangement, shown in Figure 8.7, one opening of a pair is directed upstream and the other faces crosswise to the flow. This third arrangement depends, though, on proper orientation with respect to the external flow. At least one kind of insect larva does it; some caddisflies that live in streams make appropriate tubes in the substratum and equip the tubes with catch nets. In addition, many fish that use nostrils in interconnected pairs put enough of a hood on the front one so it faces upstream. Since streams flow downhill and fish swim forward, orientation to flow presents no problem in either case.

We know a few other schemes by which nature takes advantage of flows across solid surfaces to obtain a little power. Some seabirds, for instance, repeatedly make long vertical loops without beating their wings. Gliding alternately high and low is a way to use altitudinal

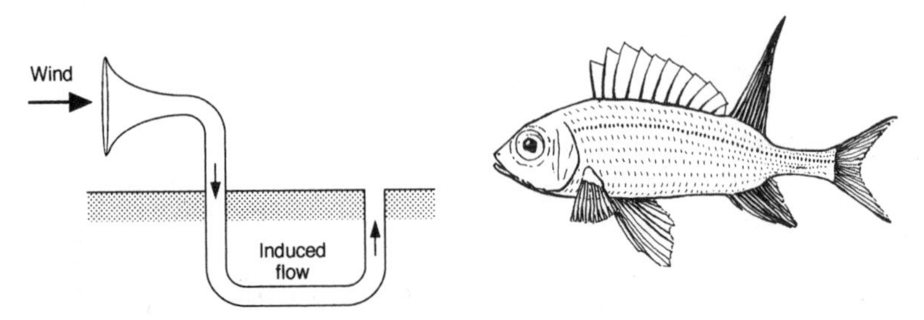

FIGURE 8.7. *A more potent scheme for using ambient flow to induce a secondary flow, but one no longer independent of the direction of the ambient flow. This one drives water in the mouth, across the gills, and out the operculum of many fish when they're swimming rapidly.*

velocity differences to stay aloft without expending energy.[18] For that matter, we sometimes ventilate simple buildings, such as Indian tepees and partially buried mound houses, with the same tricks as sponges and prairie dogs. Mine ventilators often use reversible fans to take advantage of any wind-induced flows. Such flows will occur in all tunnels that have multiple openings except where the external wind is the same over all openings, the openings are geometrically identical, and the land is perfectly flat.[19]

All these cases use wind or water currents to drive internal flows of air or water. One flow induces another, which takes a minimum of transducing machinery. However interesting as bits of natural history, they're insignificant next to photosynthesis as a way by which nature gets energy from the environment. Moreover, none provides energy in a storable form, the most important feature of photosynthesis.

MUSCLES AND CILIA

By this point the reader may wonder whether nature always comes off second best. For one type of engine, though, nature has the only team on the field. I refer here to what are coming to be called molecular motors. Human technology works large, dealing with materials in bulk and using such processes as casting, pressing, slicing, dicing, and crushing. We struggle mightily to make ever tinier components for our complex electronic devices so they can be powerful without getting dysfunctionally large. Nature works the other way; hers is a microscopic, even submicroscopic technology aggregated with some minimal integration into larger systems. Molecular motors, if a value judgment is permissible, show

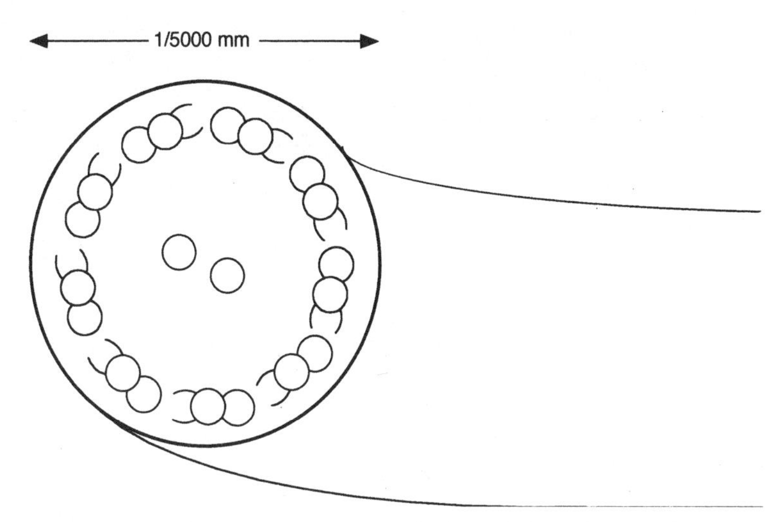

FIGURE 8.8. *A cilium cut across (imagine a sliced strand of cooked spaghetti), showing the typical set of nine doublet microtubules surrounding a pair of singlet microtubules.*

nature both at her finest and at her most distinctive. Two of these—muscles and cilia—drive her fastest motions.

Cilia (and flagella, any distinction being unimportant here) of a fairly stereotyped kind occur in a wide variety of animals and plants.[20] They're tiny hairlike machines that push stuff around by waving, beating, or wiggling, either singly or in coordinated arrays. Some stick out of single cells—such as our sperm—and other tiny creatures, propelling them along. Others line surfaces (clam gills, for instance) and pipes, acting as pumps and filters. In fallopian tubes, they propel eggs toward the uterus. All use chemical energy in the form of adenosine triphosphate (ATP), as does muscle. Figure 8.8 gives a simplified look at the structure of a cilium. Of particular relevance to us are two proteins of their internal structure. A polymer of one, tubulin, makes up the nine lengthwise doublet tubules; a polymer of the other, dynein, forms a pair of arms on each of those doublets. Tubulin and dynein slide along each other to generate motion; the dynein arms cyclically attach to and detach from sequential sites along the adjacent tubules. Bending of the overall cilium results when sliding occurs only on one side. In short, the basic motor moves by ratcheting along somewhat the way a person does when poling a boat up a shallow stream.

Muscle uses the same chemically powered ratcheting motion, however

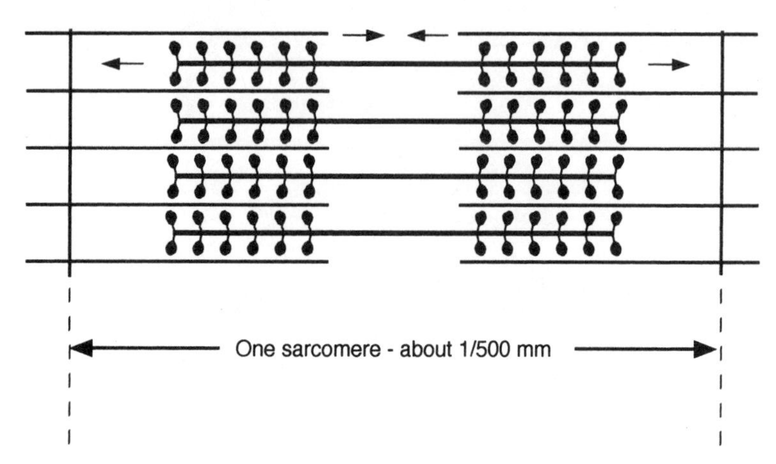

FIGURE 8.9. *The basic contractile unit of ordinary muscle. Cross bridges on the thick myosin filaments alternately attach and detach from sequential sites on the thin actin filaments. That makes the two sets of filaments interdigitate farther so the whole apparatus gets shorter.*

that differs from what seems to happen when you lift something by shortening your biceps. Shortening occurs because the basic filaments increasingly interdigitate as they ratchet along each other, as when you slide the fingers of one hand between those of the other, as in Figure 8.9. The fundamental motion may resemble that of cilia, but the main proteins of muscle, myosin and actin, have evolved separately from dynein and tubulin. Muscle can also tighten without actually getting shorter; it can just develop an end-to-end pull. So either microscopically or macroscopically, we're misled a little by the term "contraction" for what muscle does. A muscle develops force, which may or may not (depending on the load) bring its ends closer together and make its belly fatter.[21]

Muscle's need to use energy to develop force, even when it doesn't shorten and thus do work, generates much of our common confusion between the terms "force" and "work." The peculiarity is peculiar to muscle. The chandelier's chain exerts enough force to hold up the chandelier, and it does so year after year with no fuel supply whatsoever. (Oddly, one kind of muscle can maintain a force without additional expenditure of energy. It's the muscle that holds together the half shells of clams, mentioned in Chapter 2. But offsetting that cheap, steady pull is a slowness of action perhaps tolerable only by an unhurried clam.) Muscle has another slight disability compared with the engines of our technology. Active contraction is completely irreversible. Not only can't muscle actively reexpand, but stretching it doesn't make it produce chemical energy. If you turn the shaft of an electric motor, you can get electricity back out; you've made a generator. Forcing pistons back and forth or spinning the rotors of turbines makes them work as pumps. A loudspeaker can be used as a microphone. But you can't regain energy by running downhill or by putting a motor on your exercise bike to drive your legs.

In at least one way, though, muscles behave like most of our internal-combustion engines. The maximum power that either kind of engine can produce depends on how long it has to sustain that power. We get higher power by briefly supercharging (pressurizing the intake) when an aircraft takes off or by briefly tolerating high heat production in the starter motors of our cars. Nature goes us one better, tailoring muscle for specific durations of use. For good sustained output, muscle sacrifices some of its contractile fibers (and thus power) to provide space for more oxygen-processing metabolic machinery. Dark meat on a bird or fish is the sustained output muscle; light meat is the more fiber-rich, higher output, intermittently active version. Lots of so-called white muscle (a little lighter in color) makes for success in a hundred-yard dash; red muscle produces less power but wins in the long events. How much of each kind you develop depends on your training regimen. The strategy for brief bouts of high output may be the same for heat engines and muscle, but the tactics differ. Our heat engines increase their peak output by laying on extra fuel and oxidant. Muscle, by contrast, changes the way it uses its fuel, temporarily doing without oxidant and accumulating (to our discomfort) lactic acid instead of carbon dioxide.

Musclelike systems occur elsewhere among organisms. All nonbacterial cells contain the basic motor proteins, actin and myosin. Such things as the movements of the contents of plant cells and the locomotion of amoebas seem to be driven by ratcheting interactions of actin and myosin filaments. Similarly, when a cell divides, tubulin appears involved in moving the chromosomes around, although probably without the same kind of ratcheting as in cilia. (We find it much easier to discover what proteins are present in a system than to understand how they operate.)

While organisms are a diverse lot, these proteinaceous motors aren't. Nature invented a few versions early on and stuck with them. Perhaps that's not too surprising. Enzymes, virtually all of them proteins, are nature's chemical machines. But they do their chemistry in a mechanical way: They work by grabbing, manipulating, and releasing other molecules. Both myosin and dynein are enzymes; they just move other proteins, actin and tubulin, instead of performing more conventional chemical transformations.

Muscle does something very different from what any human machine can manage. This presents a terrible problem for biomedical engineers as they try to contrive prostheses for muscular organs. A heart is quite a simple thing, compared, say, with a kidney or a liver, but it's mainly muscle. We can make fine valves (although as replacements pig valves still have certain points in their favor), but we can't yet make a full prosthesis that's anything like one's original equipment.

OTHER NATURAL ENGINES

Besides cilia, muscles, and a few other schemes in which proteins ratchet along one another, nature, as noted earlier, has several other kinds of engines up her sleeve. While they're slower and thus not so obvious to us impetuous animals, they're undeniably powerful. They also have few parallels in our technology, and they're of particular interest because they work without solid parts that move. Three are especially common.

- An ordinary corn plant lifts about four quarts of water each day from the soil. Lifting is work, so a corn plant must have an engine, as must almost all terrestrial plants. The main engine is simple but strange: a direct-acting solar-powered evaporative engine. If (1) pipes filled with water run continuously from the roots up into the leaves, if (2) water can evaporate into the atmosphere from the leaves without air leaking in, and if (3) the pipes are stiff enough so they don't collapse, then water lost to evaporation will be replaced by water ascending from the soil by way of roots, trunk, and branches. The necessary conditions appear to be met, and most of the ascent of water in a corn plant or a tree is due to this pull from the top.[22] Evaporation of course takes a lot of solar energy; evaporating a gram of water at room temperature takes a bit more energy even than boiling off a gram in a hot pot. We'll return in Chapter 10 to this remarkable machine that may pull against pressures of more than one hundred atmospheres (nearly a ton per square inch) without any moving parts.

- Most carbohydrates (starch, mainly) and proteins are hydrophilic; they avidly attract water. Cornstarch and gelatin do just that in the kitchen, which is why we use them as thickening agents. Put a lump of either starch or protein (many kinds of protein, at least) in water, and the lump will swell almost irresistibly. If dry seeds beneath concrete pavement get wet, their expansion can crack the pavement. Germinating seeds commonly use this engine to split their seed coats and to penetrate soil. One case of this so-called imbibition is known in animals: The male mosquito hydrates pads of protein to raise the hairs on its antennae in order to sniff out the location of an odoriferous, receptive female.[23]

- Like all other molecules in gases and liquids, water molecules diffuse around—that is, they continuously move around at random, mixing themselves. Work can be extracted from this random motion in the following way. If water is more concentrated in one place than in an adjacent place, more water will move from the concentrated place to the dilute place than the other way around simply because there *is* more water in the concentrated place. How to do it? Say we dilute some water by adding a solute such as a salt or sugar. We then place the diluted water in a compartment that's separated from another compartment filled with pure water. As a barrier between the com-

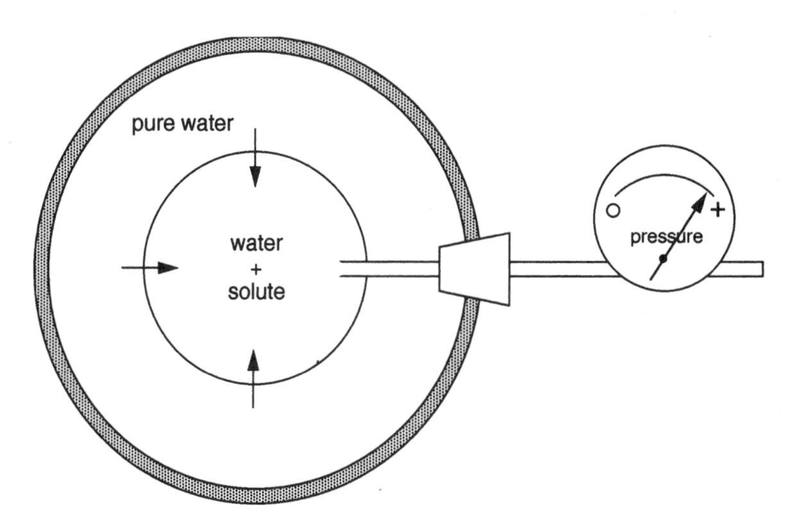

FIGURE 8.10. *The basic pressure-generating osmotic engine. Water enters a compartment in which it is less concentrated, passing through a barrier that's impermeable to the solute that dilutes the water.*

partments we use something through which only water can pass, as in Figure 8.10. The compartment with the added solute will swell up as more water enters than leaves it. While it may swell slowly, it will do so very forcefully, working as yet another expansion engine. Pressure generated by such a scheme ejects the stingers (nematocysts) of jellyfish from their parent cells. Roots absorb water from soil and pump it into stems with a version of this engine. Movements of leaves and other parts of plants depend on pressure changes inside nonrigid cells; such pressure changes are generated by adjusting the concentration of solutes in the cells and letting diffusion do the rest.

COMPARING EFFICIENCY

Both power relative to weight and energetic efficiency measure the quality of engines. But even together they don't provide a full yardstick, and for some engines, such as the evaporative, imbibition, and osmotic ones just described, neither measure is particularly revealing. Still, energetic efficiency merits at least the attention that weight economy got in Table 8.1. For electric engines one ought to include the generator's efficiency as a factor. For none of the engines is it easy to peg the costs of obtaining and processing fuel, whether by mining or by digestion. And the costs of transporting fuel vary a lot. Nor can we easily account the use of "waste" heat for thermoregulation in warm-blooded animals (and some large, active fish and flying insects) and for space heating of our building. Bearing these limitations in mind, let's look at a few figures for energetic efficiency.[24]

Piston engines have a lot of virtues, but efficiency is not high among them. At best automobile engines reach about 25 percent, but we usually run them far below their optimal output level. Diesel engines do somewhat better, but their efficiency also is often compromised by the ways we use them. In steam piston engines where the steam expands in successive cylinders, efficiency got up to about 17 percent by 1900 and can reach about 19 percent currently. (Watt's single-stage engines did about 2 percent.)

Turbines do better and have superseded piston engines for power plants, large ships, and most aircraft. A coal-burning steam power plant might have an efficiency of 40 percent, a nuclear steam plant (using, for safety, slightly cooler steam) an efficiency of 32 percent, or a gas (internal-combustion) turbine a 26 percent efficiency.

Consider an electrical system composed of a turbine, a generator, and a motor. The turbine has an efficiency of 40 percent or less, so it represents a huge power loss. Large generators are better than 95 percent efficient, so very little is lost at this step. Electric motors are widely variable, ranging from 20 percent for the motor of a small electric fan to a most impressive 90 percent for a hundred-horsepower polyphase induction motor optimally loaded. So the overall efficiency of only the best systems will exceed 30 percent. One does better with hydroelectricity—water turbines pass around 90 percent of the energy of the stream to the generator—but since one has to have a decent head of water on tap, the benefits are not universally available.

What about muscle? It turns out to be flabby, if functional. Even ignoring losses in, for instance, locomotory machinery, it's still less than 25 percent efficient, and sometimes it's a lot less. The shortening machinery isn't too bad, but the chemical operations that transfer energy from sugar to adenosine triphosphate and thence to myosin suffer substantial losses. Also, in muscles (as in other engines), efficiency depends a lot on speed of operation, something that varies a great deal from muscle to muscle, animal to animal, and moment to moment. Muscle may win no prize, but on the other hand, it's no disaster either.

Both nature and human technologies give energetic efficiency a fairly high priority in their respective designs. The evidence of this for natural technology may be indirect, but it's persuasive: Most structural and behavioral aspects of organisms make sense only if energy is considered precious. For human technology we have good historical records, whether we look at improvements in the harnesses of draft animals or at the historical development of steam engines. The similarity of engine efficiencies is curious, even if accidental—in the 20 to 30 percent range for gas turbines, steam or gasoline piston engines, electrical systems, and muscle. If either technology has any slight edge, it's our human one. Also, we're still improving, as nature probably isn't.

Beyond all the detail, what might we say about the good and bad of all these engines? Heat engines and engines that expand or rotate dominate one technology; constant-temperature engines, most of which shorten or shear, dominate the other. One technology transports energy over long distances, either as electricity or as chemical fuel; the other transports energy for only short distances and uses electricity only for signaling. While human technology may be marginally better in energetic efficiency, we definitely win in power output relative to weight; nothing comes close to modern aircraft engines.

Nonetheless, we can still look with envy on muscle, nature's preeminent large and fast engine. An individual muscle of a tiny insect might weigh a microgram; a large muscle of a big whale may approach a hundred kilograms—several hundred pounds. Those masses are a hundred billion times different, 10^{11}-fold, and performance doesn't deteriorate noticeably at either extreme. Very little in either technology does so well over such a gigantic size range.

Besides, muscle is good to eat. This is no mere Parthian shot. Expeditions that used beasts of burden sometimes exercised the option of consuming them. The Lewis and Clark expedition to the Pacific Northwest of the United States in extremis ate horses, and a century later Amundsen's expedition to the South Pole ate its dogs on a predetermined schedule. Try that with your internal-combustion engine.

Chapter 9

PUTTING ENGINES TO WORK

Making an engine run efficiently doesn't make it run usefully; it must be persuaded to do its assigned task. Once in a while sagacity or sheer luck lets us couple an engine directly to a machine, the way the shafts of their motors bear the blades of lawn mowers, fans, and blenders. More often machines need things between motor and output device to couple the two—transmissions. While the machines of our devising usually need transmissions, muscle-powered machines *always* need them. That's because our rotary engines, at least, make full cycles under their own power. Muscles can't do this; a muscle does its work by shortening, but it always needs something to restore it to its original length. In human technology transmissions are ubiquitous, but in nature they're universal.

For engines the important factors are power produced relative to power consumed, power produced relative to weight, and how power varies as speed changes. For transmissions, the only relevant power is the maximum that can be handled without failure. Weights run well below and efficiencies well above those of engines. Two things, though, matter a lot for transmissions. First is the particular trade-off of force and distance (for a given amount of work) or force and speed (for a given power level). You start a car moving with the transmission in a low range, meaning a lot of force (to accelerate) while the car's speed is low. You then shift (or the car shifts) to a higher range, with less force but a higher speed. Meanwhile the engine only has to operate over a narrow range of speeds (RPMs). Second is the match of motion between engines and final components, what's called the kinematics of machines—the way the hipbone is connected to the thighbone, and so forth. Proper kinematic design lets a piston going up and down make wheels turn and lets the rotor of an electric motor make a saw blade go back and forth.

In both technologies the diversity of transmissions exceeds that of engines. But each technology has its bag of favorites, with less overlap between the two than one might guess. Transmissions in fact contrast almost as sharply as do engines. That's partly driven by differences in the components from which they're crafted, partly by differences in the initial engines, and partly by differences in what the machines finally do.

LEVERAGE

Probably no mechanical device is older than the lever; simple and versatile, it's no doubt older than we humans. Prying with a lever is something that must be done at least occasionally by the finch of the Galápagos Islands, which uses a thorn to extract insects from bark, or by a chimp dealing with a moundful of delectable termites. Levers are powerful devices—Archimedes exaggerated only a little in claiming he could move the earth if given a fulcrum for his lever—and ubiquitous devices; without levers how could you pry the cap off a bottle, puncture a can, tighten a nut, switch on a light . . . ?

A lever at its most basic consists of a stiff rod along which three locations can be recognized. One is the fulcrum, the spot around which the lever turns. Another is the load, where what's being worked on makes contact. Finally there's the effort, where the driving force is applied. Levers come in two precisely opposite versions, both shown in Figure 9.1. In one version, a force amplifier, the distance between fulcrum and effort is greater than the distance from fulcrum to load. In the other, a distance or speed amplifier, the distance between fulcrum and effort is less than that from fulcrum to load.[1] We'll defer talking about speed for a few pages.

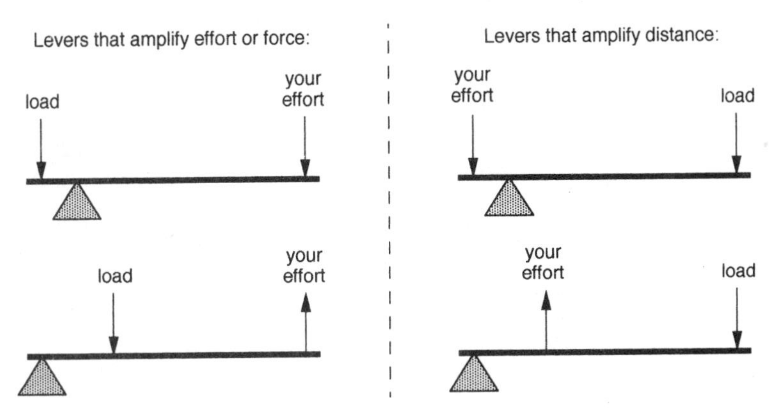

FIGURE 9.1. *Two basic kinds of levers. The crucial feature that distinguishes them is the relative distance of load and effort from the fulcrum.*

A lever is about the simplest possible transmission. It does no work, which we have defined as force times distance. It just changes the particular mix of force and distance of an engine (muscle or motor) into a more useful mix of force and distance. Levers allow one to apply an effort force less than the loading force (as with a prying bar). They also allow moving the effort a shorter distance than it moves the load (as when one swings a golf club or baseball bat).

Levers are just the simplest of a diversity of devices that trade force against distance and find use as transmissions. All give the same choice between force amplification and distance amplification. Figure 9.2 shows a few that find use in human technology: a windlass in which ropes run around coupled wheels of different sizes, a so-called block and tackle in which a rope goes around several pulley wheels, and a belt drive that uses pulleys of different sizes. The windlass resembles a simple lever in both structure and operation; for the others one shouldn't let structural differences obscure the functional equivalence. In natural technology, levers are common enough, but lever analogs like the block and tackle aren't quite so obvious.

In this distinction between force amplifiers and distance amplifiers lies an interesting difference between natural technology and the gadgets we humans use. We have a strong preference for force-amplifying devices, a few of which are shown in Figure 9.3. Your appendages move far, but they don't move very forcefully, so you use force amplifiers to get, as we say, "leverage." Look around your kitchen. Almost all the hand-operated

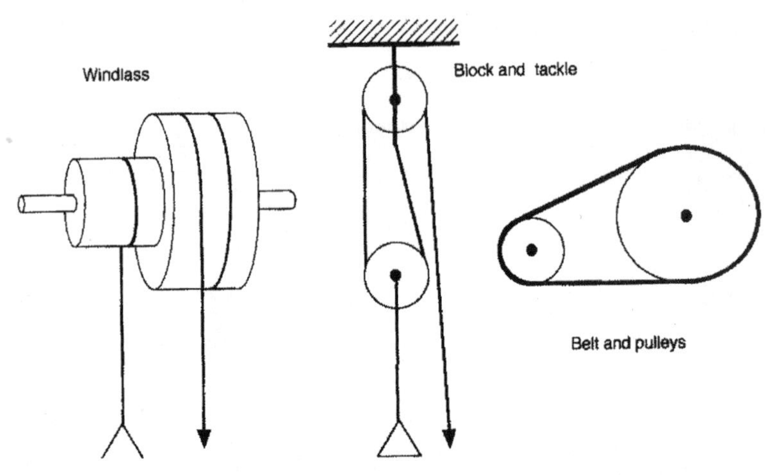

FIGURE 9.2. *Devices that get "leverage," shifting the relative magnitudes of force and distance or speed.*

tools are force amplifiers. In our kitchen the only distance-amplifying devices are a salad server, hinged between short handles and long tines, and a pair of wooden tongs, hinged at one end, that hoists English muffins from the depths of the toaster. Among the hand tools on our workbench are screwdrivers, pry bar, wire snips, metal shears, socket wrenches, and pliers—force amplifiers all. Among our garden tools the only distance-amplifying device is the pair of snips used to trim grass that's too close to barriers for the power mower to reach; all others are force amplifiers.

We use hand-operated force amplifiers whenever we crank something.[2] If you haul a boat onto a trailer, you use the crank handle of the winch or windlass so a lot of arm motion by the crank exerts a lot of force on the wire cable. Door handles and faucet levers are just small-scale cranks. Our bias toward force amplifiers shows in the way we use the term "mechanical advantage" for "force advantage"; evidently distance advantage isn't as advantageous. Thus we say that a lever in which the effort-to-fulcrum distance is twice the load-to-fulcrum distance has a mechanical advantage of two, not one-half.

A biologist would have made the opposite choice. In nature, distance amplifiers rule the roost. They do so because muscles are short-stroke engines; they make lots of force but shorten a relatively short distance. A muscle does its greatest work (again, load times distance) when it short-

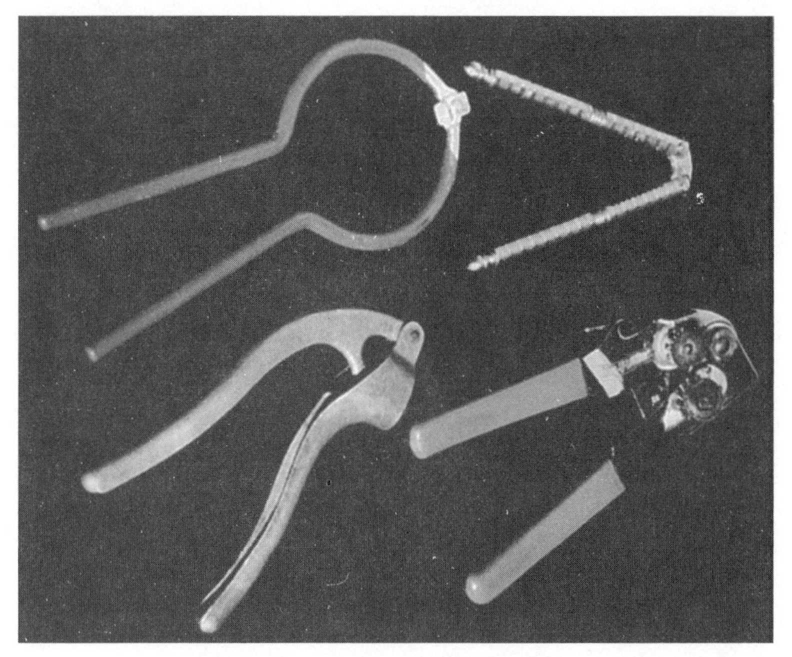

FIGURE 9.3. *Household devices using levers that increase force at the expense of distance or speed: jar lid loosener, nutcracker, garlic press, and can opener.*

ens by only about 10 percent of its length, even though most muscles can shorten as much as 30 percent with lighter loads. To make arms and legs swing through substantial arcs, to make their far ends move substantial distances, to permit muscles to run close in alongside the bones that they move—such tasks require distance amplifiers. Consider the muscles and bones that raise and lower your forearm, shown in Figure 9.4. Both the biceps muscle of the front of your upper arm and the triceps behind your humerus are substantial distance amplifiers.[3]

So our muscles are force specialists, and our bodies compensate with distance-amplifying internal levers of tendons and bones. That makes our appendages distance specialists, for which our technology compensates with force-amplifying hand tools—our can piercers and pliers. Paradoxical, even irrational perhaps, but we do prefer to use our tools without first being surgically adjusted.

When substantial force is still needed, less drastic distance amplifiers are used, as with the attachments of its muscles to the forearms of a mole,

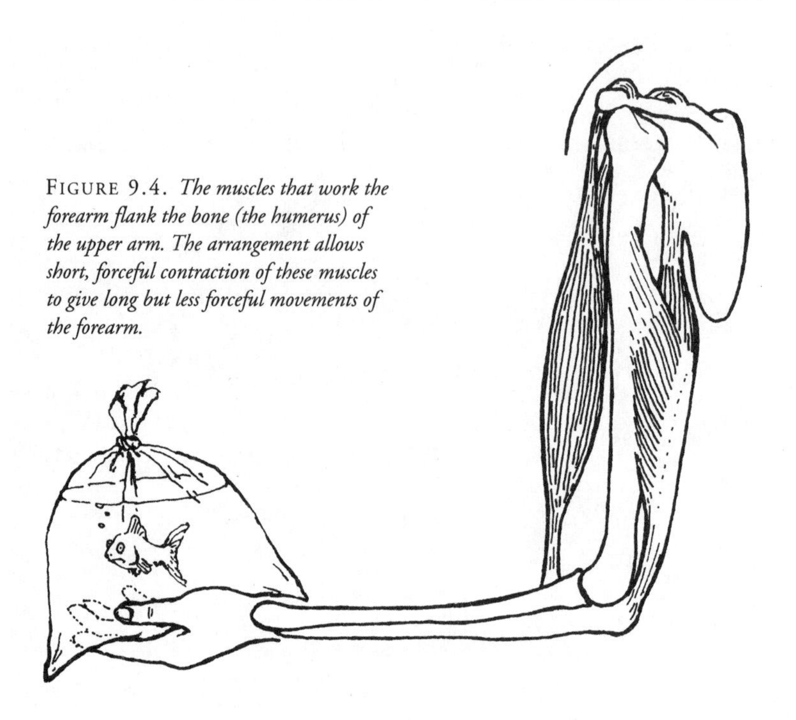

FIGURE 9.4. *The muscles that work the forearm flank the bone (the humerus) of the upper arm. The arrangement allows short, forceful contraction of these muscles to give long but less forceful movements of the forearm.*

which burrows through soil. Its bones are shorter and the muscles attach farther out from the joints than do our equivalent ones. The same choice sets the relative positions of jaw muscles and teeth in mammals and reptiles.[4] The front teeth of a protruding jaw snap at prey, with lots of distance but little force, while the rear teeth, closer to the jaw-closing muscles, chew and grind with lots of force. Still, both the mole's forearms and the lion's molars remain distance amplifiers, just less drastic ones than our forearms and the lion's incisors. Conversely, where great distance (or speed) is needed, distance amplifiers get really extreme. The flight muscles of many insects shorten less than 5 percent of their lengths—such a short stroke may be necessary for insects to achieve their high wingbeat frequencies—and since the muscles are tucked inside the insects' thoraxes, they have to be short. Insect wing tips may move as much as a hundred times farther than their muscles shorten.

To move body parts either farther or faster, nature often raises distance advantages by hooking up muscles in a peculiar way. Instead of running straight from the tendon on one end to the tendon on the other, the fibers are short and oblique, usually running from a pair of outer tendons

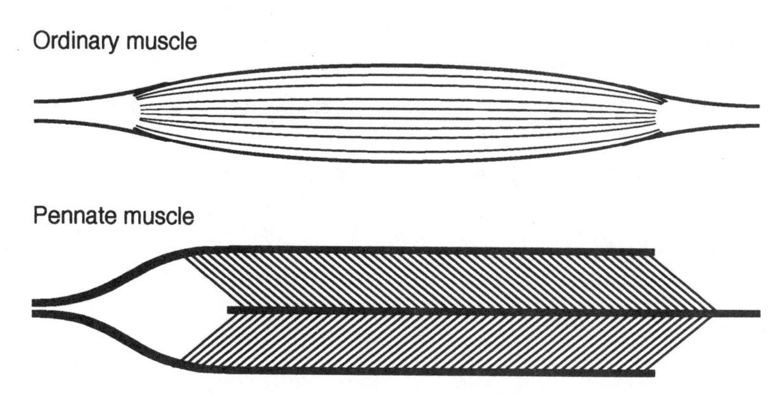

FIGURE 9.5. *An ordinary muscle, such as one of those of the previous figure, and the really drastic high-force, low-distance arrangement of a pennate muscle.*

to a central one, as in Figure 9.5. The outer tendons attach to one skeletal element, the central one to another. The arrangement greatly limits the muscle fibers' shortening distance, but it much enlarges the total cross-sectional area of the fibers, on which the force depends. The arrangement is especially common in insects and crustaceans, which need it because their muscles run inside their tubular skeletons. With muscles inside, getting a substantial sideways distance between the hinge of a joint and the attachment of a muscle is nearly impossible. Muscles simply can't shorten very much, so they have to produce a lot of force and then turn to distance amplification. A lobster claw is a fine case, closing its pincer with one of these pennate (penlike) muscles and opening it with another. A look at the relevant anatomy might provide adequate excuse to eat one;[5] otherwise Figure 9.5 will have to suffice. Another example is the fattest part of the hind leg of a grasshopper; fiber directions correspond to ridges visible on the outside.

Several other arrangements let nature move things farther, faster, and less forcibly than does a direct muscular hookup. Near the end of Chapter 7 a class of hydrostatic devices called muscular hydrostats was mentioned; examples included the tentacles and arms of squid, various tongues, and the trunks of elephants. If a squid's tentacle is made entirely of muscle, as it very nearly is, and if it's long and thin, as it certainly is, then a slight decrease in diameter ought to give a great increase in length. If muscle contraction reduces the diameter of an incompressible cylinder by 10 percent, then the cylinder will lengthen by almost 24 percent. But a tentacle is already long and skinny. If its length is twenty-five times its diameter, then in absolute, not percentage, terms a one-unit decrease in diameter will give an almost sixty-unit increase in length. Force, of course, will work the other way, with a sixtyfold decrease in force from input to output. Squid snag moving prey by extending tentacles far and fast but not very forcibly.[6]

A more widespread and personally relevant way to make the most of limited shortening is to wrap a thick coat of incompressible muscle around a spherical chamber, as in a heart. A little calculation reveals that if the volume of the wall is twice that of the chamber inside, shortening the muscle fibers by 6 percent will drive out half the contents of the chamber, and shortening by 13 percent will fully empty the chamber. Six percent and 13 percent span muscle's efficient range of shortening.[7] Similar calculations can be done (the reader might be inclined to try) for muscle-wrapped cylinders, as in our intestine. The wall is thinner, a little greater shortening is needed, and we use a different kind of muscle, but the principle is the same.

Nor is the problem of converting high force to high distance unique to muscle-driven machines. The proteins that ratchet past one another in a cilium are near enough to the center of the cilium so a little ratcheting gives the whole thing a substantial bend. And the wilting of the leaf of a plant from horizontally outward to vertically downward takes only a small loss of volume of a few cells in its stem, as shown in Figure 9.6.

The engines of our technology are more diverse in their force-distance behavior than are muscles, cilia, and swelling cells. But in most instances their moving parts travel far but less forcibly—quite the opposite of living engines. What particularly characterizes engines of recent vintage—whether electrical or combustion—is that they move fast. High speed has consequences similar to long distance: Power output is force times speed just as work output is force times distance. Higher speeds let us use smaller and lighter engines. For instance, we use particularly high-speed motors to power hand-held power tools and aircraft, cases in which size and weight must be minimized. In effect, higher speed means that a given engine is used more often. Higher speed also means that we often have to slow things down between engine and application, and a speed-reducing transmission is a force-amplifying one.

In short, muscles most often need distance-amplifying things like tendons and bones to couple them to their tasks, while rotary motors commonly need force-amplifying gearboxes to drive useful machines.

Still, distance and speed aren't quite the same, and our arms and legs

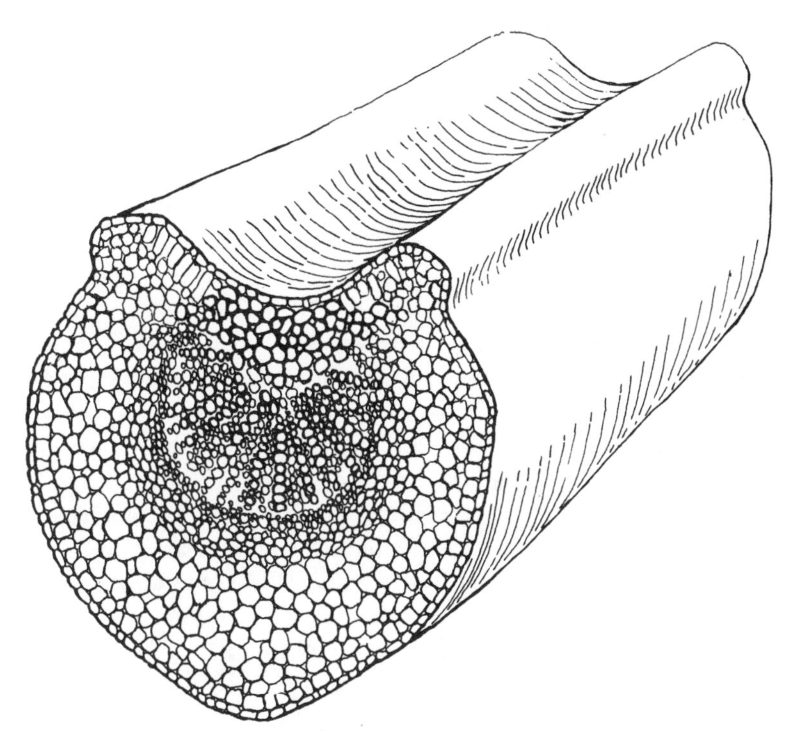

FIGURE 9.6. *Osmosis can generate spectacularly high pressures. Slight but forceful swelling and shrinking of the large cells in the lower part of this leaf stem (petiole) raise and lower the leaf's blade as sunlight and water supply change.*

move relatively far while the engines of human technology move relatively fast. As a result, motorizing a hand-operated machine rarely works well. Years ago almost every household had a hand grinder designed to be turned about once each second by a willing arm. Most of its tasks now get done by a food processor, a radically different design with a directly connected electric motor. Less common is a grinder of the old design but with a motor. We have one, and its performance is unimpressive.

Where simple motorization works, though, the comparison is instructive. Consider the hand-operated and electric forms of meat grinders and ice-cream freezers. For hand operation a long crank is used, the usual force amplifier undoing the distance amplification between arm muscles and hand. The motorized versions replace that long lever with motors that turn between fifty and a hundred times each second, so a problem of excessive speed replaces one of excessive distance. Lest the motor stall or the contents be immediately liquefied, a set of gears must be used to reduce the speed and increase the force. We make textile thread with machines not at all like old spinning wheels, spinning blade power lawn mowers have displaced the motorized reel type of mower, and the most effective bits for use in electric drills differ from those that work best in a hand-cranked drill.

WHEELS

Human technology turns on wheels. Except for the odd sled or sledge, all our land vehicles ride on them. Our ships are driven by rotating paddle wheels or propellers. Our aircraft use rotary turbines with or without rotary propellers. Snow blowers, trench diggers, conveyor belts, and chain saws go in circles around axes. Internally we use crankshafts, rotary electric motors, rotating pulleys, gears, capstans, hinges, cams, windlasses, ratchet wheels, roller bearings, and spindles—just to mention the more obvious ones. Nature, with only one exceptional case, doesn't go around using wheels. Only metal use provides as dramatic a contrast.

When I was a student, the issue was simply put. "Nature has never invented the wheel," went the textbooks. But science progresses, and we now know that a perfectly fine and true wheel and axle does occur in nature. It's new only to us since the organisms equipped with wheels are of enormous antiquity. The discovery of this wheel—by Howard Berg and his associates during the 1970s[8]—demands that we ask not why nature hasn't invented the wheel but why she uses it in only one instance. First, of course, we have to examine that instance.

The talk about cilia and flagella in the last chapter quietly excluded bacterial flagella. They're much smaller than the standard appurtenance of higher organisms, and they lack the "normal" internal gear with which cilia actively wave around. In the high magnification of an electron micrograph, a bacterial flagellum usually looks like a carefully drawn set of very regular waves—suspiciously like a rigid structure. It turns out to be a rigid helix much like a corkscrew. Instead of passing one wave of bending after another along its length, it spins around, ten to one hundred times each second, as in Figure 9.7. The base of the flagellum forms a driveshaft that passes through the cell membrane, connecting it to a rotary engine. And the membrane works like a proper set of bearings. The engine bears a curious similarity in both appearance and operation to our electric motors. It's even reversible. The whole thing—engine and

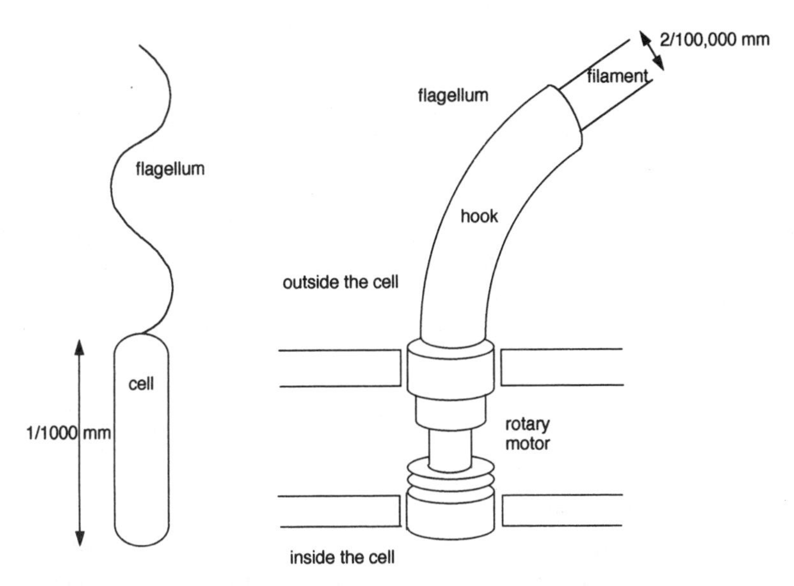

FIGURE 9.7. *A bacterium with its flagellum, and the base of the flagellum in more detail as we presently understand it. The magnification here is extremely high, about three hundred thousand times, so specific details represent interpretations of smudges. How this rotary engine works is far from clear.*

corkscrew—either singly or in groups, pushes or pulls a bacterium around much the same way a propeller pushes a ship or pulls an airplane.

How well does the bacterial flagellum perform? In terms of power output per unit weight it's more than fifty times better than muscle, better even than a gas turbine. Still, even protozoa, some only a little larger than a typical bacterium, swim with conventional flagella. These latter rely on the much less potent tubulin-dynein engine described earlier. This situation is powerfully puzzling. Could it be that key bits of information simply never got passed on to the nonbacterial world? Odder yet, higher organisms have adopted other bits of bacterial machinery by the drastic step of symbiotically expropriating the bacteria themselves, as noted back in Chapter 2.[9] Perhaps for some reason the bacterial engine can't be enlarged: impracticality of making larger bearings, difficulty in electrical transmission over any longer distance in a nonmetallic world, or something else entirely.

By wheels we mean proper wheel and axle devices that can rotate without limit with respect to the rest of a machine. If you roll down a hill, your whole body may be a wheel, but you're no wheel and axle. So we're not talking about tumbleweeds, or about the tiny turds that dung beetles roll homeward for their grubs, or about a few crustaceans that get around by rolling as a whole. Nor are we worrying about how far we can rotate our fists around our arms or our heads on our shoulders. By "rotation" we also mean something fairly specific. When you draw a circle on a piece of paper, do you rotate your hand? You may move it in a circle, but you don't truly rotate it; after all, your hand at all times points in the same compass direction. Human dances make elaborate use of such circular but nonrotational motion, most likely because it doesn't make us dizzy. Not all dances, of course; waltzes are rotational and, one suspects, intentionally vertiginous. The wheels of a bicycle rotate; your feet and the pedals just move around in circular paths. The Ferris wheel rotates as a whole, but the seats and people just go in circles. In this precise sense—excluding both rolling as a whole and merely going in circles—the only known instance of a wheel and axle in nature is the bacterial flagellum.

The classic view is that wheels are terrific, but that nature (sweeping the bacteria under the rug) just can't figure out how to make them do decent service. Stephen Jay Gould made just this point,[10] noting the difficulty of moving nutrients into structures with sliding connections to the rest of an organism. That suggestion fits nicely with one of the main points here; the different contexts of human and natural technologies. He mentioned, additionally, the problem of evolutionary continuity. How could an incompletely evolved wheel have conferred benefit on a creature? The argument, like the equivalent one made here for metals, may be attractive but isn't fully persuasive. Gould's whole discussion, though, hinges on the absolute superiority of wheeled transport.

As with so many issues, close scrutiny uncovers complications. At least two people, Michael LaBarbera and George Basalla, a biologist and a historian,[11] have taken exception to Gould's freewheeling assumption of superiority. They admit that wheels may give cheaper transport than legs. After all, bicycling is more efficient than walking or running; adding twenty-five pounds of passive machinery reduces the cost of transport severalfold. But they point out that wheels realize their advantage only on smooth surfaces—on roads or floors. A wheel on a cart can go over bumps no higher than a quarter of its diameter, and even those bump up the cost of transport. Prairie schooners had enormous wheels, much larger than the ones on ordinary on-the-road wagons. To use wheeled transport routinely and productively, a culture must be sufficiently settled and organized to do a lot of civil engineering. For most organisms, much smaller than we, the natural world is an even bumpier place.

Wheels, however wide our use of them, don't define human technology. They seem to have appeared first in the Middle East, about five thousand years ago, in two applications: wheeled vehicles and potters' wheels. Whether the two were linked or the wheel was invented on more than one occasion remains unclear. But they were absent from the pre-Columbian New World; pack animals and sledges were used for transport, and even axisymmetrical pottery was built up by coiling long, thin cylinders of wet clay rather than by turning on potters' wheels. A visit to a museum of Incan or Mesoamerican artifacts should persuade anyone that these weren't technologically primitive cultures. Moreover, they knew about wheels; Mayan and Aztec toys have been found in which animals had wheels on the ends of their legs.[12] Nor was the Western Hemisphere the only site of non-wheel-ridden cultures. The inhabitants of sub-Saharan Africa, Southeast Asia, and Australia got along without them as well.

For wheeled transport to make sense, streets must be wide enough for a cart and draft animal to turn around since the combination can't easily back up. Draft animals of sufficient size must be domesticated, and that seems not to have happened in pre-Columbian America. Furthermore, roads must be hard. Shag or well-padded carpet almost completely immobilizes a rider-propelled wheelchair, and government standards prohibit thick carpeting in public buildings. Wheeled vehicles were given up in North Africa and the Middle East between the third and seventh centuries (C.E.) and not used there again for a thousand years. Domesticated camels that carried rather than pulled apparently provided a better alternative under the circumstances.[13]

So one may be skeptical of claims of wheeled transport's absolute superiority. But the arguments don't shed much light on other uses of wheels in technology. Here I think the case for the broad utility of wheels and axles is much clearer. Wheel-based transmissions are highly versatile and efficient. Pairs of simple gears (Figure 9.8) routinely pass power from one shaft to another with better than 99 percent efficiency. The efficiencies of such things as bevel gears (which connect shafts at right angles), belting, and roller chains range between 95 and 99 percent. In worm gearing, a long, turning driver (the worm) slides through the teeth of the gear it turns (the wheel), but even this turns out to be about 80 percent efficient.[14]

Roads may limit the use of wheels for vehicular transport, but bear-

FIGURE 9.8. *Several kinds of gears and such, a far from exhaustive selection (clockwise from the upper left): spur gears, a cam, a worm and wheel speed reducer, and bevel gears.*

ings limit their general use as mechanical elements. Doing without bearings is possible only by using such unattractive schemes as placing free rollers (such as tree trunks) under an object and moving the rollers as they come free from the rear to the front of the object. Large stone blocks were certainly moved this way by the builders of ancient monuments, and I've found it a useful way to move long, unsplit logs for short distances. Alternatively, of course, the whole vehicle can simply roll. But in properly upstanding vehicles some parts must slide against one another, the sliding surfaces must bear loads, and friction between sliding surfaces costs power. So a cart needs smoothly turning bearings, either between the wheels and their axles or between the axles and the frame.

So everything turns on bearings and on the twin problems of friction and wear. These must have made big trouble until metal lathes of decent precision were developed a few hundred years ago. The underlying problem is a nasty one. A thicker axle gives good support with low stresses on its material, but its bearing surfaces move around one another at higher speeds and thus with more heat and wear. The opposing surfaces of a thinner axle move more slowly, generating less heat and wear, but stresses are higher, and the axle is more likely to break. Good bearings, like electric wires, are mundane but critical items whose present excellence we take for granted—except, of course, when they seize up or screech thirstily for lubricants. Or during wartime, when ball-bearing factories become targets of the highest priority.

Even though she uses no wheels, nature makes some impressive bearings. On the ends of our bones are layers of porous cartilage. A lubricant (synovial fluid) oozes through them, so adjoining bones never actually touch each other. The resulting friction is about as low as anything in engineering practice—unless you suffer from arthritis or bursitis. One can get a feel for the impressive slipperiness of our joints by dissecting the muscle and connective tissue away from the knee or hip of a lamb and then moving the bones on either side of the joint while pressing them together as hard as one can. At least perceptually, the joint is frictionless.[15] Nature's bearings won't bear blame for the absence of the wheel and axle.

In using wheel-based devices, humans can't copy nature. Our widespread use of them proves their technological value better than any verbal argument or calculation. Their scarcity in nature, though, isn't so simply explained. At least we don't understand what mix of factors might matter. Are they less useful to an unstiff technology in a bumpy world? Are they hard to evolve and maintain? Had they worn out their welcome when organisms got much over a thousandth of a millimeter long?

MATING MOTORS AND ROTORS

The lack of wheels and rotational motion among organisms, on one hand, and their commonness among human devices, on the other, raise a peculiar problem: how to turn a rotary machine. If, as in every human- or animal-powered machine, the engine doesn't rotate, then linear or reciprocating motion must be converted to rotary motion. We commit mechanical miscegenation when our pushes and pulls work rotating machines. But the conversion problem doesn't face only muscular drivers.

How can a push-pull engine, whether it has muscles or moving pistons, make rotation? The engines of our cars use piston rods to turn crankshafts, and some of us recall the slider cranks attached to the driving wheels of steam locomotives. But these devices weren't common until late in the eighteenth century, when James Watt introduced a version of the crankshaft so his stationary steam engines could turn wheels.[16] Nor was

Watt the first to get rotation from a reciprocating engine. Rotary motion had sometimes been produced with Newcomen beam engines, using a heroic but inefficient scheme: The engine drove a pump that raised water that then dropped onto an overshot waterwheel.

This problem of getting repeated rotation out of muscle can be circumvented if the engine itself goes in circles, as happens when an animal tethered to a long cranking arm walks in monotonous circles. Otherwise one needs cranks like those of our piston engines. Modern cranks may have been invented by Archimedes, about 200 B.C.E.; they're first mentioned in Chinese literature in 31 C.E., surely as an independent development. But the idea goes back to the ancient Egyptians, although their version required practice and skill to use effectively. The difference between theirs and ours lies in how much freedom of motion the main shaft is allowed. Most of our cranks use fully guided rotors in which a set of bearings keeps the main shaft from doing anything other than turning about its axis. The operator's hand or foot is guided around a predetermined circle, as in an old-fashioned meat grinder or a bicycle. Using the opposite extreme, an unguided rotor such as a spinning lasso, takes real dedication.

Intermediate is the partially guided rotor, which is what some Egyptian relief sculptures seem to show. As in Figure 9.9, one end is fixed and the other swings in a circular path. It's all too prone to wobble; with only one end fixed the operator has to keep the shaft from swinging to one side or the other. The scheme does make a perfectly serviceable drill, although one that takes a little practice to operate and works best for shallow holes, where a little wobble is acceptable.[17] We still use a few partially guided rotors. A conventional brace and bit require that the operator maintain the alignment of the drill shaft. But alignment only requires that the aligning hand act as a stationary upper bearing rather than move in a specific circle. On the other hand, the brace and bit occupies both of the operator's hands, while the old Egyptian device left a hand to hold the object that was being worked on. A hand-operated eggbeater is another partially guided rotor, and like a brace and bit, it demands a little skill.

A device that's now fully obsolete, the bow drill, may be used to make a muscle-powered arm drive a rotor. As in Figure 9.10, a flexible cord connects the ends of a bow; the cord is looped around the shaft to be spun. The lower end of the shaft holds a drill or is otherwise pointed, while the upper end turns against a hand-held thrust bearing. Pulling the bow back and forth makes the shaft revolve first one way and then the

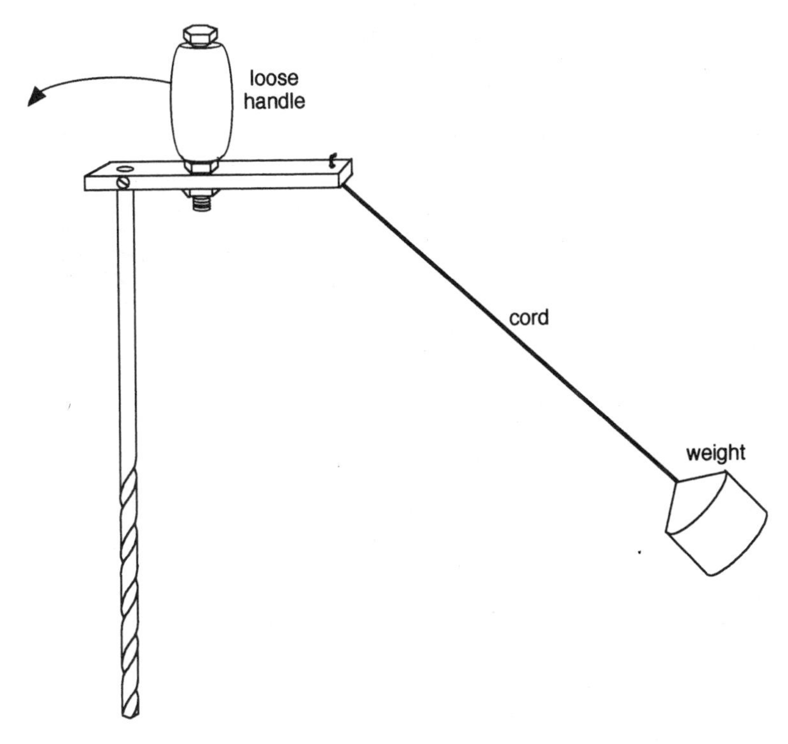

FIGURE 9.9. *A partially guided crank, in this case an interpretation of the mechanics of an ancient Egyptian drill. The details come from my test version rather than actual antiquity. The operator grasps the handle and moves it in a circle, watching to make sure the drill stays vertical and doesn't wobble. The weight swings out and around on its own; it smooths the motion and provides downward force on the drill.*

other. True conversion to rotary motion isn't strictly involved, since periodically reversing the shaft balances the clockwise and counterclockwise turns. Still, it realizes most of the advantages of rotation for tasks such as drilling. Bow drills are ancient and multicultural. North American Indians, who used no other obviously rotary devices, used them to generate sufficient heat (through friction with the lower bearing block) to start fires, a fine, nonobvious (as a patent office would put it) invention.

HYDRAULIC CONNECTIONS

In a hydraulic device something solid gets pushed by a pressurized liquid. (Where a gas gives the push, it's a pneumatic device, but the principles are

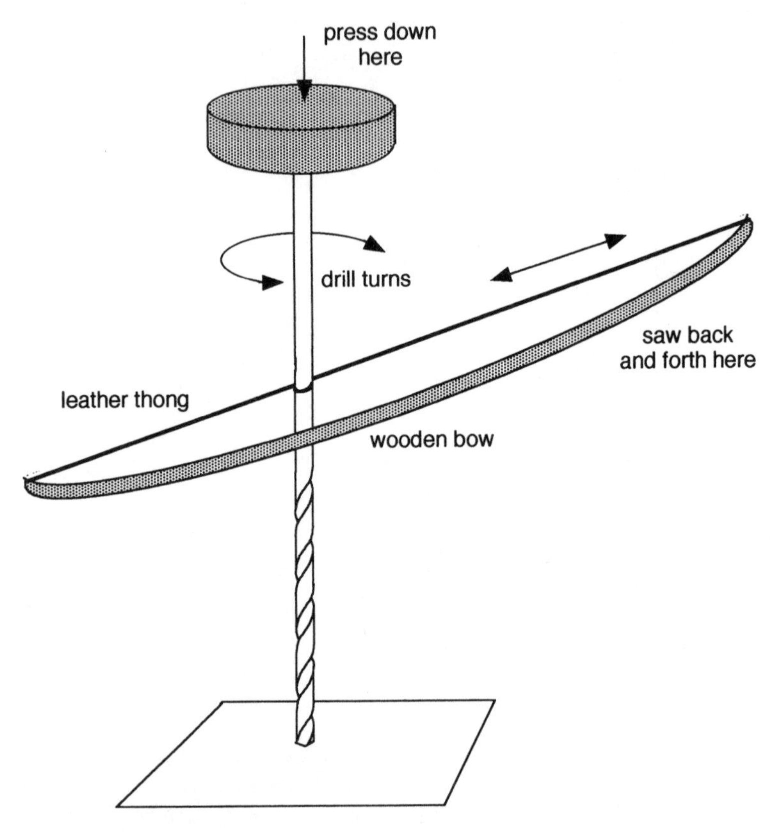

FIGURE 9.10. *A bow drill, shown with a modern bit. The upper, hand-held block has a small concavity on its lower side to guide the drill. Making and using one of these are easy, although you'll need to roughen the bit's surface so the string or leather thong doesn't slip.*

the same.) We've already run across such devices in various guises. In Chapter 7 we looked at hydrostatic and aerostatic supportive systems, such as worms and blimps; in Chapter 8 we considered water transport, water absorption, and evaporative engines (mainly in plants); and earlier in this chapter we talked about muscular hydrostats like a squid's tentacle. Except for a few aerostatic blimps and buildings, all the examples involved organisms, and they ranged from unicells to trees and sharks. Nature apparently finds this an "easy" technology. That shouldn't surprise us since nature pumps a lot of fluid around for other purposes and since a muscle encircling a spherical chamber or a cylindrical tube suffices for pressurization.

The brakes of cars are our most familiar hydraulic machinery. You push on a pedal that pushes a piston farther into a cylinder. That in turn forces brake fluid out of the cylinder, through some pipes, and into another cylinder, where the fluid pushes another piston outward. The second piston then pushes the brake lining against the brake drum (or disk), creating lots of car-slowing friction. The system works because a pressure applied at any point in a closed fluid-filled system appears everywhere else in that system; pressure is transmitted almost undiminished and instantly.

The handiness of hydraulics comes from its basic principle. The pressure in the hydraulic fluid is the force on the piston times the area of its face, and that pressure remains the same throughout the system. Thus a wide range of forces can be produced by doing nothing more devious than changing the face area of the pistons. If, as in Figure 9.11, you push inward on a piston with a force of ten pounds in a cylinder one square inch in cross section (a pressure of ten pounds per square inch), a piston in a cylinder of four square inches will be pushed outward with a force of no less than forty pounds. Although this sounds like getting something for nothing, work and energy in fact are properly conserved. The piston in the big cylinder may move more forcibly, but it won't move as far. If you push the small piston in an inch, the large one will move outward only a quarter of an inch. In short, the device behaves just like a lever. What a nice way to move force, work, or power from one place to another, trading force and distance (or force and speed) whenever convenient!

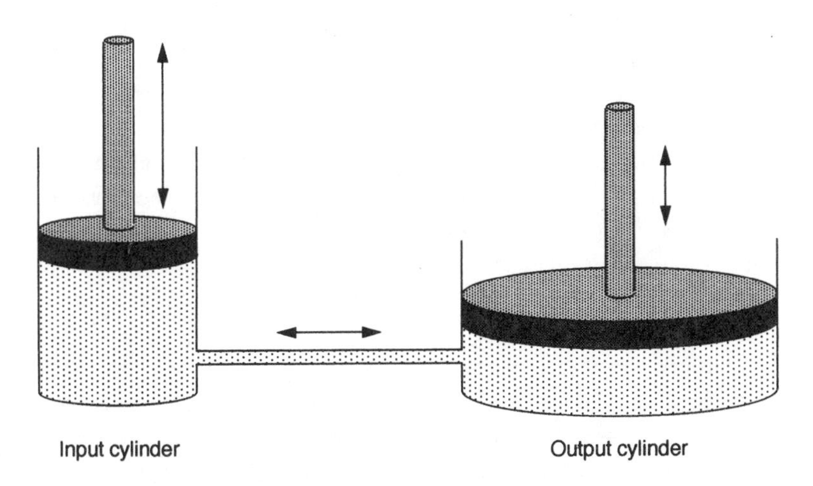

FIGURE 9.11. *A hydraulic connection that achieves a fourfold force amplification.*

Where does nature use hydraulic transmissions? Put a freshly cut flower in a vase of water, and the water is hydraulically sucked up the stem from the vase. A starfish moves around on a thousand or so tiny tube feet, hydrostatic devices interconnected in a low-pressure hydraulic system. A worm adjusts the pressures in its sequential compartments so that with the aid of some rearward-pointing bristles, it can penetrate soil. A spider flexes its legs with an ordinary set of muscles, but instead of using extensor muscles, it moves its legs outward hydraulically; it squeezes together the top and bottom of its body (the cephalothorax), increasing its blood pressure and extending its legs. On emerging from its pupal skin, a butterfly briefly pulls its abdominal segments inward to increase its blood pressure, inflating the veins of the wings, and expanding their membranes. We run the filtration equipment (glomeruli) in our kidneys hydraulically, using our high arterial blood pressure to force plasma (but not blood cells) through tiny pores as the first step in making urine. That's partly why keeping proper fluid balance depends on having a good heart. We (male humans and some but not all other male mammals) use the same blood pressure to erect our penises. At least we do so in part, with valves and local muscle activity then pushing the pressure up to much higher levels. Hydraulic devices are nothing if not widespread among organisms.

But for human technology, hydraulics and pneumatics have come harder. We have no handy squeezer like muscle, and most of our manufactured pipes are rigid. We make pressure with pistons; indeed we use pistons for little else. But pistons must fit snugly and smoothly in their cylinders if they're to slide freely yet not leak. (Detecting such leakage is the purpose of a compression test on an automobile engine.) The ancients could do none of this, and hydraulic devices were limited to siphons and other pistonless schemes that worked at close to atmospheric pressure. Leakage between pistons and cylinders limited the pressures that could be used in early steam engines. Now that leakage is a manageable problem, hydraulic and pneumatic devices find an ever wider variety of uses. The technology may have presented great difficulties early on, but it turns out to be so marvelously handy. But while our applications resemble nature's, our machinery looks thoroughly different.[18]

Consider, again, the brakes of a car. These days the brake pedal works not one but two cylinders, a safety feature to make the system leak-tolerant. If your car has power-assisted brakes, the reduced pressure associated with the engine's intake works as a pneumatic element to generate the hydraulic pressure. If the system leaks a little, you add extra brake fluid to the reservoirs or remove air or froth from the system ("bleeding" the brakes). Some cars use hydraulic connections to run their clutches as well; the device simplifies construction of models available with either right- or left-hand drive. Farm tractors usually have a hydraulic system that raises and lowers various attachments, with the hydraulic pressure produced by an engine-driven pump. Control surfaces—rudder, elevator, ailerons, and flaps—and retractable landing gear on aircraft are often hydraulically operated. Heavy construction equipment steadily increases its use of hydraulic power transmission. Closer to home, door closers, shock absorbers, and hand-operated squirt or spray bottles are all hydraulic or pneumatic devices.

A particularly subtle and clever use of hydraulic power transfer is basic to almost every automatic automobile transmission. These fluid couplings aren't especially efficient—early automatic transmissions had to be water-cooled to dispose of their waste heat—but they have the nice property of slipping badly (or usefully) when run slowly. That means that the driver can let the engine idle impotently without mechanically disconnecting it from the wheels (as a manual clutch does). The basic scheme of a fluid coupling is shown in Figure 9.12. It works in the following way. First, any mass prefers to go straight rather than to turn in a circle; if you turn around while holding a string with a weight at the other

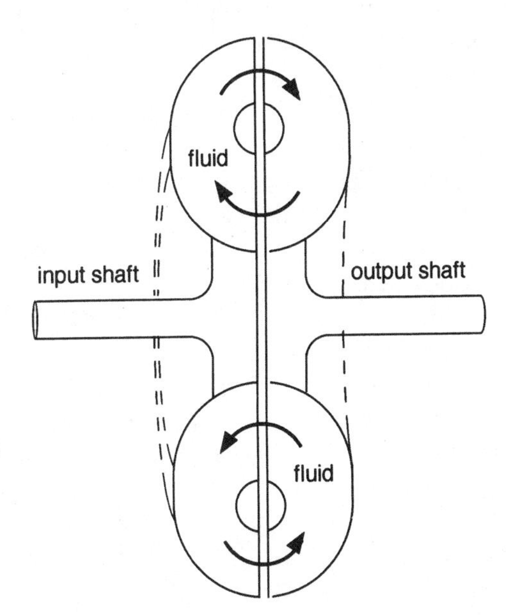

FIGURE 9.12. *A fluid coupling. Input and output shafts each have hollow half toroids running around their circumference. The half toroids, filled with oil and baffling, are sealed together but are free to slide against each other. Imagine the well-buttered sliced surfaces of a halved bagel pressed together.*

end, the weight flies outward and pulls the string taut. Liquids have mass, so when spun, the liquid in a chamber tries to move outward, experiencing what's ordinarily called centrifugal force. Second, moving mass in toward the axis of rotation makes a system rotate faster, a notion familiar from watching skaters increase their spinning rates by pressing their arms closer to their bodies. (To look up the details, try "conservation of angular momentum.") If fluid moves outward as a result of the spin of an input shaft and moves inward again in an adjacent chamber attached to an output shaft, the output shaft will speed up. Rotary motion and force (together, torque) have thus been transferred from input to output. Within limits, the faster the input shaft is spun, the more efficient the coupling.[19] So when you speed up the engine, the wheels recall that they have an engine attached.

BRIEF BATTERIES

Speeding up or slowing down anything that has mass takes force. To exert a force over a distance takes work, and that implies expenditure of energy. Yet a pendulum repeatedly accelerates downward and decelerates upward, needing only slight periodic pushes to keep it going. Where does the pendulum's energy come from? What it does is shift its energy investment four times for each full swing. As it rises and slows, it gets out of energy of motion (kinetic energy) and into gravitational energy. Only a small amount (offset by the occasional push) is diverted to heat. As the pendulum swings earthward again, the gravitational investment is cashed in and reappears as increased speed. It then swings up the other way and does two more energy transfers. In effect, the pendulum repeatedly stores kinetic energy in a gravitational battery.

Besides using gravity, how else can energy be stored? The next most obvious agency is elastic resilience, the reversible deformation of material—what the springs of our cars do. In addition, rechargeable electric batteries are common in cars, portable appliances, the clocks of computers, and elsewhere. Finally energy can be stored inertially, by making a flywheel spin; a machine can then be run by slowing that spin. Some old potters' wheels took only an occasional push, maintaining their motion by doling out the energy of that push.

Human technology uses all four storage modes: gravitational, elastic, electrical, and kinetic. Pendulums have long served as speed regulators for mechanical clocks, and counterweights within the frames of windows and in the shafts of elevators allow us to raise heavy things with the energy from previous lowerings. In a medieval catapult, energy was slowly put in by raising a counterweight; letting it fall released the energy, which was used to throw a missile. We occasionally use gravitational storage on a large scale when we generate electricity; the scheme is called pumped storage. Our use of electricity varies a lot—but predictably so—with the time of day, so utilities must be able to adjust production. Nuclear plants aren't much help because they cost so much to build and so little to run that they're best operated near full capacity. Coal-burning plants are better for meeting intermittent demand, but they leave a utility with a lot of unused capacity for much of each day. And electrical batteries to smooth out the demand would be uneconomically large. A pumped storage system makes no new electrical energy. It just uses excess capacity when demand is low to pump water into an uphill reservoir; it then operates as a hydroelectric plant when demand is high. But it requires a handy mountain and local people willing to tolerate a reservoir whose level fluctuates daily.

Local acceptability can't be brushed aside. Pumped storage works best in undeveloped hilly country near centers of population—a formula for maximizing offense to people who value access to scenic nature. During the 1960s New York City's main electrical utility proposed building such a plant about fifty miles up the Hudson River, near West Point, in the beautiful and historic Highlands. A mighty outcry eventually put an end to the matter.[20] As a biologist who grew up only a few miles from the site I am perhaps predictably biased: Keeping the Highlands unsullied merits a little energy conservation in the lower Hudson. One must remember that, their image makers notwithstanding, utility companies are in business to sell, not save, power.

More common is elastic storage, although most of the springs in our cars and appliances are there more for our comfort or because they permit simpler designs than for storing great amounts of energy. Mechanical typewriters used large numbers of springs, and every dishwasher, tape player, and camera has a few. They provide more serious storage in clocks, watches, and wind-up devices, such as toys and an occasional lawn mower starter. As mentioned a few chapters back, metals have high resilience and accept a wide range of deformations: tensile, compressive, flexural, and torsional. In addition, springs may be nonmetallic (rubber bands, for instance) or even gaseous. Squeeze a gas such as air, and it can reexpand with almost no loss of energy. Elastic storage was of great importance in weaponry before explosive charges came into use: An arrow is propelled by energy stored in the bent bow, and the great rock-throwing ballistae of classical antiquity used twisted cow tendon. Still, the use of elastic storage has declined relative to the next scheme.

Until recently most rechargeable batteries stored electricity to restart internal-combustion engines. Such batteries now grace all kinds of small appliances, even though they're heavy for the energy they store; electric vehicles run into just that drawback. In addition, batteries deteriorate far more rapidly than springs and pendulums. Fashion may play a role; fine wind-up mechanical shavers, devices that could be left in car or briefcase and didn't care at all about chargers or local voltage, were available a few years ago.

Finally, we use an occasional flywheel, nothing more than a large mass rotating in a wide circle. These simplest of short-term batteries may also be the most ancient; intermittently driven potters' wheels have been used for more than five thousand years. We use flywheels to smooth out the action of our engines, both piston engines and (in turntables and tape recorders) electric motors, and we use them in toys, such as tops, Yo-Yos, and tiny model cars. Occasionally one hears suggestions that automobiles might employ flywheels along with smaller internal-combustion engines that run at more efficiently high and steady speeds; the flywheel would intermittently supply energy for acceleration and hill climbing.

Two of these four schemes, gravitational and elastic energy storage, find wide application in nature, probably more than in human technology. Just about every way that animals move around—walking, running, jumping, flying, swimming—uses one or the other. Three of nature's peculiar disabilities cry out for short-term energy storage. One is her lack of wheels and consequent reliance on reciprocating or pulsating devices like legs and wings. A second is muscle's inability to reexpand itself and the resulting need for some easy way to undo its contraction; an opposing muscle is bulky and expensive. The third is that muscle contraction isn't instantaneous, and chemical explosives aren't part of nature's ordinary armamentarium, yet motion with high initial acceleration has obvious advantages: Predators from alligators to cats get their food by springing forward. Through slow accumulation and sudden release, energy storage permits pulses of extreme power output.

Elastic storage must be more common than gravitational storage simply because the latter works only for terrestrial organisms of significant weight and the bulk of the earth's creatures are aquatic or tiny or both. No absolute choice, though, need be made; a single animal may use both.

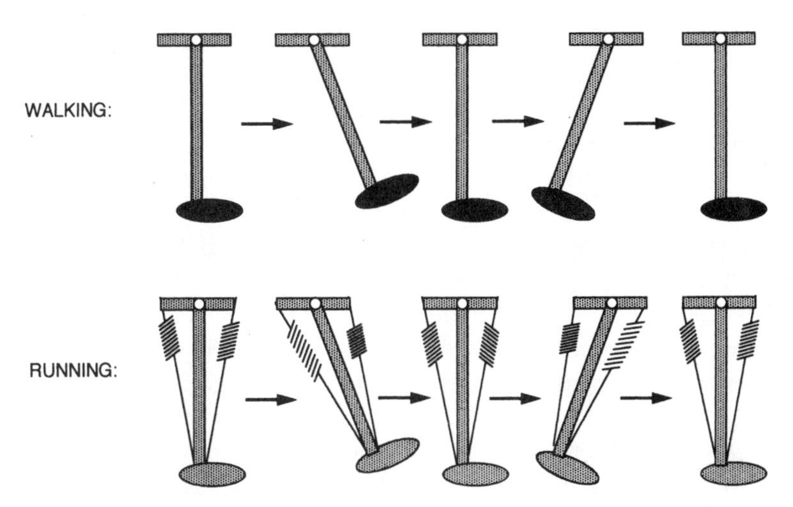

FIGURE 9.13. *The essential difference between walking and running. In the former what matters is the weight of the appendage since storage between strides is gravitational. In the latter the elasticity of tendons provides the equivalent reservoir for absorbing and releasing energy.*

Consider how we get around on our hind legs. Legged locomotion isn't especially efficient—recall that the cost of getting from one place to another can be substantially reduced by attaching a bicycle to oneself—but getting around on legs would be worse without energy storage. Gravitational or elastic? As shown in Figure 9.13, we use both, switching back and forth effortlessly and almost mindlessly. We cheapen our walking gait with a little gravitational, pendulumlike storage between strides. As we walk faster, we tend to swing our legs farther rather than more frequently. Just as with a pendulum, one frequency is "natural" or most efficient for a given leg length. Eventually, of course, your anatomy resists any greater amplitude, and you switch to a gentle jog, another gait altogether. For a typical human, that happens at about a twelve-minute mile.

R. McNeill Alexander, of Leeds University, who has done the best work ever on the mechanics of walking and running, found the rule that sets the switch point. Except for the value of a constant, the formula for the gait transition point is the same one that gives the period of a pendulum. Whether you're a crow or kangaroo shifting from walking to hopping, or a human or dog (or even an insect) switching from walking to trotting, you follow Alexander's rule. Switching happens when the square of your speed is around half of gravitational acceleration times the distance between your hip and the ground. For a middle-size human, again, that's about a twelve-minute mile or five miles per hour. For a smaller animal, the transition speed is lower; you walk while a small child, dog, or cat must trot to keep pace.[21]

While gravitational storage isn't of much value above the transition speed, energy storage is still very much in the picture. When jogging or running, you use elastic rather than gravitational energy storage, stretching tendons rather than swinging legs up and down. The storage mode differs, and the two gaits are unmistakably distinct. In a way, gravitational storage is the special case, used only as legs rise and fall in walking. All the other common gaits—trotting, galloping, hopping, and so forth—depend on elastic storage. A hopping kangaroo, for instance, regains about 40 percent of the energy absorbed in landing when it bounces up again. In storing the energy by stretching tendon, it's using the protein collagen as its battery—the same stuff the ancients took from cows for their ballistae and the main material of our own tendons. Collagen has a resilience of about 93 percent—that is, 7 percent of the energy put in when it's stretched fails to reappear in mechanical form when it springs back. That's not bad, better than the ordinary rubber we make from the sap of rubber trees. But where collagen brings home the bacon is in the amount of energy it can store relative to its weight—nearly twenty times that of spring steel.

The problem of appendages that repeatedly change directions isn't limited to terrestrial locomotion with legs. Insects beat their wings up and down at frequencies up to almost (in the smallest ones) a thousand times a second. We've known about (and wondered at) these remarkable rates for a long time. During the 1940s an unusually gifted Finnish investigator, Olavi Sotavalta, compiled a compendium of wingbeat frequencies simply by listening to them. Or perhaps not so simply: Sotavalta not only had perfect pitch but had trained himself to distinguish fundamental notes from overtones, avoiding, as he put it, the "soprano-tenor error." Tone-deaf people like me resort to microphones, recorders, and other bits of electronic assistance. (In my youth I tethered fruit flies by fine wires to phonograph needles.) Sotavalta's data have always proved reliable. Incidentally, one can force the frequency even higher by unloading the system—snipping the ends off the wings. The record is 2,218 beats per second, held (and checked by others from a recording) by the same Olavi Sotavalta;[22] that's by far the most rapid back-and-forth movement made by any organism.

As with stride frequencies in walking, wingbeat frequencies don't vary much for a given insect, and changes in wingbeat amplitude and other variables handle adjustments of flying speed. A sharp optimal frequency characterizes almost any gravitational and elastic energy storage system—why we use either pendulums or tiny springs to regulate our mechanical timepieces.

Flying insects use elastic storage, and in their wing hinges are pads of the best elastic polymer known to either technology, resilin, which got brief mention in Chapter 2. Resilin was discovered about 1960 by a great Danish scientist, Torkel Weis-Fogh, who measured its resilience as an astonishing 97 percent. Resilin thus loses only 3 percent of the energy put in rather than the 7 percent of our collagen. The difference in power economy, though, is probably trivial, for 97 percent is only a little better than 93 percent. What must matter more is the mischief caused by any energy that's lost: It turns into heat, and 3 percent is less than half of 7 percent, which means that the muscular flight motor can be run a lot harder without cooking itself.

Reextension of muscle was mentioned as another place where short-term energy storage is common. We use some gravitational storage every time we raise an arm. But we mostly use muscles in pairs or groups so one muscle can be used to reextend another. For instance (as in Figure 9.4), the biceps on the front of your upper arm both raises your forearm and extends the triceps on the other side of your humerus. That triceps both lowers your forearm and extends the biceps. A more important role for storage occurs in bivalve mollusks, such as clams and scallops. The two half shells are brought together and held together by muscles; elastic hinge ligaments reopen the shells and reextend the muscles. The process happens rapidly in scallops, which repeatedly clap their half shells together and squirt out water in brief bouts of swimming. Their hinge ligaments, of the protein abductin, have a respectable 91 percent resilience. The 9 percent loss to heat shouldn't matter much for a water-cooled machine that runs for only a few seconds at a time.

An important case of elastic energy storage came up in Chapter 5. Our hearts beat; they're pulsating pumps. Part of a heart's work pushes blood directly around our circulatory circuits. But part of that work, using the blood as a hydraulic fluid, pushes out—stretches—the walls of our arteries. In between beats the arteries then rebound, constricting elastically and pushing blood onward. In this way the elasticity of your arterial walls reduces the extreme blood pressure fluctuations generated by your beating heart and makes blood flow more smoothly through the capillaries and other small vessels downstream.

Finally, energy storage can be used to get high acceleration, something of especial importance to small creatures. To go any decent distance, a projectile needs a high initial speed—no matter whether it's jumping, being kicked, or being shot. A projectile's speed is at its highest when it leaves its pusher. The smaller the creature, the shorter the distance in which that speed ("muzzle velocity" for a firearm) must be reached. So a shorter distance demands a greater acceleration to reach the same final speed. A flea, a grasshopper, and a kangaroo have about the same takeoff speeds, but the flea must have about a hundred times and the grasshopper about ten times the acceleration of the kangaroo. The kangaroo can jump mostly with direct muscle power, but fleas and grasshoppers use their muscle to store up energy by deforming resilient material—resilin pads for fleas and chitinous cuticle for grasshoppers. They then use trigger mechanisms to release the accumulated energy in short order. Many plants do the same thing in order to throw their seeds. Their devices form a wonderfully diverse lot. One way or another, all crank energy into elastic material; triggering is variously done by drops of water, by drying to a critical point, by brushing against animals, and so forth.

We commonly use four different ways of storing mechanical energy for brief periods: gravitational, elastic, electrical, and inertial. Nature uses only two: gravitational and elastic. Here again, ours looks like the more versatile technology. Nonetheless, short-term energy storage matters more for nature. Why? Simply because most of nature's brief batteries solve problems that using rotational devices, explosives, and other such tricks, we don't ordinarily encounter.

Two general points about the engines and transmissions of the two technologies. First, the very intricacy of nature's alternatives perhaps best points up the usefulness of rotational machinery. Second, the behavior of engines interacts with the character of transmissions. Each technology makes a great diversity of transmissions perhaps because each, in its own way, calls on engines that are few in number and fairly similar in performance to do a wide range of tasks.

Chapter 10

ABOUT PUMPS, JETS, AND SHIPS

How can human technology help us understand the nonhuman world? Let's ask the question of more complex devices than we've worried about so far, turning to whole systems rather than individual components. We'll look, in particular, at three cases where we can make sense of what nature does by combining physical rules with the practical experience of human designers. By "make sense" I mean seeing order amid diversity and discerning rules that transcend mere accidents of ancestry. The three cases are pumps for moving fluids around, propulsion by means of jet engines, and swimming on the surface of bodies of water. Notice that all are physical rather than biological categories. Getting the most mileage out of these comparisons requires us to begin with physical phenomena and consider pump, jet, and surface swimming rather than heart, squid, and duck.

PUMPS

Our blood circulates; sap rises; a clam filters food; a squid jets. These share a crucial common feature: In each case a pump moves a fluid. Still, hints of commonality get hidden in the dazzle of diversity. Perhaps that diversity should come as no surprise; beyond the diversity of pumping creatures, the pumps manage a ten million–fold range of pressure.[1]

What does a pump—any pump—do? It uses power to raise the pressure of fluid flowing through it. So three things matter: power output, increase in pressure, and rate of flow. Power is the increase in pressure times how fast the fluid flows (volume flowing per time, not the speed of a bit of fluid). Put another way, a pump makes fluid go through some system (load) that it wouldn't go through by itself. If the load resistance is high, as when fluid has to flow through a long and skinny pipe, then a lot of pressure produces only a modest trickle. Conversely, if the load resistance is low—as with a short, fat pipe—a little pressure gives a great gush of fluid. Obviously a pump needs sufficient power to do its intended job. A bit less obviously a pump ought to be appropriately matched to the resistance of its intended load.

The designer faces a choice. One pump might produce a great pressure but deal with a relatively low flow rate. Another pump might invest its power in a high flow rate but give only a small increase in pressure. If you use a pump that's a good pressure producer for an application that mainly needs a high flow, things will go awry. I say this with the conviction that comes from learning the hard way. Long ago, so long that the embarrassment has faded, I built a recirculating flow tank (or flume) that used a two-horsepower centrifugal pump to pump its water around. That pump, the largest that could be plugged into the room's electrical outlet, moved only two gallons per second. A few years later I built a flow tank that instead used a marine propeller to push the water around; a half-horsepower motor now moved over thirty gallons per second. So I got sixty times as much flow relative to power expended, and flow is what matters in a flow tank.

My first pump would have been a good choice for lifting water ten or twenty feet or making it squirt through a nozzle; it produced unnecessarily high pressure and too little flow. The propeller, by contrast, did good service as a low-pressure, high-volume pumper. At least I wasn't the first to pick a pump that gave the wrong combination of pressure and volume. Before the Normandy invasion in 1944 harbor components were prefabricated and then, to prevent storm damage, flooded with water and sunk off southern England. The pumps expected to empty them for floating across the English Channel, while big enough, were the wrong sort and couldn't generate enough pressure. An alert

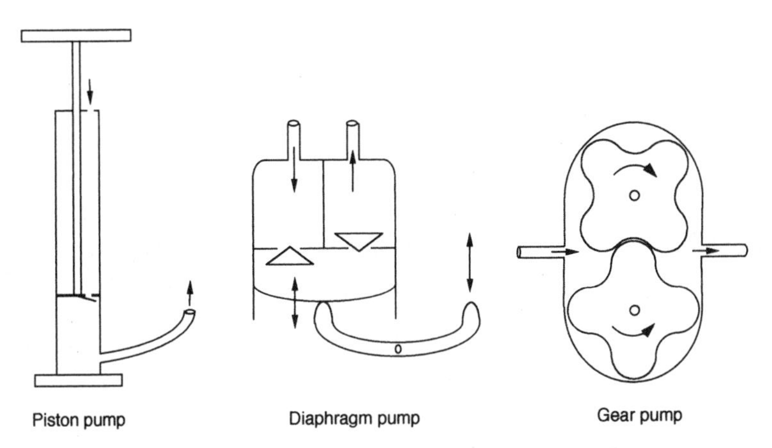

FIGURE 10.1. *Several positive displacement pumps in common use. The piston pump is of course a tire inflator. The diaphragm pump is often used to push gasoline into a car's engine, and the gear pump moves its lubricating oil.*

naval officer forced his superiors to face the problem by staging a persuasively unsuccessful demonstration. To save the day, the pumps of the London fire department were borrowed, no trivial accommodation at a time of air raids.[2]

Two general classes of pumps can be recognized by their particular trade-offs between pressure and flow. The first are so-called positive displacement or fluid static devices; see the examples in Figure 10.1. They include the piston pumps we commonly use as hand-operated tire inflators, the diaphragm pumps that serve as fuel pumps in most automobiles, and the vane or gear pumps that move a car's lubricating oil around. In all these, either the volume of a chamber is decreased, pushing fluid out some intended opening, or else the position of a chamber is moved, and the fluid it contains moves with it.

Those of the second class are called fluid dynamic, or rotodynamic, or kinetic pumps; several are shown in Figure 10.2. Some, just enclosed versions of aircraft propellers, push fluid lengthwise down pipes. Others (like my centrifugal pump) spin the fluid and thus fling it outward. Still others use the motion of one stream of fluid to move another stream of fluid, without having any moving parts themselves. One version of the latter is a jet pump. It squirts fluid from a nozzle, drawing other fluid along with it. Another is an aspirator. Here, as in automobile carburetors, the pressure reduction caused by the motion of one fluid can draw in

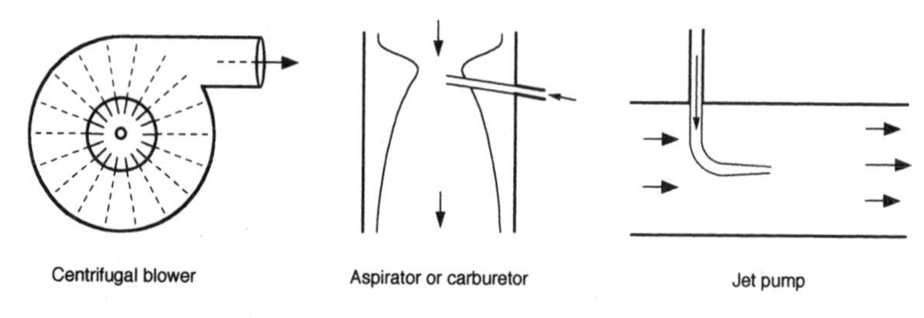

FIGURE 10.2. *Several fluid dynamic pumps. Centrifugal blowers are used for tiny cooling fans and for the large circulators of forced air furnaces. In an aspirator (once in every automobile) fuel is drawn in by motion of air past an orifice. A jet pump works in the opposite manner, with a high-speed flow out of a small pipe drawing with it a flow in the large pipe.*

another fluid—what in Chapter 8 we saw in sponges, prairie dog burrows, and similar systems.

While "fluid dynamic" sounds superior to "fluid static," in fact neither one wins in any general sense; each is suitable for a particular range of applications. Fluid static or positive displacement pumps produce high pressures and low flows, doing best with loads of high resistance. You need one, for instance, if you're to raise water up from a deep well. Fluid dynamic pumps produce higher flows and lower pressures. Within each class, pumps cover a wide range of pressures; thus the pump that was too strongly biased toward pressure for my old flow tank was still a fluid dynamic one. Also, functionally competent exceptions exist. For instance, air entering a jet engine first goes through an axial compressor (Figure 10.5), a fluid dynamic pump that manages to reach high pressures. It does so step by step, compressing air as it passes through a long series of alternately rotating and stationary blades.

What of living pumps? Can the distinction that works for the pumps we build help us recognize common features among nature's? Nature's pumps certainly look nothing like ours, but they do the same job of pushing liquids and gases and do it over an equivalently wide range of operating conditions. Table 10.1 summarizes the character and performance of several biological pumps.

Without much doubt,[3] the biological pump operating against the highest resistance is the evaporative sap lifter of trees and vines. It's about as different as can be from any common machine, but it's a true positive displacement pump. Recall how it works. A continuous column of liquid water extends from roots to leaves through conduits (xylems) less than a

TYPE	CATEGORY	SYSTEM RESISTANCE	EXAMPLES
Evaporative	positive displ.	highest	leaf sap sucker
Osmotic	positive displ.	very high	root sap pusher
Valve and chamber	positive displ.	high	heart, bird lungs, squid jet
Peristaltic	positive displ.	high	intestine, some hearts
Piston	positive displ.	medium	some tubicolous worms
Vane or gear	positive displ.	medium	other tubicolous worms
Valveless chamber	positive displ.	medium	jellyfish jet, mammalian lung
Paddles	fluid dynamic	medium	crustaceans in burrows
Propellers	fluid dynamic	low	hive ventilating honeybees
Ciliary layer	fluid dynamic	low	bivalve mollusk gills
Flagellar	fluid dynamic	low	sponge pumping cells
Venturi aspirator	fluid dynamic	very low	prairie dog burrow

TABLE 10.1. *Various kinds of pumps used in biological systems, arranged in descending order of the system resistance they encounter.*

millimeter in diameter. Evaporation through the fibrous meshwork of the walls of cells in the leaves reduces the volume of sap in them, and that in turn pulls water up from below. A column of water a little more than thirty feet, or ten meters, high exerts a pressure of one atmosphere on its base. So for every ten meters of height, a tree has to pump water against a gravitational pressure of one atmosphere. In fact, it has to work against two other sources of resistance. The skinny conduits offer a hydrodynamic resistance to flow about as high as their gravitational resistance. In addition, the soil around the roots (especially if nearly dry) may hold its water with great tenacity. Separating water from soil is much like wringing out a wet towel; it's harder and harder to extract water as the supply decreases. Faced with the problem, we inevitably switch schemes and resort to a state change—evaporation—to extract the final bit and dry the towel. But the tree can only raise liquid water so its roots can't make that switch. Desert plants, in the driest soil, have to pull the hardest; to raise water, they may work against pressures of as much as a hundred atmospheres.[4] That's the pressure that would be exerted on the base of a column of water thirty-three hundred feet high or the pressure on the hull of a submarine thirty-three hundred feet down.

Another positive displacement device is an osmotic pump, also one of the engines described in Chapter 8. Like the evaporative pump, it involves no moving parts and strikes us as a strange machine. Offsetting relatively ordinary concentration differences takes pressures of tens of atmospheres,[5] so an osmotic pump can work against a high load resistance. Osmotic pumps find their greatest use in small-scale systems of one or a few cells, where very high pressures are associated with only modest tensile stresses (recall Laplace's law from Chapter 4). But they're important in water secretion in our pancreases, in water absorption in roots, and in pushing sap upward in stems—the latter a lower-pressure process complementary to the evaporative sap lifter.

The best-known positive displacement pump is the valve and chamber heart, shown as a generic device in Figure 10.3. A muscular chamber between two conduits forms the basic element, with a one-way valve where each conduit connects to it. By allowing only one-way flow, the valves ensure that one conduit leads fluid in and the other leads it out. These pumps often produce higher overall pressures by using several pumping chambers in sequence—atria and ventricles, for instance—but none comes anywhere near the pressures generated by the sap lifter. Our hearts, for instance, manage no more than a quarter of an atmosphere, not tens or a hundred. The highest pressures, reached by giraffe hearts and squid jets, are still less than half an atmosphere. The gills of many fishes and the lung inflators of frogs also use valve and chamber pumps. All the jet engines of animals appear to be positive displacement devices. Some have valves (squid, for instance); others (such as jellyfish and the anal jets of dragonfly larvae) have single, valveless conduits through which both squirting and refilling happen, as in a kitchen baster. Nature's valve and chamber pumps resemble our piston and diaphragm pumps despite the absence of sliding parts like piston rings

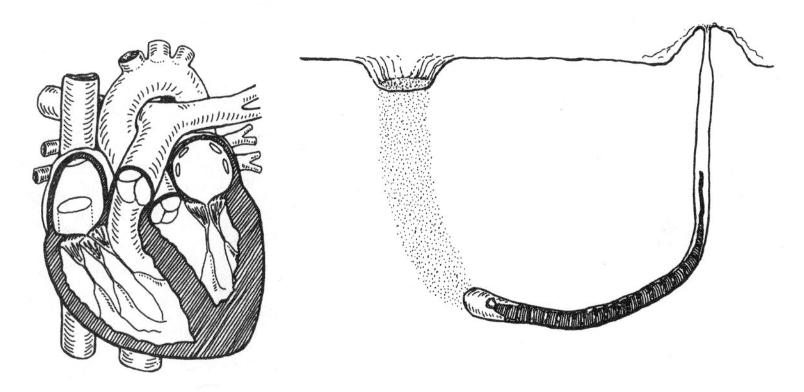

FIGURE 10.3. *Positive displacement pumps in nature: a human heart, with its valves and chambers, and a marine worm,* Arenicola, *that lives in subtidal sand and, by passing waves of contraction down its body, pumps water down through the sand and up through its burrow.*

in nature and of any actively contractile element like muscle in our own devices.

Yet another kind of positive displacement pump propels digesting food down our intestines. In one of these peristaltic pumps, waves of constriction pass down a muscular tube and push along blobs of fluids or slurries. Many worms have hearts—or large blood vessels, since the distinction gets blurry—that pump peristaltically. The ugly lugworm, *Arenicola* (shown in the figure), moves water through its partially sand-filled (and thus high-resistance) burrow by passing peristaltic waves down its body. Peristaltic pumps find only limited technological use (they'll come up again in Chapter 13, and one is shown in Figure 13.1), mainly where we want to keep the fluid in the tube that conveys it and where the inefficiency of the devices (lacking a decent analog of muscle) isn't prohibitive.

In our technology fluid dynamic pumps mostly use wheels and axles, so nature's versions (a couple are shown in Figure 10.4) don't look much like ours. Furthermore, our fluid dynamic pumps are big, fast things, so pushers, such as fan blades, work well. By contrast, most of nature's fluid dynamic pumps are smaller and slower, so pushing with propellerlike blades is hydrodynamically less effective.[6] Instead the basic push works better if done in the fashion of a submerged oar: by alternately moving something downstream in a high-drag orientation and then upstream again in a low-drag position. Human technology hasn't used that scheme much since the demise of paddle wheels; propellers blow away oars if they're large. But

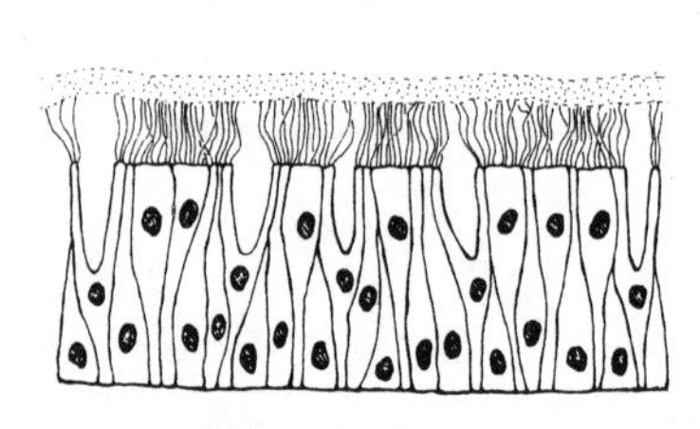

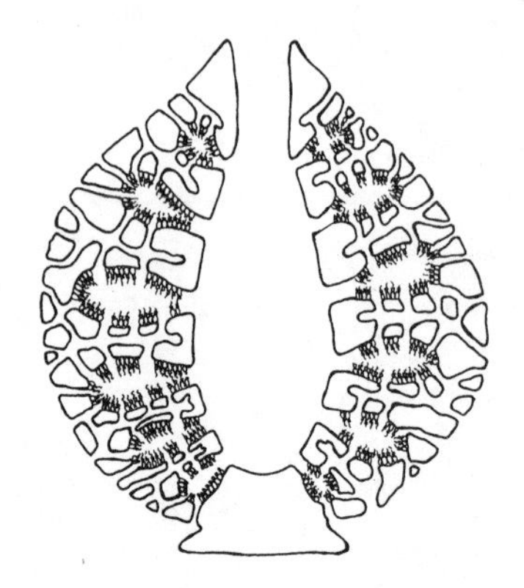

FIGURE 10.4. *Fluid dynamic pumps in nature: a ciliated epithelium (as lines much of our respiratory tracts) pushing a layer of mucus across itself, and a sliced sponge, in which chambers of flagellated cells draw water in through the general surface and out into the central cavity and apical opening. This sponge is highly diagrammatic since in reality the chambers and passages are far too small to be visible in an overall view.*

despite such divergent appearance and operation, the basic principles and the applications of natural and human-made fluid dynamic pumps are every bit as close as those for positive displacement pumps.

Living or not, fluid dynamic pumps deal with low-resistance systems. Some of the lowest resistances are handled by devices that we don't think of as pumps at all. These move large volumes of an external rather than an internal fluid. We're talking here about swimming and flying by moving some appendages, whether by beating wings or paddles or by oscillating cilia or flagella. All such locomotion requires processing huge volumes of fluid but imparting only slight pressure increases. Slightly higher, but still relatively low, resistances face the fluid dynamic pumps (mainly ciliated surfaces) that push mucuses around. Most of the rest of nature's fluid dynamic pumps are involved in suspension feeding. Far more animals than most people imagine make livings in this quiet but demanding way—from the simplest sponges to the largest whales. Most natural waters contain edible particles, but they don't often reach high concentrations. So a lot of water has to be processed to get a little food. For the nutritional benefit to exceed the processing cost, a separation system can't afford great pressure differences. Thus the task calls for fluid dynamic pumps. For suspension feeders, life becomes a competitive battle of extraction efficiencies, with the winner the one that can put the most energy into growth and reproduction and the least into food collecting.

We have lots of data on suspension feeding pumps, mainly because they're used by bivalve mollusks, of both ecological importance and culinary appeal. Soft-shell clams and mussels, pushing hard, can press about one two-thousandths of an atmosphere, but in normal operation they (and sponges too) work at about one hundred-thousandths of an atmosphere. But while these pressures sound trivial, they push impressively high volumes; water equal to as much as half of an animal's body volume gets processed *per second.* (Our circulatory systems don't come close to these volume flows; even during vigorous exercise the human heart pumps less than 1 percent of body volume each second, fifty times less.) Similarly low pressures are involved when sponges use local water currents to augment their pumping and prairie dogs ventilate their burrows; only very low resistance systems can take advantage of ambient flows. Yet another fluid dynamic pump working in the same pressure range is the honeybee hive ventilator, a pump consisting of stationary bees beating their wings at the hive's entrance. This last can be a multistage pump, like the compressor of a jet engine, since the bees often stand one behind another. Lacking decent ductwork, though, it's suitable for only low-resistance work.

Thus the distinction we make between our two general classes of mechanical pumps helps us see order in some diverse natural devices; the same resistance-based difference in performance drives the choices made by both technologies. And we need help if we're to understand nature. Just assuming that natural selection leads to good design doesn't take us very far. Not only do organisms differ dramatically in size and anatomy, but their different lineages form separate technological microcosms. The sap lifter isn't available for use by a vertebrate heart, nor can a plant order up any muscular device. But a few thousand years of human technology let us see the forest through the trees. At the least it counteracts the unnatural division of natural systems among people who study different kinds of organisms, who consider different biological functions, who publish in different journals, and who write different chapters in different textbooks.

The analysis also helps us understand why certain arrangements don't occur. Our circulatory system uses tiny pipes—capillaries—to exchange material between blood and cells, and it uses large pipes—arteries and veins—to interconnect different capillary beds. Pumping of course is done with muscular hearts joining the largest vessels. Why not build a circulatory system that uses ciliated capillaries instead of muscular hearts?

After all, blood flows through capillaries at speeds consistent with ciliary pumping, and clams pump vast amounts of water with cilia. Some years ago Michael LaBarbera and I guessed that cilia were simply so much less efficient than muscles that the scheme would be impractically costly—whatever the advantages of decentralization of function.[7] We should have argued that cilia, as fluid dynamic pumps, don't match the high resistances of circulatory systems. Keeping resistance low enough for ciliary pumps would demand huge interconnecting vessels containing a great volume of blood, along with extremely short capillaries—if the system could fit into the body at all.

One final note on nature's pumps for fluids, a point that might not occur to us except through a comparison with human technology. The biologist studies what occurs, not what doesn't occur, since what doesn't occur isn't available for examination.[8] But the comparisons in this book have repeatedly drawn our attention to oddly glaring omissions in the natural world. Each of these has either told a tale or raised some provocative question. What follows is another of nature's peculiar omissions.

Potential mismatches such as using cilia to drive blood through a mammal can be fixed with a device that trades off pressure change against volume flow. We're invoking nothing more radical than what levers do with force and distance or electrical transformers do with voltage and current. Our technology uses many such converters and has done so since antiquity. A device of the ancient Mediterranean world called a noria is shown in Figure 10.5. A noria used the flow of a stream either to drive paddles that ran a revolving chain of buckets or to drive an undershot waterwheel equipped with buckets. Either way, a lot of slightly descending flow in the stream moved a lesser volume to a much greater height—high pressure and low flow from low pressure and high flow. A so-called hydraulic ram, still in occasional use, accomplishes the same conversion in a different way. A stream flowing downhill makes a small amount of water flow considerably farther uphill—here by capturing the energy released when some of the moving water stops. In effect, the sudden end of a surge of flow drives a smaller volume to a higher level—repeatedly. A best-selling book of the late 1940s, *The Egg and I*, talks about how much life on the farm improved when the family got its ram; the modern urban reader might wonder about the odd ovine allusion. All of nature's valve and chamber pumps produce pulsating flows, but no hydraulic ram has yet been described in nature.

We reduce pressures and increase volumes as well as the opposite. A

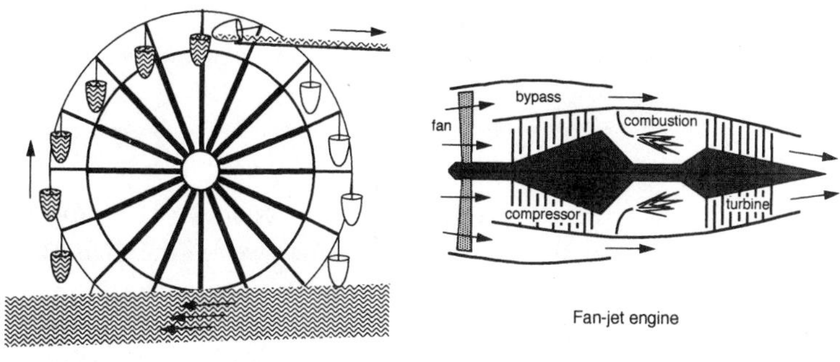

FIGURE 10.5. *Old and new ways to trade pressure against volume flow. A noria run by an undershot waterwheel increases pressure and decreases flow, so the motion of a stream can lift water into an irrigation system. A fan-jet engine decreases pressure and increases flow to improve efficiency at subsonic speeds. Air enters the front; fuel is pumped directly into the combustion chamber.*

ducted fan-jet engine, as in Figure 10.5, uses a big fan in front to move additional air through ducts that go around the engine itself. It thus gets greater volume flow with less overall pressure change. (We'll face the "why bother" shortly). Animals commonly make one flow induce another, but no squid or jellyfish uses such a converter. The combination of long fixed wings and small propellers on an airplane does the same conversion. The propeller produces its thrust by giving the small stream passing through it a large pressure increase. The long wing uses some of that horizontal thrust to make a lot of air flow just a little bit downward. So it converts a high-pressure, low-volume flow to a low-pressure, high-volume flow in a different direction. But among flying animals only beetles make any use of this separation of propeller and wing.

Nature uses many devices that make flows go faster or slower by changing the sizes of pipes. The nozzle on the squid's jet speeds the water, working just like the nozzle on a garden hose. But all these trade speed for area, not pressure for volume flow. The scarcity of pressure flow converters in nature is a puzzle. Perhaps we're missing something or looking at these systems in the wrong way. Perhaps the extreme range of pressures spanned by nature's pumps reduces her need for converters. But that makes the improbable assumption that natural selection has a wide choice of pumps in each specific instance. Or perhaps the small size of organisms is an insurmountable burden. For small, slow flows, viscosity places a severe tax on the efficiency of fluid mechanical devices, so maybe nature simply can't build converters efficient enough to be worthwhile.

JET PROPULSION

What could be simpler? Eject a fluid in one direction, and get propelled in the other. If ambient fluid (usually air) is used, we call the engine a jet; if the fluid comes entirely from within, we call it a rocket. The distinction doesn't matter much here, so we'll refer to what all such reaction engines do as jet propulsion. The process can drive an airplane, a surface ship, or a submarine; it can even work in a vacuum. What could be more natural for an animal? One need only wrap a muscle around a bag of fluid and give the bag a squeeze; out will come fluid through any hole, deliberate or fortuitous. Someplace or other, it's done by almost every animal big enough for us to see. Squeezing our hearts sends blood outbound; contracting our leg muscles helps send blood inbound again; a traveling squeeze of the esophagus pushes food stomachward, while a similar squeeze of the intestine propels the processed slurry farther aft. Expel rather than just propel the fluid, and you have a jet engine.[9] Furthermore, the most ordinary fluid, liquid water, is nicely dense, flows easily, and thus serves admirably. Plastic water rockets are splendid toys; just half fill the rocket with water and pump air into the remaining space. When the rocket is launched, reexpansion of the air expels the water downward. Recoil sends the rocket up a hundred feet, and the experienced operator gets only minimally dampened by the effluent.

Nature has many such engines. Perhaps no other locomotory scheme has independently evolved in so many different lineages. Jellyfish do it. And the cephalopods—squid, octopus, and cuttlefish—do it. A scallop swims in short bursts using a pair of jets on either side of the hinge of its shell. A young dragonfly swims in a pond by squirting water out its anus. Frogfish maneuver by squirting the water that has passed over their gills out through nozzles placed more or less amidships. And on and on.

All these jets work like toy water rockets. A squid, certainly nature's champion jetter, has a water-filled cavity between an outer muscular sheath (which got notice earlier as a hydrostatic skeleton) and its various viscera. When a squid tightens that sheath, the water inside is forced out through a nozzle, and the squid can get elsewhere in a hurry; fifteen to twenty miles per hour is impressive for an aquatic animal less than a foot long. Avoiding an approaching mouth, a squid can briefly outdistance all but cetaceans or the fastest fish. It can shoot sixteen feet up from the ocean's surface or take an arcing aerial trajectory as much as fifty feet long.

So jets sound terrific. But while jets may be simple and common, no jet-driven animal ever goes both fast and far. Squid can't maintain top speed for more than a few pulses. Jellyfish may swim steadily, but they reach only about a quarter of a mile per hour. Frogfish do only about twice that, terribly slow by piscine standards. Dragonfly larvae swim at a little more than one mile per hour, and scallops up to one and a half miles per hour; both, like squid, perform only intermittently. Where's the rub? Is the limitation biological or fluid mechanical?

How we ourselves use jets is revealing. Jet and rocket engines may be the oldest devices that burn fuel to make motion; in principle they're the simplest of all heat engines. The earliest-known steam engine was a direct jetting device; it had only one moving part and needed little precision in its construction. This was the famous engine of Hero of Alexandria, in the first century C.E., shown (with necessary artistic license) in Figure 10.6. A fire heated water in a spherical chamber mounted on an axle. The chamber had two tangentially pointing nozzles through which steam emerged; the jets of steam thus spun the chamber. But Hero's engine never made anything better than a heroic steam whistle for reasons that, as we'll see, cast no aspersions on Roman culture.[10] The steam engines that drove the Industrial Revolution were based on a completely different scheme.

Even after fifty years of development we mostly use these reaction engines for very fast aircraft and almost never for cars or trains or boats. Why do both natural and human technologies take this peculiar arm's-length attitude toward what seems an attractively simple and powerful scheme? Because the scheme has a major and fundamental flaw.

The crux of the problem is the low efficiency of jets under all too many circumstances. Here's how it arises. By pushing fluid rearward—ejecting momentum—a jet engine produces a forward force. One might take a lot of fluid and increase its rearward speed just a little. Or one might take a small amount of fluid and greatly increase its rearward motion. All that matters is mass flow rate multiplied by velocity. Air or water passing at one kilogram per second and given an extra speed of two meters per second yields the same thrust as two kilograms per second given one extra meter per second.

But the efficiency of the device cares a lot about the particular mix of amount and speed of expelled fluid. For a given flow rate, the rearward thrust you get depends directly on the speed you give the passing fluid.

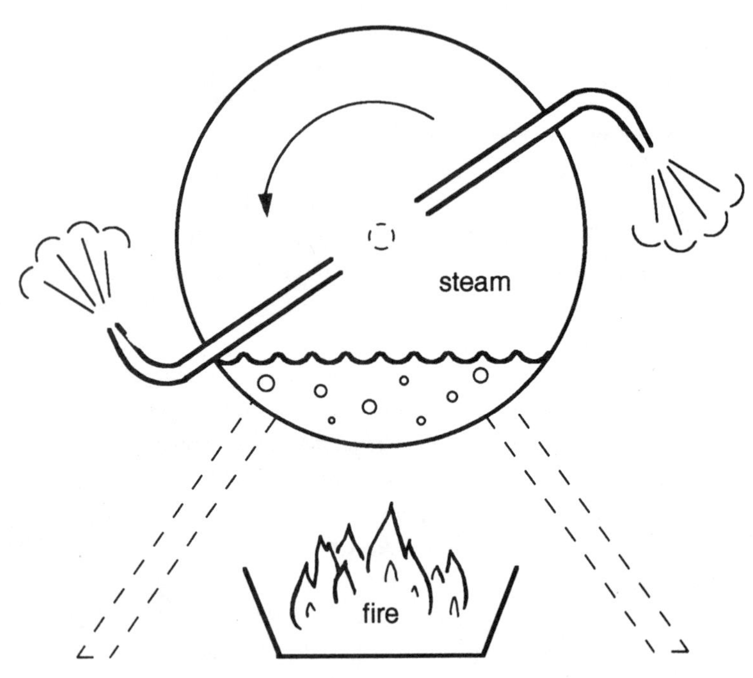

FIGURE 10.6. *The steam engine attributed to Hero (or Heron) of Alexandria, first century C.E. A fire beneath a metal sphere spins the sphere by causing a mix of water and steam to squirt out the jets.*

But the cost in energy needed to produce that push increases not directly with the speed but with the square of the speed. Double the speed, and you double the thrust, but doubling the speed costs you four times as much energy. So you want to keep the speed of the expelled fluid as low as possible; minimizing cost means using lots of fluid and giving it only a little rearward push. There's the rub. That's just the opposite of what jets and rockets do: They squirt small but rapid streams.

How can an engine be made to use lots of fluid? Definitely not by making all the fluid pass through its middle and out a nozzle; too little fluid can get through, so what goes through has to go too fast. You need to take the opposite approach. Instead of forcing a little fluid through the engine, you ought to force a lot of fluid around the engine. Better to make the engine go through the fluid than to make the fluid go through the engine. How to do this? Attach long, moving appendages to the engine—fins, wings, paddles, or propeller blades. Any of these will beat the jet in propulsion efficiency. All, though, are more complicated than Hero's engine or the simple squirting machine of a jellyfish. Nonetheless, the complication pays off. A trout, waving its body and tail, deals with about ten times as much water per unit time as a jetting squid of the same size. As a result, the trout takes half the power to go twice as fast.

Should jets be dismissed as primitive because of their simplicity and inefficiency? Are they just the bad hands dealt out by an animal's ancestry? Wrong as well as unfair. To swim or fly by pushing the local fluid backward, a machine (living or not) must push it faster than the rate at which the craft goes forward.[11] Consequently, the speed of fluid emerging from its propulsion device sets the craft's maximum forward speed. If going fast is what matters, then jet engines look a lot more attractive. That's how we use jet aircraft: We make both small and large ones, but we don't make slow ones. A squid does much the same. With small fins on its rear, it has a choice of systems. For slow swimming, as when feeding, it mainly uses its fins. But when pursued by a fish or cetacean, it turns on its jet and zips away. In a brief life-or-death maneuver, biological fitness doesn't turn on energetic efficiency!

For that matter, the jet engines of commercial aircraft have changed over the past few decades, processing ever larger amounts of air and reducing their average output speeds. They've added turbine-driven ducted fans, evolving from pure jets to ducted fan jets (Figure 10.5), as noted when we considered pressure flow converters. Early jets had small intake openings, and all the air coming in entered a fanlike compressor prior to receiving its charge of fuel. Fan jets have big intakes with large, conspicuous entrance fans. Engine designers have worked hard to increase the amount of air going around relative to that going through the combustion chamber—the bypass ratio—to improve efficiency and thus get lower fuel consumption and greater range and payload.

Poor propulsion efficiency makes jet aircraft impractical for flight at the sizes and speeds of animals. We know of no living jet aircraft—almost certainly nature never made one—some engaging bits of science fiction (and flatulent humor) notwithstanding. But oddly enough, jets aren't all that bad for slow swimming. In steady swimming, the thrust produced equals the drag incurred. Low speeds generate disproportionately low drag, so balancing it takes very little thrust. Some squid make lengthy migrations using their jets, but they do so at speeds of around a body length per second, less than a mile per hour. Evading drag simply by being slow probably underlies the success of slow jetters such as jellyfish.

While the same fluid mechanical rules govern motion through air and water, flying is a lot harder than swimming. The lower density of air may help a craft go forward, but it demands an additional force to keep the craft aloft. This additional component consumes (except for buoyant blimps and balloons) a lot of power, whether or not the craft goes forward at all. For flight, never mind moving slowly to take advantage of low drag. To the cost of going anywhere must be added the cost of staying aloft, and slower flight means longer periods aloft. In short, the drag-evading slow swimming of jellyfish or frogfish provides no model for any heavier-than-air flying machine.[12]

To stay aloft, an aircraft must push air downward. Some air must be given an increase in speed, but now the speed that matters is downward. Again the choice is between giving a large amount a small speed increase and giving a small amount a large speed increase. But for staying aloft by making air flow downward, the relevant speed of the craft is how fast it's ascending. If (as is usual) it's simply holding altitude, then ascent speed is zero. So the lower the speed of the downward airflow, the better. Concomitantly, the greater the volume of air moved downward, the better. Thus directing a small, high-speed stream of air—a jet—downward is a particularly inefficient way to stay up. A small downward-facing propeller works only a little better.

That's why helicopters have long rotors and why tilting the engines of an ordinary propeller plane from horizontal to vertical produces a very inefficient hoverer. One military aircraft, the Harrier, is a jet that can hover, but it gulps fuel when it does. The opposite extreme would be a helicopter with blades infinitely long; it could hover at no cost at all!

This issue of propulsion efficiency explains a basic difference between flying animals and ordinary airplanes. Birds, bats, and insects get both thrust, to go forward, and lift, to stay aloft by beating their wings. Their wings ordinarily beat not just up and down but to some extent fore and aft. Slower flight usually comes with less up-and-down and more fore-and-aft motion; a creature just tilts the plane of beating backward when it slows down, as in Figure 10.7. When hovering at flower or feeder, a hummingbird has its head up and tail down, and its wings mainly move back and forth. What a helicopter does when hovering is almost identical. Like a hummingbird's wings, a helicopter rotor moves in a horizontal plane when it hovers, tilting the plane front down a bit for forward flight. But ordinary airplanes use propellers to go forward and fixed wings to stay aloft, successful airplanes preceded successful helicopters by about thirty-

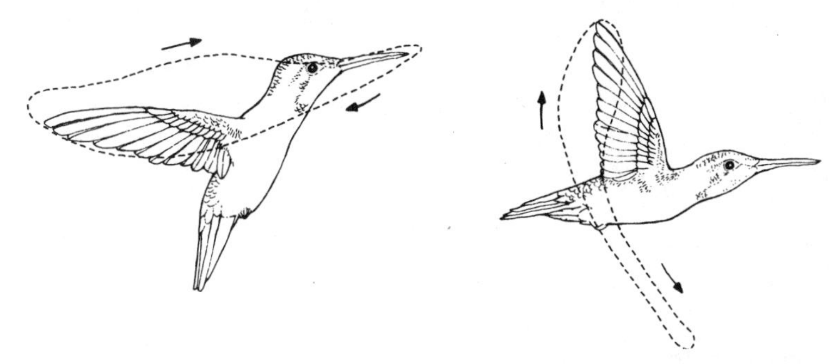

FIGURE 10.7. *A hummingbird switches from hovering to rapid forward flight mainly by changing the plane in which its wings beat from horizontal (back and forth) to vertical (up and down).*

five years, and helicopters are still profligate fuel consumers. We managed to build airplanes when we gave up the bird or helicopter arrangement and began using airfoils in two different ways on the same craft. (A crosswise slice of a propeller reveals a shape just like that of a wing; they're the same kind of device.)

The fixed wing airplane produces its thrust by moving air backward at a speed greater than that of the plane's forward motion, which is usually quite rapid. It produces its lift with long, fixed wings that give a large amount of air a little downward push. What's subtle is the power source for operating fixed wings. It can only be the propeller since that's the only thing that has an engine attached. What happens is that the power to stay aloft is felt by the propeller and engine as an additional drag that requires more thrust; when producing lift, a wing (unless it's infinitely long!) has more drag than when simply sticking out into the moving air. In effect, the airplane produces forward thrust by using small airfoils or jets operating at high horizontal speeds. It converts some of that thrust to lift by using large wings operating at near-zero vertical speeds, as we noted when talking earlier about pressure flow converters. The arrangement is efficient, which is why, long after fine helicopters first became available, we persistently build planes that take long runways to reach high takeoff speeds.

How then do birds, bats, and insects manage so well with only one set of airfoils? The argument, alluded to in Chapter 3, turns on the relative sizes of the flying machines of the two technologies. A wing produces lift in proportion to its area. But an aircraft requires lift in proportion to its weight. After all, for steady, level flight, lift and weight must balance just as do thrust and drag. Consider what happens if the size of a craft doubles, with no change of shape or density. All lengths—length, wingspan, etc.—will double. All areas—total external surface, wing area, etc.—will go up fourfold. Volume and weight, though, will increase no less than eightfold. So weight relative to area will double; the bigger craft will have twice the weight relative to its wing area. Thus it will be relatively lift-deprived, something not at all propitious for flight.

Two evasions suggest themselves. The first is to use disproportionately large wings for larger craft. The original Wright Flyer of 1903 had huge wings as does the ultrasophisticated Gossamer series of human-powered aircraft.[13] But both are intentionally slow, disdaining the second evasion, which is to go faster. Double the flying speed, and the lift of a wing goes up fourfold. (So a doubling of weight can be balanced by a 1.4-fold increase in speed.) Ordinary aircraft may weigh more relative to their wing areas than do birds, but they fly faster. Only a few large birds can exceed fifty miles per hour in level flight, while only a few specialized airplanes can fly so slowly. Larger size not only permits higher speed (less total surface to incur drag relative to volume) but practically requires it (less lifting surface relative to volume). Higher speed, in turn, requires a smaller and faster air pusher. Airplanes gain efficiency by separating wings and propellers. Birds, smaller and thus slower, find larger and slower air pushers practical, and they have little cause to use fixed wings for lift and flapping ones for thrust.[14]

But even flying animals aren't completely out of the woods. Big animals weigh more, relative to their wing areas, than do small ones. Big birds may fly faster than small ones, but they don't have enough wing area to hover effectively; their propulsive systems move too little air too fast for decent efficiency in hovering. Large aquatic birds often need to taxi furiously to reach takeoff speed. A pigeon can hover for a few seconds, but it can't maintain sufficient power output to keep it up for long. Only hummingbirds are proper avian hovercraft. But for small flying insects hovering is routine. These small, slow creatures have a high ratio of lift-producing wing area to lift-requiring body weight. With all that wing area they move a lot of air downward, so they don't have to move it downward very fast.

A pair of lessons: First, something absent in both, jet engines propelling craft of moderate size at moderate speeds, draws attention to the unromantic but inescapable issue of propulsion efficiency. That in turn provides proper skepticism of proposals for jet backpacks, jet boats, and the like. Second comes a point made repeatedly. Comparing situations where fixed wings work best with those where flapping wings prove practical turns on the subtle, pervasive, perhaps even pernicious influence of size. Size differences all too often confound judgmental comparisons between what we make and what nature makes.

SWIMMING AT THE SURFACE

Two kinds of machines, surface ships and submarines, swim, and both technologies make both. That every submarine we've built can also swim at the surface shouldn't be allowed to obscure basic differences between the two kinds of swimming. One might guess that surface swimming is easier; only at the surface can a thrust-producing appendage enjoy a recovery phase in low-density air. Rowboats, canoes, and paddle-wheel steamers use aerial recovery, but it's ignored by all our better boats as well as by nature's. So aerial recovery can't be all that valuable. A more important difference, one with the opposite effect, emerges from the various ways water resists a swimmer's motion. Beneath the surface lurks ordinary drag, most of which can be avoided by careful streamlining. The surface ship must also contend with drag from surface waves. With less of its hull submerged, the surface ship avoids some ordinary drag, but the saving is less than the ship pays in wave-induced drag. A submarine can go faster with less fuel than a surface ship of the same size.

Both technologies build both kinds of swimmers, just as both build positive displacement and fluid dynamic pumps and just as both build jet engines. Here, though, we see a major bias. Nature's swimmers are almost all submarines, with only an occasional duck, muskrat, or water strider moving at the surface. Our swimmers are nearly always surface ships, with self-propelled submarine technology of any kind barely two centuries old. Even now, submarines are limited to damn-the-cost military use. So we have two questions: What limits submarines for us, and what's wrong with surface swimming for living boats?

Answering the first question is easier but not as interesting as the second. For one thing, we like to breathe air at sea-level atmospheric pressure, so we build our submarines with rigid, pressure-resisting hulls. Such hulls have to be rigid because with air inside, a submarine faces a peculiar instability. As it dives deeper, the rising pressure tries to compress hull and air. That makes the overall density of the submarine rise, so it gets less buoyant and all too eager to go deeper yet. (Whale and skin diver contain air only in the lungs, although compression of air in the lungs and the resulting transfer of nitrogen gas from lungs into blood do present a serious problem.) In addition, most of our engines breathe air, and they demand more oxygen than the humans inside the craft. So fully submerged boats were first hand-cranked, then driven by electric motors and rechargeable batteries and most recently by nuclear reactors driving steam turbines.

If nuclear engines were commercially available, we might use submarines at least for transporting bulky and incompressible cargoes such as oil for long distances. Fuel efficiency would be better, vulnerability to storms might be reduced, and the whale-shaped ships themselves would be more compact than present supertankers. A small pressure hull would of course be needed for the crew.

The trickier issue is why nature's swimmers so rarely keep their heads above water. Avoiding the surface is remarkably widespread. Ducks swim on the surface, but they do so slowly; they're far faster either submerged or in flight. Air breathers, such as penguins, seals, and cetaceans, do most of their swimming underwater even though they have to come up to breathe. Nature's submarines vary in size from the smallest microorganisms to the largest whales, while her surface ships encompass a size range from just below a centimeter to less than a meter in length. By contrast, think how ancient and successful are our surface ships, how varied in materials and designs, how diverse the cultures that have built them, how many places on earth were settled long ago by people who came by boat. In short, why do surface ships present a problem for nature, or does she simply build such good submarines that surface ships are superfluous?

The culprits, waves, pose a worse problem for swimming animals than for boats. The waves that give the most trouble aren't the ones that winds generate but the ones that surface swimmers themselves make. Neither duck nor ocean liner can avoid making waves as it moves along. As in Figure 10.8, a floating craft moving along the surface creates a pair of waves separated by roughly its own waterline length—so-called bow wave and stern wave. That waterline length (hull length) thus sets the distance between waves, the wavelength. Waves always move, and their wavelengths set their speeds. As has been known for at least a century, wave speed increases with the square root of wavelength; double the wavelength, and you increase the wave speed about 1.4 times. By making waves of longer length, the bigger boat makes waves that travel faster. With a string, a stopwatch, and some toy boats (add weight if they ride too high) you can test the assertion in any swimming pool.

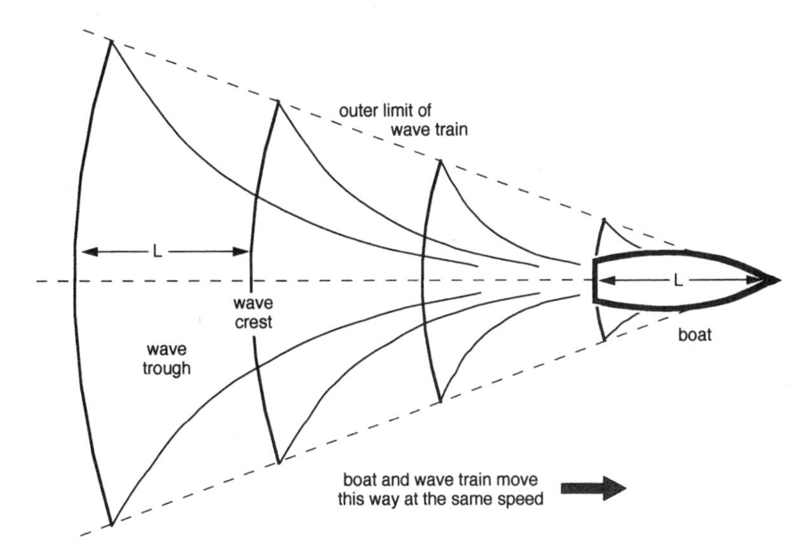

FIGURE 10.8. *An ordinary boat generates a pair of wave crests as it moves—a bow wave and a stern wave—separated by about the length of its hull (here marked "L").*

Trouble comes because an ordinary floating ship or a swimmer finds it hard to go faster than its waves. If it tries, a hill of water faces it, and it has either to cut its way through or go perpetually uphill—up the down escalator as it were, as in Figure 10.9. So a surface ship faces a practical speed limit, what's called its hull speed. Above its hull speed its drag increases suddenly and severely. Since hull speed depends on hull length, the larger ship can go faster before hitting that limit. Tinkering with the shape of the hull can have some effect on the speed and suddenness of that drag increase, but the underlying problem can't easily be evaded. We build fairly large ships, and a hundred-foot-long ship reaches hull speed at about a respectable fifteen or so miles per hour. Hull speed for a ten-foot-long boat, about as small as we ever use, is about five miles per hour. For the foot-long duck hull, it's only one and a half miles per hour—for an animal that can fly at thirty. For a muskrat the speed limit is even lower. Even swimming humans, far from streamlined, can go faster underwater than on the surface, despite giving up any aerial recovery phase for their arms. So much better is underwater swimming for us that excessive use of it has been banned for races lest fanatic competitors injure themselves; it's healthy to breathe when working maximally.

FIGURE 10.9. *A small boat hits hull speed at a low speed; for this rubber duck being towed in a tank it's uphill all the way above about one mile per hour. Notice the slight tilt of the duck at the highest speed shown.*

One can evade the limit by hydroplaning, scooting across the surface, which we do in small to medium-size motorboats in fairly calm water. A few birds, such as razorbills and loons, hydroplane for short distances. But it's just not a practical way to carry large masses for respectable distances. The problem then is one of size. Surface ship technology is more attractive if you're big than if you're small. So the profound divergence turns on nothing more obscure than the difference in scale between the two technologies.

As if in the interest of symmetry or fair play, quite a different way of getting around on a liquid's surface works better for nature than for human technology. Whirligig beetles, small black ovals (Figure 10.10), use it to dash around on streams and ponds. An ordinary surface wave involves two competing phenomena. The inertia of the moving water keeps the surface wavy[15] while gravity tries to flatten the surface, and the interaction of these phenomena sets the speed of a wave. Gravity, though, isn't the only agency trying to flatten the surface; it's just the important one for large waves. The phenomenon that makes droplets round up, that causes water to bead on a waxy surface, that permits us to rest a clean needle on the surface of a pan of water, that we minimize by using soaps and detergents—surface tension—also flattens the water's surface. Gravity works at large scales, while surface tension becomes significant when things are small. So the interaction of inertia and surface tension rather than inertia and gravity rules waves shorter than about two-thirds of an inch.

If only gravity were relevant, then a centimeter-long hull could make

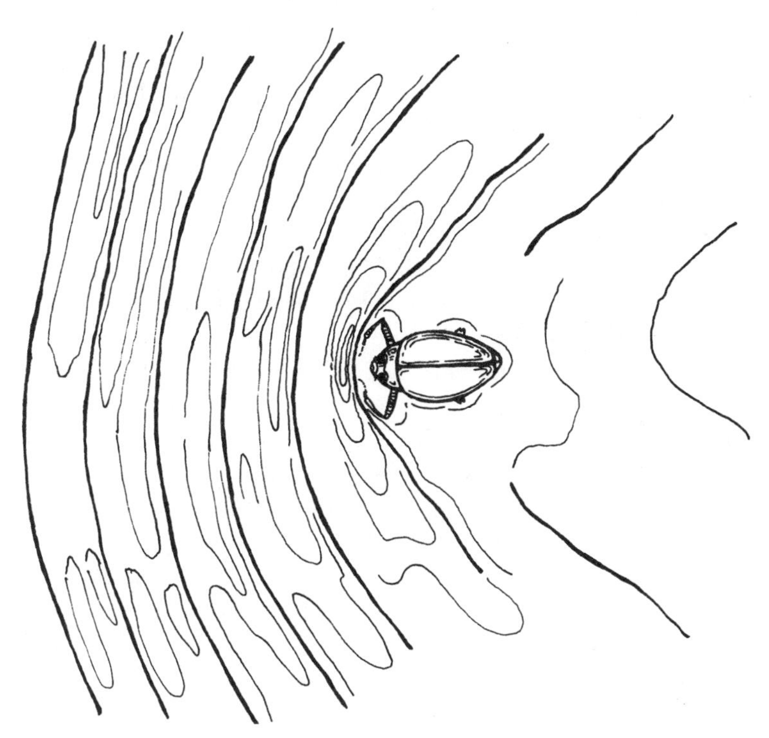

FIGURE 10.10. *A whirligig beetle on a pond is a surface ship so small that it encounters waves whose speeds are determined more by surface tension than by gravity.*

only about five inches per second. But when surface tension is taken into account, the speed limit reaches almost ten inches per second. For a half-centimeter hull, surface tension makes even more of a difference: twelve inches per second instead of three and a half. So the speed limit is by no means as bad as we might have guessed for creatures in the millimeter to centimeter range of hull length. What's more, in this peculiar domain smaller is faster, not slower.

Nonetheless, whirligig beetles are among the few inhabitants of this world of small surface ships. (Water striders, springtails, and a few other creatures walk on the surface rather than swim in the present sense.) Using surface tension requires fairly calm water.[16] Also, it's not a friendly world for animals less than a few millimeters long. For these still-smaller animals the speed limit isn't the problem. Instead surface tension itself becomes a trap, as you may have noticed when a tiny fly falls onto the surface of a pond or puddle. Sprinkle a little talcum powder in a bowl of water, and you'll see the problem. Talc is several times denser than water, but the particles get caught on the surface and stay there persistently.

Pumps, jet propulsion, and surface swimming are each used by both technologies. What do we learn from the comparisons? Both technologies make elaborate use of pumps of diverse designs; for both the same rules link basic design with practical applications. Both make substantial use of jet propulsion, but each gives the scheme something less than a clear and sweeping endorsement—for the same reason. Jets get used when something transcends their inefficiency, something such as the desire for high speed, the need for a brief burst of speed, the acceptability of very low speed, or the simplicity with which preexisting structures can be modified. Both technologies use surface swimming, but human technology finds it far more attractive than does nature. The problem for nature is one of size, the difficulty that the mechanics of wave propagation poses for small ships trying to move rapidly. Looking at pumps, jets, and ships, we're once again reminded that however divergent their aims and appearances, the underlying constraints and imperatives are very much the same for human builders and nature.

Chapter 11

MAKING WIDGETS

Before they work, things must be made; to keep them working, they must be repaired. We've looked extensively—exhaustively, the reader might declare—at what natural and human technology make, but we've yet to ask about how each produces and services its things. In no aspect do the two diverge farther. The primary story is the biological one; being competent organisms gives us no intrinsic sense of how we operate! But human technology needs a little attention as well; participating in a modern industrial society doesn't instill much understanding of that either.

THE LIVING FACTORY

We large creatures operate in two distinct domains. On one hand, our pieces make up a single product, an organism tested against other organisms in the marketplace of natural selection, with reproductive success as criterion. On the other hand, the factory that makes our pieces is the cell. A tiny organism may be just a cell, but the animals and plants we see about us are decidedly supercellular. A human contains about 100,000,000,000,000 cells (a hundred trillion, or 10^{14}). Assuming a typical diameter of a hundredth of a millimeter, in single file such cells would stretch a million kilometers, a thin line encircling the earth about twenty-five times. These microscopic production units vary little to either side of their typical length of four ten-thousandths of an inch. It's no accident that the ancient icon of the biologist, the light microscope, views structures at this scale. Biology textbooks discuss levels of organization such as organs and communities, but only the species approaches in importance the cell and the organism.

If this dualism of cell versus organism strikes you as theological or metaphysical, expunge such heretical thoughts; this is reporting, not New Age analysis. Never mind, at this point, why a cellular scheme of organization was retained almost every time macroscopic organisms evolved. What matters here is the strangeness of a manufacturing system whose products are larger than its factories. This is cottage industry with a vengeance, and it has important consequences for how nature makes and maintains her things and for what particular things she makes.

Building Up, Not Down. The basic production units, cells, make things bigger than themselves, organisms, from parts smaller than themselves, molecules. The nearest equivalent of a single cut, cast, or pressed part is a single protein molecule. If a cell were simply filled with such molecules and nothing more, it would hold roughly 10,000,000,000 of them (ten billion, or 10^{10}). Despite their small size, these parts are complex. A protein may be a polymer of a few hundred monomeric chemical units strung together, but it differs in two important ways from any polymer produced industrially. The monomeric units (amino acids, if you need a name) are nonidentical, and the sequence in which different ones are strung together is all-important. Not just the proportions but the specific sequence matter. Complexity starts within the molecules themselves.

Making proteins is the most important thing that cells do. These proteins either get used within cells or get exported, mainly to form intercellular structures or to flow in intercellular fluids. As a manufacturing system a cell should find it easiest to make things smaller than itself. Of course assembling subcellular things made of several kinds of proteins isn't necessarily automatic. But it should be a lot less tricky for a cell than making things larger than itself. The evidence of microfossils in the oldest rocks accords with this view; for most of the time that life has existed on

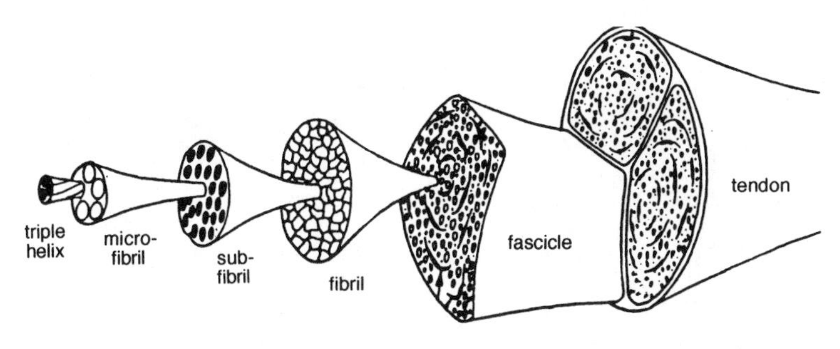

FIGURE 11.1. *The hierarchical structure of a tendon.*

earth, supercellular structure was limited to whatever gluey stuff stuck a few cells together.[1]

The larger things made by organisms reflect this process of building up from the molecular level. A steel beam is uniformly composed of steel, while the wooden beam taken from a defunct tree has level beneath level of structure. In making metals and plastics, we try to achieve material homogeneity, while living systems seem strongly biased against homogeneous materials. Homogeneity isn't reached until one gets well below cellular dimensions. A tendon (Figure 11.1), for instance, isn't just a stiff elastic band; it's divided into fascicles, which are made up of fibrils, which are formed of subfibrils, themselves composed of microfibrils, the latter mostly bundled triple helices of amino acid chains. Even hair, of a single protein, keratin, has a hierarchical (or as Julian Vincent shamelessly quips, hairarchical) structure.[2] Nature routinely and easily introduces organizational complexity at a microscopic level, at least compared with the barriers that face our own fabricational techniques.

Nature uses composites for all her rigid materials while we crow a bit when we make a composite that's competitive in a nonmilitary marketplace. Perhaps nature just does what comes naturally, taking advantage, though, of some coincidence in how cell size coincides with the best size for the components of good composites. Only a little coordination among cells yields highly anisotropic composites—composites with regularly rather than randomly arranged components, composites whose properties depend on the direction of loading. We do that with fiberglass, making sheets in which the glass fibers run in the same plane and rods in which all the fibers are parallel. But even the fanciest fiberglass is monotonous next to wood or bone, as in Figure 11.2.

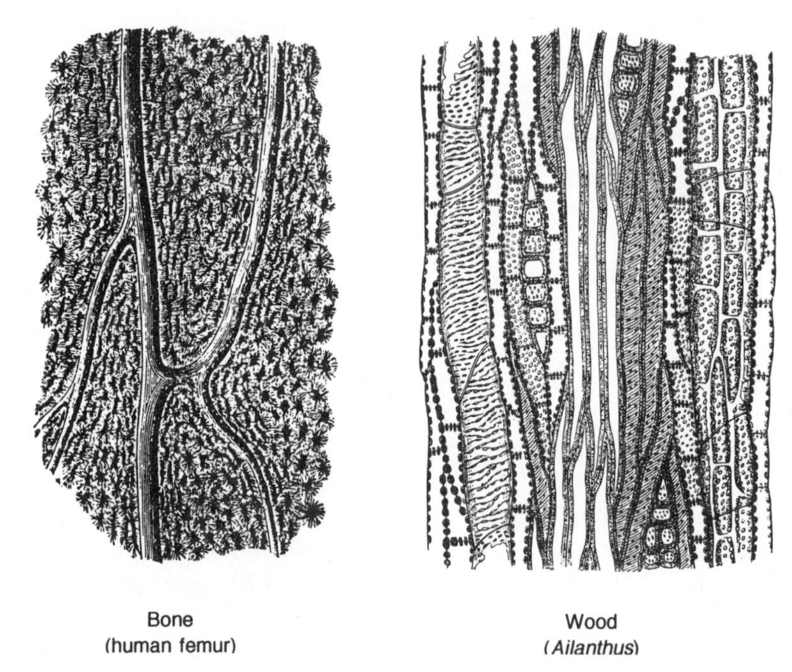

FIGURE 11.2. *Recognition of the complex composite character of bone and wood is no new thing. Here are micrographs of lengthwise, glancing (tangential) sections of the two, from the 1878 edition of Gray's human anatomy and Sachs's plant physiology of 1882.*

Making a biological composite amalgamates the outputs of a multitude of microfactories to good advantage. For other biological products, any superiority of cellular over macroscopic synthesis is less clear. What can't be doubted, though, is that cellular synthesis has consequences far beyond the internal arrangements of structural materials. Consider muscle: It differs from an electric motor in a way not yet mentioned. The motor is a single machine; omit almost any part, and it will fail to work, and no portion of the motor can run by itself. The muscle, by contrast, is an amalgam of small, identical units, the sarcomeres (Figure 8.9), each about two micrometers, about one ten-thousandths of an inch, long. Whether one operates doesn't depend at all on whether the others can do so. We put that independence to use, adjusting how hard a muscle works by changing the number of motor units—somewhat larger operational elements than sarcomeres—that are active.

Nor is muscle the only living machine in which larger simply means more of the basic units. A kidney and an intestine and a liver look substantial enough, with their functions, as organs, described in textbooks. But the functions of these organs are just the functions of identical elements made up of individual cells or small groups of cells. Each of us carries a strong sense of organismic individuality, and none of us holds special affection for our cells as individuals, yet we're minimal confederations of these tiny elements. In even the most complex organism, far more information moves within cells than between cells. Even our brains are bit players compared with the rate at which information from the genetic material directs the synthesis of protein, an entirely intracellular activity.

Put another way, nature achieved something glorious when, something more than half a billion years ago, she invented well-integrated, multicellular, macroscopic organisms.[3] Nature builds up more often than she miniaturizes. With rare exceptions, large size is the specialization, whether in how creatures work, how they mature as individuals, or how they evolve. Evolution has engendered big things many times, but major evolutionary change happens mostly in small organisms.[4] In short, big creatures usually descend from small ancestors, not the other way around.

Recipe or Blueprint? "Bake until golden brown." "Boil, stirring constantly, until the temperature reaches 230 degrees." "Reduce heat until just barely simmering." Each instruction specifies an end point rather than a specific course. Each demands that you observe something and then alter a process on the basis of that observation. Each employs feedback, called that because you *feed* information about results *back* to control a process.[5] In the first two instances you terminate the process, while in the third you adjust something. By analogy with cookbooks, we'll call these kinds of instructions recipes, even if recipes for cooking don't inevitably involve feedback. By contrast, instructions may be outcome-independent and lack such an informational loop; we'll call these latter blueprints. The level of detail doesn't matter; what's relevant is whether or not the instructions depend on the results. Figure 11.3 puts a result-dependent scheme in more general form.

Feedback control may be automatic, with no human link in the feedback loop. The earliest clear cases of automatic controls, according to Mayr,[6] were a few float valves of classical and early Islamic civilizations. These worked like the ones we use to regulate the water level in the tanks of household toilets: High water raises a float and turns off the water supply, as in Figure 11.4. Early in the seventeenth century Cornelis Drebbel, a remarkably prolific Dutch inventor about whom we rarely hear, invent-

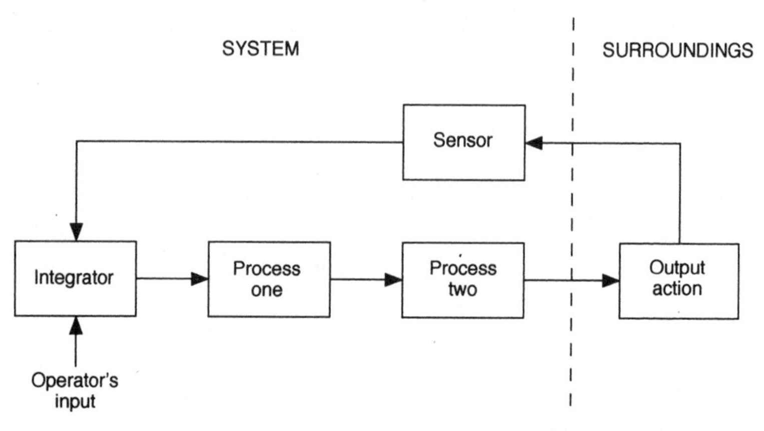

FIGURE 11.3. *The crux of a feedback system is the ability to adjust what it does depending on conditions outside itself, where those conditions include the results of its own actions. In a strictly mechanical and formal sense, it has self-awareness.*

ed thermal controls for ovens and incubators. In them, high temperature reduced the rate of fuel combustion. One of James Watt's finest accomplishments was a governor for a steam engine; an increase in the speed of the output shaft cut back the steam entering the engine. Feedback controls now occur everywhere: temperature controls on ovens, furnaces, and refrigerators; error-correcting protocols in modems; load-dependent speed controls on motors; and on and on. Nor have feedback controls with human links become obsolete or superfluous. You close a feedback loop when you steer a car. If it drifts toward the left, you turn the front wheels a little rightward; when it drifts too far rightward, you turn toward the left. No road runs so straight that you can take aim and then close your eyes.

In organisms, feedback controls in physiological systems get most attention. Such things as the tension produced by a leg muscle, the output of the heart, the rate of breathing, and the diameter of the pupil of each eye are result-regulated. Lift one leg off the ground, and the muscles that stiffen the other leg detect the increase in load and pull harder in compensation; you don't sag down toward the ground, as would a system that wasn't keeping track of what it was accomplishing. Less widely appreciated is how extensively feedback loops lace together the process of making an organism in the first place. Its genetic material, DNA, is often

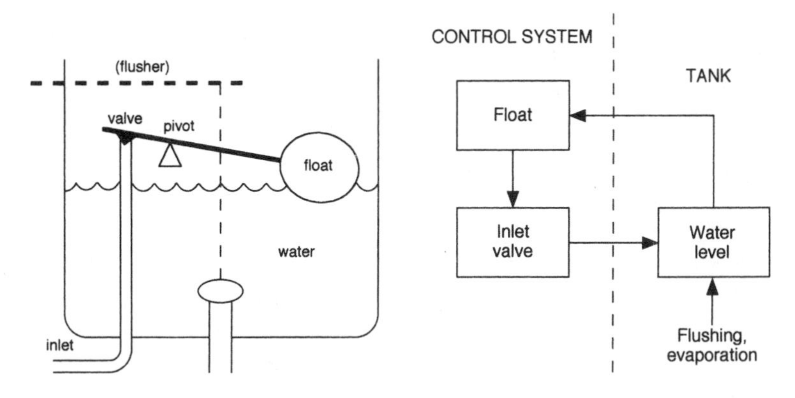

FIGURE 11.4. *A specific feedback control system, as machinery and as a formal scheme—the water-level control of a household toilet. Not only will it restore the level after a flush, but it will compensate for evaporation from the tank and is unaffected by changes in water pressure.*

referred to as the blueprint for making an organism. If "blueprint" implies a detailed and complete set of instructions that need only be read out by the synthetic machinery, then the word misleads.[7]

DNA in fact functions as a recipe in the present sense. No instructions could be precise enough to make something as complex as an organism without adjusting course on the basis of how things were going. You might program a car to drive a hundred miles without running off the road, but only if you equipped it with lane-detecting sensors whose outputs feed back to the steering apparatus. Otherwise, if there's even the tiniest alteration in tire pressure, if a passing cloud causes the road temperature to drop ever so slightly, if a crosswind comes up, you're done for. Chemical processes, such as those central to the development of an organism, depend both on the concentration of reactants and on temperature, and different reactions differ in their exact dependence. A minor change in environmental temperature, in the concentration of some minor ion, or in any of a host of other variables would fatally derail development. Without elaborate use of feedback, the possibility of success would be remote.

The realization of how self-regulating were the processes that generated organisms caused shocks that rumbled through biology and philosophy. In 1891 Hans Driesch, a German embryologist, let a fertilized sea urchin egg divide into two cells. He then separated the cells and watched the further development of each. To his surprise (frog eggs had given the opposite result for Wilhelm Roux a few years earlier) each cell went on to make a perfect, albeit smaller, larva. That any cell could so reorganize its developmental program converted Driesch from his mechanistic view of life into a staunch supporter of vitalism—the idea that life involves something not explicable in terms of physics and chemistry. Retrospectively, Driesch's problem was the unimaginability of feedback control at a cellular level to the biology of his time.

Up to this point the development of an organism doesn't sound all that different from modern industrial practice, where machines and operators provide lots of nice error-correcting feedback control. But nature takes things a step farther, and the word "recipe" gets even more apt. To a large extent,[8] two products of a modern manufacturing process perform alike because they're formed alike. We most often achieve acceptably precise products by making them of equally precise components. By contrast, two organisms, even genetically identical ones, resemble each other only superficially. The closer one looks, the greater the differences one sees. We put names on the large blood vessels but not the small ones; the latter vary from individual to individual. Nature strives for fitness rather than for precision or consistency, and at every stage development is goal-regulated rather than programmed precisely. The investigator can often remove large pieces of early embryos with little or no postpartum effect. In their DNA itself, different individuals vary greatly; nature tolerates and (in an evolutionary sense) even values diversity. That of course is why we can distinguish among individuals by testing the DNA of blood samples and other bits of body. Some of the variation among individuals may be put to use in social species to facilitate recognition of one member by another; that's a fancy way to say that we'd have problems if all of us looked alike. But most of the variation in detail serves no immediate purpose; it's tolerated because it's without effect on our fitness.

Without the gene you couldn't make the structure—that's usually how we identify genes—so the gene is indisputably necessary. But while necessary, the gene isn't sufficient, even in a purely informational sense. An earlier generation of developmental biologists coined a name, equifinality, for the phenomenon they uncovered when they (starting with Driesch) saw how much abuse embryos could stand without affecting the viability of the complex organism that resulted. The word "equifinality" isn't much used anymore. Words may be poor substitutes for understanding, but this one did call attention to how nature, using elaborate feedback control, makes machines only crudely similar in structural detail yet exquisitely similar in performance.

Scheduled Maintenance. Almost no machine serves its full term without attention to its well-being. Beyond incidents and accidents, some part will wear out before the whole contraption deserves scrapping. Perhaps a temporary part, such as a filter or a flexible gasket, is handier or more economical than a permanent one. Or perhaps a part works by wearing out, as with the brake pads on a car. Or perhaps routine operation of the machine depends on cleaning filters and periodically adding lubricants. Do living machines need such scheduled maintenance? Being out of service for repair risks the continued existence of an organism; the weak, the sick, and the injured are neither effective predators nor evasive prey. Perhaps sleeping and hibernating play some role in maintenance, but most organisms do neither. Or are organisms designed so maintenance doesn't put them out of action?

What happens is truly remarkable. Organisms continuously rebuild nearly everything within themselves! So unexpected and inconspicuous is the phenomenon that it long escaped either suggestion or notice. Rudolph Schoenheimer came across it during the late 1930s, when he began feeding mice isotopically labeled compounds, which had just become available. Schoenheimer's account, entitled *The Dynamic State of Body Constituents*, is still worth reading. He missed on a few details, but most of the work has stood up well. Schoenheimer died in 1941, shortly after giving the lectures from which the book derived. He would probably have been awarded a Nobel prize had he lived; unfortunately the prizes aren't given posthumously.

Isotopic labeling allows marking atoms so they can be distinguished from other, chemically identical atoms. By feeding animals food that had labeled nitrogen atoms, Schoenheimer could look at where those new nitrogens went without being swamped by the huge number of nitrogens already in their bodies. He wanted to see how the extra nitrogen came out—which excreted compounds ended up with the labeled atoms. Specifically, he fed nongrowing adult mice protein with labeled nitrogen. Extra protein input led to extra nitrogen excretion, in the form of urea, the ordinary and expected result. But, unexpectedly, much of the labeled nitrogen failed to appear in the murine urine at all; unlabeled nitrogen took its place. That unlabeled nitrogen could have come only from the existing protein of the bodies of the mice. Since the mice weren't shrinking or otherwise deteriorating, they must have been adding a lot of new protein to their bodies, protein that exactly matched in amount and composition the old stuff that was converted to urea and excreted. The expectation was confirmed by finding, postmortem, labeled nitrogen in their body proteins.

The conclusion is inescapable. An organism isn't an engine in which food simply supplies energy and any material needed when the engine needs servicing. Instead much of the food becomes part of the living machine even when it's not growing; simultaneously an equivalent amount of machine breaks down and gets excreted. This happens almost all the time in all organisms and for almost all the material within the cells of organisms. Remarkable: Intracellular material gets continually replaced without changing the overall composition of either cell or organism. The organism retains the same structure and remains made of the same kinds of molecules, but the particular molecules residing and participating are ephemeral. Hence Schoenheimer's phrase "dynamic state of body constituents."

This dynamic state represents a profound difference between living and nonliving systems. The same molecules the ancient Egyptians hauled into place more than four thousand years ago still make up the pyramids, but you're not the same person you were a year ago. Organizationally, yes, but materially, no. The organization of the individual organism persists far beyond the material of which it's made. And the replacement process doesn't just copy what's scheduled for demolition; it's an entirely fresh synthesis. The instructions for making our most critical and complex material, protein, get continuously reread from the genetic material. Cells don't need to divide to do it, and organisms give no morphological or microscopic hint of their dynamic state.

Why bother? And why bring up protein turnover, a biochemical story, and the dynamic state of body constituents, a physiological story, in a biomechanical tale? To begin with, proteins are not the stablest of compounds, and their stability decreases (they rot faster) as the temperature goes up. Warm-blooded animals—mainly mammals and birds—keep their bodies at temperatures high enough for protein to go bad at a significant rate,[9] and even non-warm-blooded organisms face a slower version of the same kind of spontaneous deterioration. That makes maintenance mandatory. The schedule for replacement tells us a lot about an organism's problem. The more stable structural proteins are the most persistent; the less stable soluble proteins the least. It's handy to speak of "half-life" the way the physicists do when they compare how fast different radioactive isotopes break down; "half-life" here is how long it takes for half of a given kind of material to be replaced. The average half-life of the proteins of a rat's carcass—muscle, tendon, and bone—is twenty-one days, while the half-life of the proteins of its liver and blood plasma is only six. The average half-life for the proteins of an adult human is about eighty days. So perfect is the replacement process that you can recall the events of many years ago.

All intracellular proteins participate in this dynamic state. (Spontaneous changes in the genetic material, DNA, by contrast, are repaired—usually—without wholesale replacement.) Extracellular stuff, such as hair, we simply make and discard continuously. Mammalian red blood cells are discarded in their entirety, with an average life-span of 120 days. But they're not true cells; with no nuclei, they can't direct the resynthesis of their protein. So we have to remake them in their entirety.

In short, nature doesn't *make* her widget but *keeps making* her widget in an unending Sisyphean process. Once again, why bother? Probably the instability of proteins makes continuous maintenance necessary if they're to be usable within organisms. Probably because they're made up of specified sequences of amino acids, they're complex and informationally rich enough that no error detection system could do the whole maintenance job. So probably their combination of complexity and instability leaves organisms with no alternative but a wholesale replacement schedule.

Unscheduled Maintenance and Dynamic Alteration. In addition to random molecular disarrangement, organisms sustain large-scale breakage and injury. Bones fracture, skin tears, and appendages get bitten off. Sometimes controlled proliferation of adjacent cells just covers wounds with scar tissue, while sometimes injuries get repaired so precisely that no mark remains. We're battle-scarred by adulthood, despite modern medicine's intervention. How much repair and regeneration go on varies greatly among organisms and their parts, but all multicellular animals and plants manage some degree of restoration.

Repair and regeneration may be familiar and expected, but they're no trivial tasks. Even making a bit of scar tissue where a branch has broken from a tree or where a person has been cut or burned demands that the system have extraordinary self-knowledge. Cell proliferation has to be stimulated, directed appropriately, and then turned off. For full regeneration, at which we humans happen not to be especially good, the right kinds of cells have to be made in the right places at the right times, blood and nerve supplies have to be reconnected, external covering layers have to be extended, and so forth, all *as things were before.* Somehow nearly autonomous cells, acting in response to their individual instructional codes, must carry out the complex task in precise coordination. Injured vertebrates are especially vulnerable to death from infection or blood loss, but they still patch themselves up after severe tissue loss. Salamanders and lizards can completely regenerate lost tails; a salamander can re-create an entire leg. Invertebrates, though, can do much more spectacular feats. For instance, a starfish can regenerate well over half of itself. This came as an unpleasant surprise to some oyster gatherers, who had been inadvertently increasing the numbers of their adversaries by cutting oyster-eating starfish in half and tossing the halves back in the water. Some simpler flatworms and sea anemones that live on sufficiently hard surfaces deliberately confound regeneration and reproduction; one becomes two when parts simultaneously have a mind (or half a mind) to crawl off in opposite directions.

Nor are replacement and regeneration the full extent of nature's provision for perpetual care. Another is so familiar we forget how special it is. Exercise a muscle, and it gets larger; put an appendage in a cast for a month or so, and the muscles atrophy as the body recycles their constituents. Bone responds similarly, losing mineral and thus softening if out of use and increasing in density in response to repeated loading. Increase the blood flow to an organ—say, by exercise-induced proliferation of its capillaries—and the blood vessels that supply that organ widen.[10] A lot of the degeneration associated with aging may reflect disuse rather than old age itself. Lengthy spaceflights cause all kinds of serious physiological deterioration; fortunately most of the changes are reversible. The components of animals like us continuously readjust themselves in response to changes in how we use them.

Usage-tuned structural change isn't only an animal thing. Add weight to the branch of a tree, and it grows thicker. Bend a branch down to the ground, attach a weight to it, and it will (if the weight isn't excessive or the branch damaged) slowly lift the weight off the ground. Cut down one tree, and the adjacent ones grow longer branches where their illumination has been increased. Load one side of a tree, and the trunk will grow additional wood neatly positioned to offset the additional stress.[11] The evolutionary theory that preceded Darwin's, the Lamarckian idea that increased use of a feature led directly to its hereditary amplification, simply extended these familiar phenomena from individuals to lineages. The extrapolation, an entirely reasonable one, just doesn't happen to be correct for life on earth.

We now know a lot about how organisms manage to be so responsive, but the details matter less than two larger points. First, the routine character of such responses in living machines stands in sharp contrast with the absence of almost any equivalent in nonliving machines. An archaeologist decides whether a rock might have been a tool, or how a tool was used, or whether some item was decorative or utilitarian by checking for wear and examining the pattern of wear. By contrast, the same archaeologist learns how people did repetitive tasks, such as grinding grain, by studying how bones have hypertrophied—worn larger, as it were. Only extracellular parts, ones that weren't strictly living in their functional heyday, wear down from use—tooth surfaces and mollusk shells, for instance.

The second point is that this repair, regeneration, and demand-responsive alteration turns on feedback control. The system must compare what ought to be with what actually is, and that takes information about what it has or hasn't done. The whole process depends on taking action to minimize any difference between goal and present state, defining, again, feedback control. The key elements of the system are the sensors that inform it about its present state. More than anything else, they distinguish natural technology from our own. We understand quite well the sensors used in controlling the positions of our appendages or the rates at which our hearts beat—parts of high-speed neuromuscular and neuroendocrine feedback systems. We also know a lot about the information-carrying links, things such as chemicals that diffuse away from a sensor or else move in blood or other internal fluids. One rarely thinks of sensory equipment in plants. But growth away from the earth depends on gravity detectors, and growth toward sunlight can't happen without photoreceptors—just to point to a few.

Organisms control themselves at an even more basic level. With few exceptions, every cell contains the full set of genetic instructions—the complete recipe book. Most of the time, though, most of the information in most cells stays on the disk, even under the dramatic demands of regeneration. Except in a few instances—cloning being the most notorious—we can't call the information back into functioning software. Organisms are highly inhibited, elaborately repressed. To keep everything from being attempted all at once, a lid must be kept on almost all the possibilities. The repression may fail, as when cells that should know better produce a cancerous growth in a reproductive orgy. In a certain sense this restrained totipotency isn't uniquely biological—only predominantly so. Long ago the neurobiologist Sir Charles Sherrington pointed out (by analogy with the brain, as it happens) that when you use a telephone, the important thing isn't so much getting the right connection as avoiding the entire world of wrong numbers.

In nature, then, the processes of production and maintenance intermingle so thoroughly that we can't make any tidy distinction. Some individual organisms (like ourselves) may grow to fixed sizes while others (such as many trees and fish) may keep growing however long they live, but for neither does development ever finish.

How Good Is Good Enough? Living machines are a competitive bunch, doing overt or subtle battle for nutrients, for energy, for space, and for mates. Competition produces winners and losers, those more fit and those less fit. The conflicts resemble those of unrestrained capitalistic economies enough to have generated the nasty doctrine of social Darwinism. The latter is basically the claim that well-to-do individuals, races, or nations are well off because they are biologically fitter and that while the not so well off might deserve sympathy, attempts to alleviate their plight will inevitably founder on their lower fitness.[12] The inapplicability to human society of fitness in anything like its evolutionary sense is fortunately a fatal flaw for the notion.

But recognition of competition in nature brings up a different parallel with human activities. How good is biological design? The assumption of good design in nature, at least within her intrinsic constraints, pervades studies of how animals work. Whether one justifies the assumption by natural selection or by divine omniscience makes little practical difference.

In recent years the assumption of good design has been variously reassessed, criticized, qualified, and subjected to quantitative analysis. Strict optimality (what we might call perfectionism) has taken a well-deserved beating from the evolutionary biologists.[13] To a large extent, though, they've criticized a straw man; good design isn't a rigorous principle but a working hypothesis of physiologists. We know that organisms aren't paragons of perfection—after all, no biological element has evolved for an unlimited time in an unconstrained context—but the assumption that we're rationally assembled without a host of useless features provides a reasonable starting point for investigating how our features function.

We do seem well tuned, however hazardous the assumption of good design. The practical problem for studying how we work is that most structures are multifunctional, and it's rarely self-evident which function was preeminent in determining the design of a structure.[14]

But well tuned or well designed doesn't imply perfectly precise production; we're just not built that way. Uncertainty and scatter in scientific data often come from inaccuracy in measurements, but in biology they as often reflect real variation among the things being measured. Constancy of the speed of light or the mass of a carbon atom may be limited only by our ability to measure them. But the diameters of human liver cells vary intrinsically. Such natural slop and scatter don't get much attention. At least from anecdotal evidence, nature's tolerance for variation itself varies from structure to structure. She builds some things to very tight specifications and other things less consistently; natural selection must target variability per se.

Many things in nature must vary less than what would be tolerable without loss of fitness. Where no extra cost comes from excessive standardization it ought to persist. For instance, many substitutions can be made in the amino acid sequences of proteins without changing how well they work.[15] But most such substitutions rarely occur; the synthetic machinery doesn't have the option of inserting one amino acid instead of another when one is more abundant and the consequences of change are insignificant. (On the other hand, individual-to-individual variation may enhance fitness. Where offspring are released into an uncertain environment—climatically or otherwise—and only a few will survive, cloning may not maximize the number of survivors. That's one of the arguments commonly advanced for the ubiquitousness of sexual reproduction.)

We can ask in another way about the quality of natural design. The sagacious engineer, mistrusting all estimates and assumptions, designs everything to be better than minimally adequate. We prefer to overbuild. Does nature? The ratio of the load that would produce failure to the greatest load expected in use is usually called the safety factor. Any safety factor above one indicates overbuilding.[16] But determining nature's safety factors turns out to be far from easy. For one thing, natural selection doesn't anticipate, and safety factor in its usual sense means anticipating possible failure rather than reacting after the event—whether by redesign by the engineer or by subsequent selection in nature. For another, we can't easily estimate the greatest expected load for a natural structure. Some things are loaded steadily and predictably while others face highly variable loads. For still another, the service life of natural products varies greatly. Worse, organisms vary in their life histories as well as in their lifespans. An oak tree that blows over in year ten loses all its fitness if it normally grows skyward for twenty years before making acorns. The vine on the tree loses a lot less if it has been putting out seeds since year five. Most human products give full service from the start, while for organisms true profit awaits the reproductive payoff.

Despite such formidable complications, some biologists have had the temerity to tackle safety factors and the cost-benefit analyses underlying them. The loss of a branch for a tree or a tail for a lizard costs little, and both losses are ordinary events. So branches and tails aren't attached with high safety factors. The buoyancy chambers of two deep-sea cephalopod mollusks, cuttlefish and nautilus, face collapse from the pressure of the water around them. Their safety factors are only about 1.4 relative to the pressures at which they live. But those pressures are especially constant and predictable. Bones and tendons have higher safety factors, running between 2 and 6, with tendons usually lower than bones. Flying animals have lighter bones than nonfliers; probably their bones have lower safety factors because for creatures that fly, excess weight carries a great penalty. Tree trunks have safety factors around 4, and the stems of annual plants around 2, although data are limited and uncertain.[17] We care a lot how plants behave since we deliberately alter them for agricultural purposes, since storms are notoriously irregular, and since whole fields of crops and whole plantations of farmed trees do blow over from time to time.

AND THE NONLIVING FACTORY

Since few of us ever set foot on a factory floor, how we make things is only a little more familiar than how nature makes them. That's unfortunate because we live by the products of division of labor, mass production, and assembly lines. Handcrafts are viable only for occasional and optional use, unless we accept drastically reduced economic conditions. Farming methods increasingly reflect the same economics of specialization, scale, and labor minimization; the purveyors of fast food bring the same imperatives to yet another domain.

Even though ancient Egypt, Rome, and China manufactured a diversity of items, mass production is recent. Many historians, especially in America, have described the evolution of the modern factory; the New World may have led the Old in the process. North America experienced a slightly different industrial revolution from northern Europe, one marked by waterpower rather than by steam engines, by small cities located where fast creeks dropped into navigable rivers rather than by large ones at rail hubs, and by the incentive to mechanize of high-cost labor. The story comes with characters whom we elevate to heroic status.[18] Thus we meet Eli Whitney and his revolutionary idea that parts could be produced precisely enough by minimally skilled labor to be interchangeable, Frederick W. Taylor and the Gilbreths with their time and motion studies that systematized the organization of the manufacturing system itself, and Henry Ford and the integration of multiple assembly lines for large products. As chauvinistic Americans we ignore major figures such as Marc Brunel, who mass-produced wooden tackle blocks for the Royal Navy during the Napoleonic Wars.

Size Trends. We build downward rather than upward. The factory—even the factory that assembles our largest airplanes, Boeing 747s—dwarfs the product. For exceptions, such as large construction projects and the global telecommunications network, one might argue that factory and product simply need redefinition. In any case, we're then not talking about mass production in the usual sense.[19] We build down in another sense as well, one that reverses Cope's rule about how size most often increases in the evolution of living lineages. Early steam engines were huge, slowly moving things, slow enough at their inception that inlet and outlet valves were manually operated; recall Newcomen's engine, in Figure 8.2. Increasingly precise manufacturing permitted higher pressure differences and faster operation, so ever smaller engines produced as much power. Early waterwheels were huge and slow; modern turbines are small and fast. The dramatic miniaturization of electronic devices started with the decrease in size and increase in number of elements within electron tubes in the 1930s, 1940s, and early 1950s; the recent elaboration of digital integrated circuits just takes to a previously unimaginable level a trend begun much earlier.

Control. Relative to the complexity of the tasks, industry probably uses more specific and detailed instructions than does nature. Conversely, we make less use of feedback, at least in the number of loops involved in carrying out a particular task. Not that feedback isn't important in manufacturing; modern machinery is unimaginable without it.[20] This use of feedback has a curious history. In a way the most extreme use of feedback antedates factories since piece-by-piece handwork depends critically on the artisan's sensitivity, visual and tactile, even auditory and olfactory, to what's being made. The machine lacks such sophisticated sensory input, so it must make do with fewer and simpler, if faster, loops. Ultimately the precision and versatility of the machine become dependent on the level of feedback—on its awareness of what it's accomplishing at each stage. Testing to determine whether the end products meet some standard becomes really an afterthought. Shifting from artisan to machine-based manufacture provides an incentive to push machines toward the level of sensory equipment and judgment of skilled operators.[21] Much of robotics is about just that: sensory equipment, fast and sophisticated calculations, and feedback loops.

Maintenance. Only in a loose sense does human technology do anything analogous to organisms' continuous turnover of material. Here and there we replace things on schedules determined by their expected safe service lives. The more technically demanding the device, the more hazardous any failure, the greater the cost of redundancy in design or of unscheduled downtime, the greater our willingness to replace bits and pieces before they've failed in the normal sense. Of course what I've described is a commercial airplane. We ask airplanes to give long service now that the basic technology has stabilized and they don't quickly become obsolete. But after ten or twenty years little except the frame may remain of the craft that first flew from the factory. Replacement without failure happens elsewhere as well. I'm told that in some large buildings light bulbs are changed on schedule whether or not they've burned out, and the discards are sold on the cheap to users with dimmer cost accountants.

The Criteria of Quality. While reproductive fitness is a uniquely biological attribute, suitability for the task at hand provides a similar general criterion. For nature, uniformity matters only when it correlates with fitness, and we encounter in the phenomenon of equifinality a world largely beyond our technological experience. For human technology, uniformity in detail takes on specific importance. Uniformity in performance usually depends on consistency in construction, and interchangeability of parts demands a similarly high level of consistency. Nature, by contrast, cares almost nothing about interchangeability and may even be actively hostile to it. To make us inhospitable to pathogens, our immune system has become so potent that we must almost destroy it in order to interchange tissues and organs among individuals. Transplanting a heart presents fewer difficulties than persuading the body to accept it. (But the rejection phenomenon isn't biologically general; insects, for instance, are tolerant of dramatic transplants of glands and other organs. And French grapes grow on American roots, roots of a species resistant to a particular pathogen.)

Quality in human technology has an aspect roughly analogous to biological fitness. "As good as possible" isn't a useful way to specify either a part or an entire machine. For the machine what's important is being good enough to do its intended task satisfactorily. Does making it a little better repay the cost, or does making it a little worse generate a true saving? For an element on an assembly line, what matters is how bad it can be, how far from some ideal it can deviate and still fit in satisfactorily—"fitness" in a different sense. An important criterion of design quality for complex devices is the tolerable range of variation of their parts. The better design is the one that can be made with sloppier parts.

Safety Factors. These of course have a long and honorable history in human technology; the biomechanic borrows the concept but must strain it almost out of recognizable shape. Whether they represent uncertainties about loads that might be encountered or uncertainties in our analysis of a design,[22] modern engineering without safety factors is unimaginable. Real truth, full certainty, and fully assured safety exist only in the domains of lawyers and theologians. The rest of us have to make do with a fallible world of purely statistical anticipations, simplified assumptions, and inattentive inspectors. We learn well from failure, but we prefer other schools of education.

One other aspect of human manufacturing ought to be noted. Historically, human technology has both reduced its direct use of natural materials and increased the degree to which natural materials are modified before use. Partly that comes from our increasing sophistication in metallurgy and polymer chemistry. Less obviously it recognizes that natural materials were naturally selected for their suitability for natural structures and not for our applications. In native form they're usually less suitable for modern manufacturing methods, which demand homogeneous composition and consistent properties. Natural materials call for the sympathetic treatment of the craftsperson rather than the rapid and uniform processing of the assembly line. Wood grows by itself, but canoes of aluminum or fiberglass are cheaper than those of wood. Stone need only be quarried while bricks must be formed and baked, but a wall of brick has a lower final cost than one of stone.

The point made at the start bears reiteration. The production methods of natural and human technologies appear different at first glance, and a more penetrating analysis and more extensive search for underlying factors find them not closer but even more divergent. They're so divergent that even a common terminology often proves elusive. Everyday terms such as "assembly," "polymer," "blueprint," "safety factor," "design," and "intended application" were meant for our production systems, and we run considerable risk of self-deception when we use them for natural systems. Similarly "selection," "fitness," "regeneration," "dynamic state," and "derepression" describe biological phenomena, and using them for human production risks real danger of inadvertent linguistic misguidance.

Chapter 12

COPYING, IN RETROSPECT

Making better widgets by copying nature—bioemulation—is no new notion. In classical mythology Daedalus and Icarus fly from captivity in Crete on wings cleverly copied from birds: "Then he fastened the feathers together with twine and wax at the middle and bottom; and, thus arranged, he bent them with a gentle curve, so they looked like real birds' wings." Then Ovid (43 B.C.E.–17 C.E.)—in the *Metamorphoses*—makes a second reference to profitable copying. After the death of Icarus (his waxy wings were definitely not FAA-approved) Daedalus takes on a twelve-year-old apprentice of real creativity: "This boy, moreover, observed the backbone of a fish and, taking it as a model, cut a row of teeth in a thin strip of iron and thus invented the saw."[1]

Legendary benefits aside, copying nature holds at least three attractions for us. Foremost is the impression of nature's superiority conveyed by the sophistication and diversity of her technology. A tall tree in a storm, a running horse, a spider's web, a flying bird, a jumping flea—their commonness doesn't obscure awesome mechanical performances.

Nor does close examination dispel one's initial sense of excellent design. Each device of nature does something beyond easy reach of our technology, and a lot more besides.

Second comes a more curious motivation, one entwined with contemporary attitudes. For most of human history, the natural and human worlds stood opposed. Nature was something to be tamed and utilized; we had the ordinary attitude of organisms toward other species. Nowadays the natural world intrudes far less but gets venerated far more. And why not? When one's meat is bought in a store, when locusts don't threaten one's corn crop, when central heating and plumbing are the norm, the aesthetics of nature hold greater appeal. We embrace a kind of pantheism or, to use E. O. Wilson's less pejorative term, "biophilia."[2] That affinity for nature drives our eleventh-hour efforts at conservation. It also drives the feeling of a natural rectitude, a moral superiority in nature's ways of doing things.

The third attraction reflects a combination of culture and economics. Support for science and technology rests on a steady supply of explicit promises at least as much as on the record of past success. Whatever the real motives of the participants, such promises work best if couched in terms of practical payoff, not intellectual or spiritual enlightenment. Several kinds of promises are especially effective: industrial profit, alleviation of ill health, and military superiority. Each of these fits well with suggestions that we might make dramatic leaps forward by copying nature.

"Copying," though, isn't the ideal motto to march behind, so better words have been coined. First came "bionics," defined about 1960 by J. E. Steele as the "science of systems whose function is based on living systems, or which have the characteristics of living systems, or which resemble these."[3] The word "systems" came naturally to those, mostly engineers, initially involved; neural systems and physiological controls formed biological parallels to human technology's cybernetics and systems theory. Pattern recognition and feedback devices got particular attention. Use of "bionics" has receded lately. "Robotics" and "artificial intelligence" now hold center stage. A more recent designation is "biomimetics," whose imperatives are more explicitly mechanical—composite materials and walking vehicles, for instance.[4]

But does it work? Not as well as every book, article, and symposium on bionics and biomimetics would have us believe. Most of the lovely allusions to past successes do no more than recognize elements of mechanical commonality. That the jet emulated the squid, that the suction cup copied the octopus sucker should not be our default explanation. A common physical context is far more likely to drive technological commonality. For that matter, we're taking a pretty dim view of human creativity when we assert that copying has been extensive and successful in the past.

I claim in fact that successful copying has been rare. But defending the claim against the more appealing affirmative claim puts me in an awkward position. My best recourse is to search hard for good cases of copying. I've therefore examined the track record with some care, playing historian and aided by the professional reference librarian to whom I'm wedded.

We set some ground rules to circumscribe our search. The present book is about mechanics, so we restricted our purview accordingly. Copying had to be both credible in concept and documented in practice. The result had to be a practical thing that has achieved fairly wide use, not a prototype or a proposal. At the end we were left with fewer than a dozen acceptable cases of bioemulation.[5] These, though, turn out to be far more interesting than mere items for a list or count.

BUCOLIC ROMANTICISM?

Let's consider, for a start, three repeatedly cited cases that wilt under scrutiny. All happen to be British and of roughly the same antiquity.

The Oak Tree and the Eddystone Light. Atop a shoal about fourteen miles from Plymouth, England, stands the Eddystone light, which has guided ships in the English Channel for three hundred years. The first Eddystone light fell in a storm, and the second (of timber) burned. Between 1756 and 1759 the first great British civil engineer, John Smeaton, built the third lighthouse (Figure 12.1) from interlocking stones prepared at Plymouth. Instead of using the rectangular cross section of its predecessors or the uniformly tapered cones now in favor, Smeaton chose the graceful taper of, as he said, "a large, spreading Oak." More of his own words, written in 1791: "Let us now consider its peculiar figure. Connected with its roots, which lie hid below ground, it rises from the surface thereof with a large, swelling base, which at the height of one diameter is generally reduced by an elegant curve, concave to the eye, to a diameter less by at least one-third, and sometimes to half its original base. From thence its taper diminishes more slowly, its sides by degrees come into a perpendicular, and for some height form a cylinder."[6]

Two problems becloud a claim of copying. First, by any engineering standards, these specifications are much too vague. "Generally reduced," "at least one-third," and "sometimes to half" are far from ample instructions—analogy or inspiration, perhaps, but not a quantitative model. The

FIGURE 12.1.
The third Eddystone lighthouse, built between 1756 and 1759 by John Smeaton.

other problem, pointed out by Alan Stevenson in 1850, is that no enlightened engineer would emulate an oak tree.[7] Its main load comes from the drag of its leaves, so it's an end-loaded rather than area-loaded beam. Moreover, it's made of light, tension-resisting wood rather than heavy, compression-resisting stone. Smeaton is simply calling to his reader's mind something that's sufficiently similar to the lighthouse to substitute for an illustration, which the account lacks. Incidentally, the lighthouse still exists, although in a different location. The rock on which it originally stood began breaking up, so in 1882 the lighthouse was removed and replaced with a bigger one a short distance away. Reassembled, the upper part now stands as a monument to Smeaton on the headland above the Plymouth waterfront.

The Shipworm and the Tunneling Shield. Early in the nineteenth century Marc Isambard Brunel bored a vehicular tunnel—still in use by the London Underground—under the Thames. Little experience was available to guide this first tunnel under a river. Preliminary borings suggested a river bottom much drier, more stable, and altogether more suitable for tunneling than actually proved the case. In fact, during the seventeen years between starting and completion, just about everything that could go wrong did: money, labor, Brunel's health, and so on. Everything, that is, except Brunel's tunneling technology. It centered on his new tunneling shield, which needed almost no modification as work proceeded.[8] The shield (Figure 12.2) allowed thirty-six workers to dig at once at the advancing face of the tunnel with a minimum of unbraced excavation.

The burrowing equipment of a shipworm supposedly provided the model for that critical shield. The shipworm isn't strictly a worm at all, but an infamous bivalve mollusk whose paired shells, much smaller than the rest of the animal, serve as the hard parts for tunneling in wood. According to a biography of Marc Brunel, written by a younger engineer who had worked with him on the tunnel:

> one day, as he himself related to me, when passing through the dockyard, his attention was attracted to an old piece of ship timber which had been perforated by that well known destroyer of timber, the Teredo navalis. He examined the perforations, and subsequently the animal. He found it equipped with a pair of shelly valves which enveloped its anterior integuments, and that with its foot as a fulcrum, a rotatory motion was given by powerful muscles to the valves,

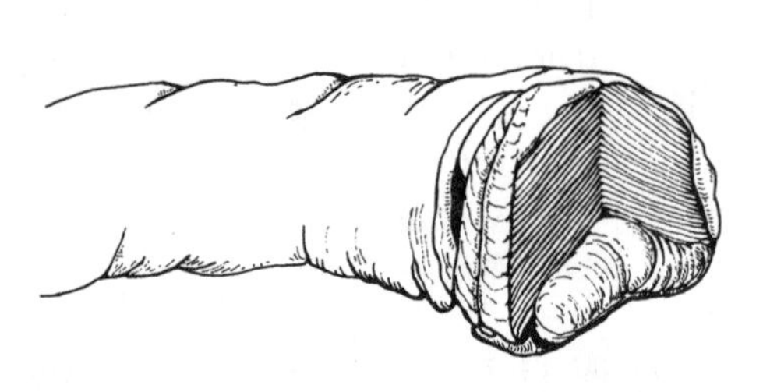

FIGURE 12.2. *The front end of the shipworm* Teredo *with its rasping shell halves, and the drawing of Marc Brunel's tunneling shield in Beamish's 1862 biography.*

> which acting on the wood like an augur, penetrated gradually but surely. . . . To imitate the action of this animal became Brunel's study.[9]

This sounds splendidly specific, but it simply can't be true. Boring by *Teredo* resists observation and was first described (after much trouble) by a zoologist a century later.[10] It doesn't happen by a rotating augurlike action at all, but instead the shells rock back and forth as rasps, scratching off bits of wood that are then ingested. Furthermore, neither what Brunel supposedly saw nor how *Teredo* really works resembles the way the tunneling shield operated. Within the shield a workman removed a single board, excavated perhaps a foot in front of it, replaced the board, and then did the same with the next one. The shield as a whole was pushed forward bit by bit with jackscrews. The shipworm must penetrate hard wood rather than soft sediment, but as a fully aquatic creature it faces no air-water interface or pressure difference. So Brunel's problem was exactly the opposite: providing access to an all too soft substratum without letting in the river. *Teredo* may have provided inspiration for tunneling rather than bridging, but the rest is mythology. Brunel deserves all the credit.

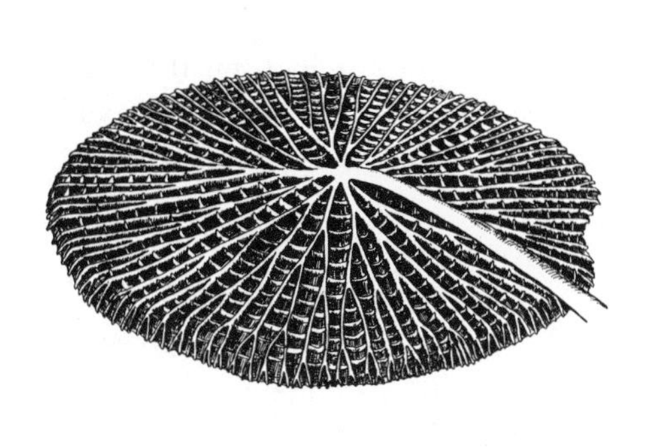

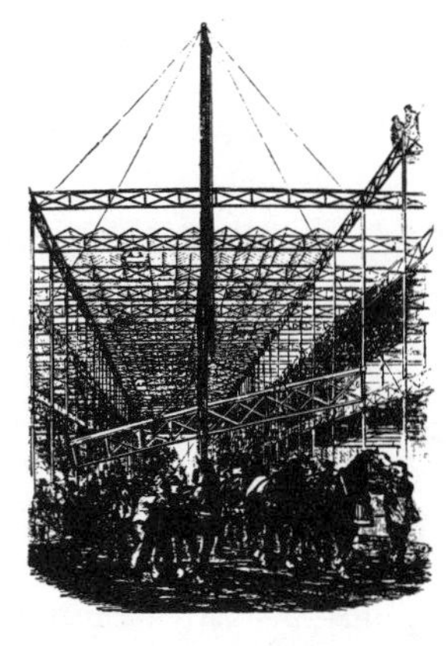

FIGURE 12.3. *The underside of the floating leaf of* Victoria amazonica *and Paxton's patented ridge and valley roofing system being installed during construction of the Crystal Palace—from the Illustrated London News, October 19, 1850.*

The Giant Water Lily and the Crystal Palace. In 1850 Joseph Paxton designed and built an enormous exhibition hall, the Crystal Palace, in London. In every aspect the structure—see Figure 12.3—was extraordinary.[11] Opened less than a year after the design was accepted, the building made unprecedented use of glass and modular components, its appearance was dramatic and unlike any contemporary style, and it was later successfully disassembled and reconstructed on another site. Paxton is often referred to as a gardener. This, although strictly true, gives a false impression of his prior accomplishments and reputation.[12] He was the preeminent innovator of his time in greenhouse construction and, among other things, patented the ridge and valley system that permitted extensive areas to be covered by a horizontal self-draining glass roof, a key feature of the Crystal Palace.[13]

This roofing system is repeatedly cited as a successful case of copying nature, in particular a giant water lily native to South America, *Victoria amazonica* (formerly *V. regia*).[14] Now this is no ordinary lily. Its leaves span as much as six feet and form boats buoyant enough to support a child. An elegant system of interconnected trusses on its undersurface stiffens each flat leaf. But it's still a floating structure, with the trusses offsetting small waves and the lateral forces of currents rather than downward gravity. Paxton was the first to raise these lilies in England, in a special structure he built for his patron the duke of Devonshire; the crucial innovation was provision of a continuous current of slowly recirculating water in its pond.

The claim of copying from nature comes right from the horse's mouth, from a speech given by Paxton to the Royal Society of Arts while the Crystal Palace was under construction. From the *Times* (London), November 14, 1850:

> It was determined in 1836 to erect a new curvilinear greenhouse, 60 feet in length and 26 feet in width. . . . This house was subsequently fitted up for the Victoria Regia, and it was here I invented a water-wheel to give motion to the water in which the plant grew; and here this singularly beautiful aquatic plant flowered for the first time in this country on November 9, 1849. [He then shows a leaf.] You will observe that nature was the engineer in this case. If you will examine this, and compare it with the drawings and models, you will perceive that nature has provided it with longitudinal and transverse girders and supporters, on the same principle that I, borrowing from it, have adopted in this building.

But Paxton's words, taken literally, don't support the usual story. *Victoria* uses a trussing system in which all the trusses, in whatever direction each runs, stay in contact with the supported surface. That's quite different from Paxton's ridge and valley system. Horizontal iron girders running in two perpendicular directions supported the ridge and valley roof of the Crystal Palace, but these girders were arranged with one set beneath the other in the ordinary way, not in the same plane, as are the lily's. And of his innovative ridges and valleys the leaf of *Victoria* shows no trace, although the large aerial leaves of some other plants are pleated into just this arrangement, as shown in Figure 4.3. But something else is wrong. "This building," the final words quoted above, refers not to the Crystal Palace of 1850 but to his greenhouse of 1836. Giving the greenhouse a roof in the pattern of its intended occupant indulged aesthetics more than it solved a problem of engineering. Finally, there's not a word about *Victoria* in the supplement to the *Illustrated London News* of a month earlier entitled "Mr Paxton's History of the Building for the Great Exhibition of 1851."[15]

In all three cases nature may have played some role, but inflating her contribution demeans splendid engineering achievements. Perhaps the legends persisted until canonized through an antitechnocratic and bucolic romanticism that came with the Industrial Revolution, something obvious enough in eighteenth- and nineteenth-century English novels, poetry, and paintings. The poet William Blake intended no praise when he called attention to "those dark Satanic mills."

WHERE COPYING HAS WORKED WELL

Having aroused the reader's skepticism, we can turn to a series of more persuasive claims. Yes, successes exist, and impressive ones at that.

Trout, Dolphins, and Streamlined Bodies. A body that travels through air or water experiences least resistance (drag) if it's rounded in the front and tapers to a rear point in the familiar, streamlined shape of tuna or whale.[16] Watch a proper marine animal—fish, seal, porpoise, or penguin—glide about underwater. No illusion: The animal moves almost effortlessly because it meets little drag, around ten times less than would a sphere or a person of the same size.

How can a rounded front and elongated, pointed rear so dramatically reduce drag? This subtle matter defied explanation until the present century. But long before, around 1809, Sir George Cayley had devised the first deliberately streamlined, low-drag shape, using the best thing at hand: animals that moved rapidly through fluids. He was explicit about what he did: "It has been found by experiment that the shape of the hinder part of the spindle is of as much importance as that of the front in diminishing resistance. . . . I fear, however, that the whole of this subject is of so dark a nature as to be more usefully investigated by experiment than by reasoning and in absence of any conclusive evidence from either, the only way that presents itself is to copy nature; accordingly I shall instance the spindles of the trout and woodcock."[17]

Cayley measured the girth of a trout at a series of points along its length. He then divided each datum by three, and he used the results as diameters to make an elongate wooden body. As Theodore von Kármán, a great twentieth-century aerodynamicist, pointed out, Cayley's streamlined body closely matches the form of the best modern low-drag airfoils and hydrofoils. So he did very well indeed. Cayley got analogous data from a dolphin (Figure 12.4), but the work on trout has received more attention.

What happened next was less serendipitous. Cayley split his wooden model lengthwise and used the shape of the resulting half body for a boat hull, saying, "We should then be deriving our boat from a better architect than man, and should probably have the real solid of least resistance."[18]

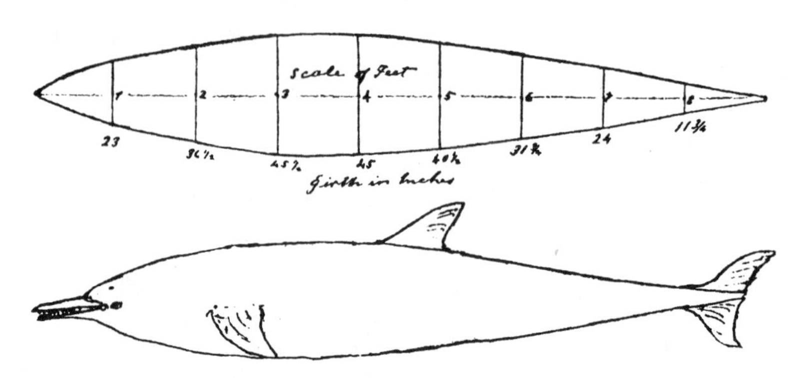

FIGURE 12.4. *A page from George Cayley's notebooks, with a dolphin and the streamlined body that he derived from measurements of its girth.*

But that approach doesn't give an auspicious hull for a shipshape boat, either for low drag or for decent rolling stability. Recall that most of the resistance met by surface ships comes from gravity waves at the surface, not from the kinds of drag faced by trout or submarine. More rationally shaped hulls came along a few decades later.

Bird Wings and Cambered Airfoils. Airplane wings have curved tops and flatter bottoms as in Figure 12.5; they're spoken of as "cambered." With that asymmetrical combination they get much more lift relative to their drag than either inclined flat plates or inclined wings symmetrical top to bottom. High lift-to-drag ratios mattered a lot in the early years of aviation, when planes flew at lower speeds with less weight-efficient engines and less refined designs. As with low-drag bodies, practical experience preceded theory; several decades elapsed between the discovery of cambering and an adequate theory.

During the 1880s two people showed the superiority of cambered airfoils over tilted flat plates. For both, bird wings provided the key—or least very important—models. In England, Horatio Phillips tested a variety of shapes, including the wing of a rook (Figure 12.6).[19] Otto Lilienthal, in Germany, made a much more extensive series of measurements. He found that plates cambered just slightly gave the best results—an upward bowing of about a twelfth (8 percent) of the distance from the front to the back of a wing. This, he noted, is the camber of the wings of the best birds. The dramatic effect of such slight curvature surprised him and probably reinforced his conviction that birds were worth close emula-

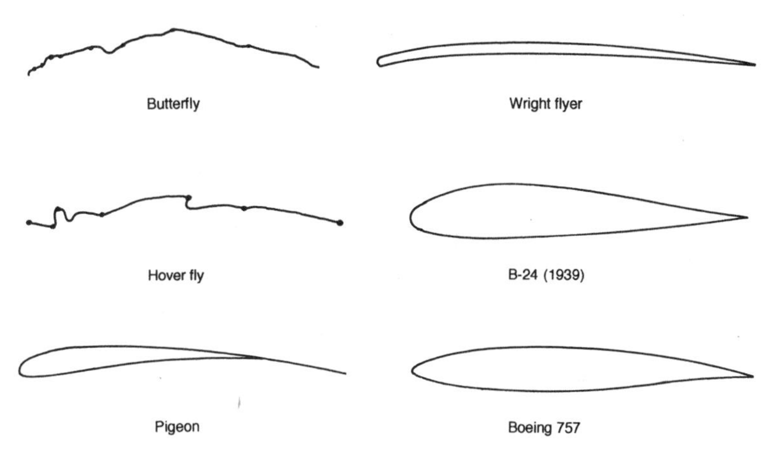

FIGURE 12.5. *Airfoils in cross section, showing the convex upper surface and (sometimes) concave lower surface.*

tion. In the years before his accidental death in 1896, Lilienthal simultaneously studied the flight of birds and constructed aircraft that we'd now call hang gliders. The ultimate aim was powered flight using a flapping wing craft. Lilienthal's main legacy is an impressive book entitled *Bird Flight as the Basis for Aviation.*[20]

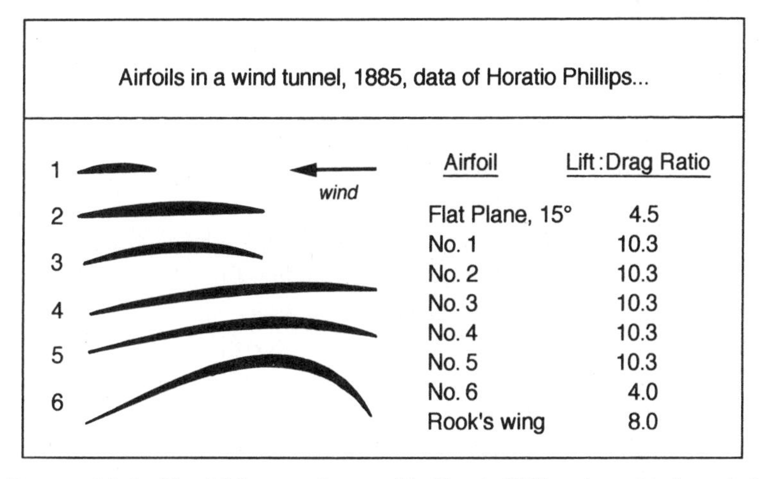

FIGURE 12.6. *The airfoil cross sections tested by Horatio Phillips, along with the results he obtained.*

(Neither Phillips nor Lilienthal but a third person who looked at curved airfoils in the 1890s, Frederick Lanchester, took the first major steps toward understanding why they work as they do.[21] The explanation of how an airfoil generates lift is based on the fact that air flows faster across the top of a wing than across the bottom. By Bernoulli's principle, that gives a lower pressure above than below, so the plane is pulled upward. But the origin of that faster flow is much more peculiar than one might guess from the polite fictions we're taught in schools and museums.)

As with Cayley's boat hull, the sequel makes the protagonists look less prescient and heroic. The Wright brothers initially trusted Lilienthal's numbers; he was, after all, a proper engineer. So they used both his airfoil shapes and his data for their first full-size glider. The glider gave too little lift, so they then built a wind tunnel and got their own data. Lilienthal's errors may have arisen because he swung wings on the end of a whirling arm rather than put them in a wind tunnel; after the first revolution on the whirling arm, the test object meets the disturbed air of the previous circuit. I'm sympathetic since I once ran afoul of the same problem.[22]

Birds and Turning Aircraft. In the two-dimensional world of automobiles and boats, steering involves nothing more complicated than changing one's heading by reaiming the front wheels or turning a rear rudder. Airplanes, though, fly in three dimensions: They can roll from side to side or pitch up and down as well as turn left and right. Early attempts to fly paid little attention to three-dimensional control. Some designs relied on familiar rudders; in others the flier was supposed to shift position like a bicycle rider so the craft would make banked turns without any specific aerodynamic adjustment.

More than any others of their pioneering generation, Wilbur and Orville Wright took control seriously. Their initial and most important patent described a system of control, and the basic scheme they worked out is still almost universally used. Bird watching helped, although much later Orville tended to minimize its contribution. But a letter written in 1900 from Wilbur to Octave Chanute contains the following: "My observation of the flight of buzzards leads me to believe that they regain their lateral balance when partly overturned by a gust of wind by a torsion of the tips of their wings. If the rear edge of the right wing is twisted upward and the left downward the bird becomes an animated windmill and instantly begins to turn, a line from its head to its tail being the axis. . . . In the apparatus I intend to employ I make use of the torsion principle."[23]

In short, the Wrights found that a bird adjusted the angles of the tips of its wing for roll control. One tip was tilted so its front was slightly upward, and that would increase its lift; the front of the other was tilted downward to decrease the lift. That asymmetry in lift would cause banking, and in a banked position the overall lift would be directed slightly sideways instead of straight upward. That sideways force (with perhaps a little compensation from the rudder) would pull the plane around in a curve. The key, then, was to twist or warp the wings, as the Wrights did with the ingenious arrangement of cables shown in Figure 12.7. You can demonstrate the change in shape, as they did, by twisting an elongate rectangular box (such as a milk carton) without ends. Of course, as Orville said about learning flight from birds, "After you once know the trick and know what to look for you see things that you did not notice when you did not know exactly what to look for." The only major change in the system since the Wrights has been replacement of warping by a pair of flaps or ailerons, one on the outer rear end of each wing. That's better for the more rigid wings of later airplanes. Ailerons, though, work the same way, producing banked turns by raising one wing and lowering the other.[24]

Wasps and Paper from Wood. While papermaking is an old art, only recently have we made use of wood fiber as our normal starting materi-

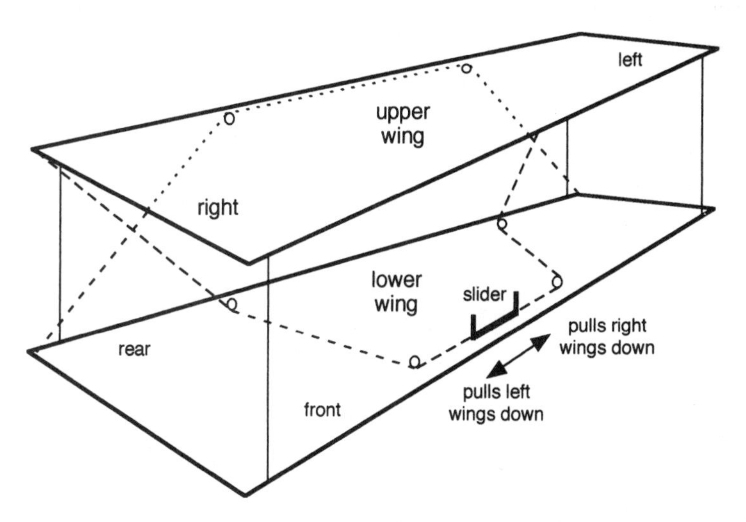

FIGURE 12.7. *The system used by the Wrights for wing warping—from* How We Invented the Airplane, *by Orville Wright.*

al.[25] Up through the eighteenth century most paper was made from cotton and linen rag, and the limited supply of rag became troublesome as increasing literacy and more complex commerce raised demand. The dead were buried in wool (in England by law) to save cotton and linen for papermaking. Around 1719 the great French entomologist and polymath René-Antoine Réaumur suggested making paper from wood, as were the nests of paper wasps (*Polistes* and related genera):

> The American wasps form very fine paper, like ours; they extract the fibers of common wood of the countries where they live. They teach us that paper can be made from the fibres of plants without the use of rags and linen, and seem to invite us to try whether we cannot make fine and good paper from the use of certain woods. . . . The rags from which we make our paper are not an economical material and every papermaker knows that this substance is becoming rare. While the consumption of paper increases every day, the production of linen remains about the same.[26]

Réaumur himself made no paper, but during the century that followed a number of people attempted to produce paper from wood, and decent evidence links Réaumur and the wasps with their efforts. The German Jacob Christian Schäffer made paper in the 1750s from a wide variety of plant material (and from wasps' nests themselves) with only a small fraction of rag. He clearly followed Réaumur's path, with conspicuous drawings of the adult wasp, its larvae, and its nest in his treatise on papermaking. In London, in 1800, Matthias Koops (an otherwise obscure figure) managed to make paper from both straw and wood with no rag at all, and his paper was suitable for printing presses. He demonstrated his achievement with a small book whose final pages were printed on his paper; the subject was—what else?—the history of papermaking. (No other area of technology seems to leave as extensive a paper record as papermaking.) In the book he cites both Réaumur and Schäffer as important predecessors with "ideas on substitutes for paper-materials."[27] Wasps are not explicitly mentioned, but the connection is plausible.

The wasps thus set the stage by showing what was possible: that cellulose fiber from wood, an almost unlimited source, could be separated from its binder, lignin, and re-created as a two-dimensional mat. They were much less forthcoming with practical guidance, and a long and arduous struggle followed the original suggestion. Koops himself went bankrupt after building a large mill; despite the fine product and the high cost of rag paper, wood paper could not yet compete. But progress thereafter was steady, and within a few decades paper mills started eating forests in quantity.

Silkworms and Extruded Textile Fibers. The same Réaumur had another suggestion that ultimately proved practical. Just before they pupate, moth larvae produce the protein we know as silk. Immediately after extrusion through a fine orifice as a liquid, the protein solidifies into a continuous fiber. The silk designed by nature for the cocoons of silkworms (the family Saturniidae) has long been appropriated by humans for beautiful and commercially valuable textile fiber. We either chop the cocoons and dissolve a gluing protein (wild silk) or, in the domesticated species of silkworm, *Bombyx mori,* unwind the fiber. Robert Hooke, in the seventeenth century, and Réaumur, in the eighteenth, suggested that a textile fiber might be manufactured by an analogous extrusion process. Hooke viewed silk as "a dried thread of glue" and casually speculated "that probably there might be a way found out, to make an artificial glutinous composition, much resembling, if not full as good, nay better, than that excrement, or whatever other substance it be out of which, the silkworm wiredraws his clew. If such a composition were found, it were certainly an easy matter to find very quick ways of drawing it out into small wires for use. I need not mention the use of such an invention. . . ."[28]

Hooke's "easy matter" proved, to say the least, wishful thinking. During the nineteenth century this possibility—extruding or "drawing out" fiber from an orifice—was explored by a number of people. Louis Schwabe, in England, extruded glass fiber as early as 1842.[29] Georges Audemars, in Switzerland, drew threads of "artificial silk" of cellulose nitrate in 1855. One M. Ozanam, in 1862, suggested that silk scrap might be reconstituted by dissolving and reextruding.[30] In the 1880s Hilaire de Chardonnet, with great labor and expense, finally developed a commercially viable process for making an extruded artificial silk, initially the same dangerously flammable cellulose nitrate. The silkworm's footprints are unmistakable. Schwabe, Audemars, and Chardonnet each were involved in the natural silk industry. Schwabe made silks in his mill for Queen Victoria, and Chardonnet worked with Louis Pasteur on diseases of silk moths—an insufficiently appreciated branch of veterinary medicine. Chardonnet later said he intended "imitating, as closely as possible, the work of the silkworm."[31]

By the start of the twentieth century both rayon and cellulose acetate fiber were being made by versions of the silkworm's (and Chardonnet's) extrusion process, and we now make many other fibers the same way. Was the silkworm a useful model? Although indirect, the evidence is persuasive. To start with, the name of the industrial extruder, spinneret, is the same as that given the silkworm's organ. Better yet, the word is thoroughly inappropriate for either, a term borrowed earlier by entomologists[32] from the spinning process long used to make long threads from short fibers. In the spinneret of neither silkworm nor fiber mill does anything spin, rotate, or go around in any sense at all. Silkworms (and spiders) don't actually "spin" their cocoons and webs, and in a fiber mill thread is spun from extruded fiber in a subsequent operation. So industry probably adopted the existing lingo of entomology.

Early mechanical extruders in fact looked very much like the analogous parts of insects—small pipes tapering down to fine apertures, as in Figure 12.8. These pipes were made of glass, and the demanding work of pulling out hot glass just enough (to a little under one tenth of a millimeter) produced a delicate product all too prone to breakage and clogging.[33] Around the turn of the century multiple-orifice precious-metal spinnerets came into use. These thimblelike devices (as in the figure) work much better, although, operating with high pressure across a flat

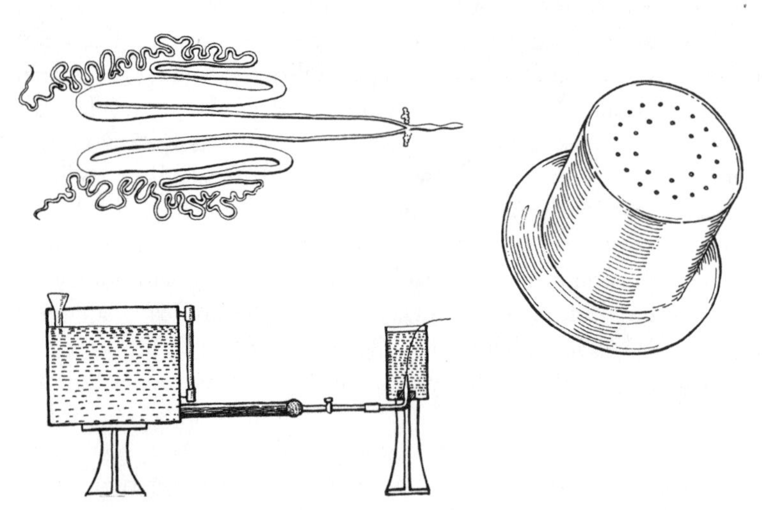

FIGURE 12.8. *The silk gland and extruder of a silkworm, an early tapered spinneret and a modern multiple-orifice spinneret.*

surface, they're not the kind of arrangement likely to occur in nature. Still, as described in 1930:

> The method of producing artificial silk [rayon, etc.] closely simulates the process by which real silk is formed. The silkworm extrudes fibroin through two apertures below its mouth, and cements the two threads together with sericin, which is ejected from the glands at the same time. In the manufacture of artificial silk, the silk glands are represented by large tanks and the apertures by fine jets. In the spinning process, i.e. the transformation of the viscous mass into threads, it is customary to make a distinction between dry spinning and wet spinning. . . . Dry spinning utilizes single spinnerets consisting of thick-walled glass tubes contracted to a fine capillary, with an internal diameter of 0.08 mm., and the separate filaments are assembled into a thread.[34]

Eardrums and the Telephone's Transducers. By the 1870s telegraphic communication was routine. But the advantage of an instant connection was offset by the need to use Morse's dots and dashes, with slow and cumbersome coding and decoding. A telegraphic circuit had only two states, connected or not connected. (Such binary coding is what our present computers use, but they can switch from one state to the other millions of times more often.) Among others, Alexander Graham Bell worked on the problem of transmitting voices instead of mere telegraph signals. One possible scheme broke a complex sound into separate frequencies that were transmitted along parallel lines; the receiver then recombined them. Bell's important insight was that such complexity was unnecessary; a single device could convert all frequencies of sound into a single electrical signal. Bell, no electrical guru, got the idea from an analogous biological device.

At the time Bell was a professor of vocal physiology at Boston University. He and his father pioneered in teaching deaf people to speak intelligibly, making sounds that they themselves had never heard, so he was immersed in the physiological and behavioral aspects of making and perceiving sounds. Bell recognized that the eardrum was a single device that handled all frequencies at once. Its movements in and out set the bones of the middle ear into motion, and these in turn communicated with the liquid-filled inner ear, where the neural equipment was located. In his own words, "It occurred to me that if a membrane as thin as tissue paper could control the vibrations of bones that were, compared to it, of

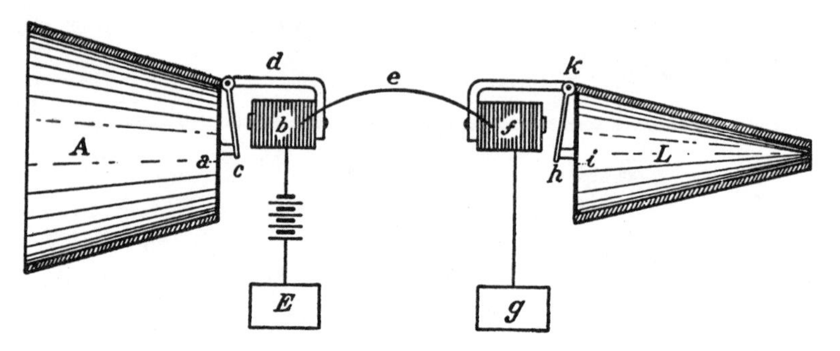

FIGURE 12.9. *Bell's diagram of the telephone transmitter (or microphone) and receiver.*

immense size and weight, why should not a larger and thicker membrane be able to vibrate a piece of iron in front of an electromagnet . . . and a simple piece of iron attached to a membrane be placed at the other end of the telegraphic circuit?"[35]

Not that all then went smoothly, but the invention of this microphone broke the back of the problem. The same device in reverse served as a receiver to get the sound back out, as in Figure 12.9, taken from what may be the most profitable patent in history.[36] Thus the device was automatically bidirectional. Although Bell's transmitter was soon replaced by Thomas Edison's more sensitive carbon microphone, his receiver survives in earphones and (at least its basic principle) in loudspeakers.

Modern life would be impossible without aircraft, cheap paper, artificial fabrics, and telecommunication of speech—major matters, all of them. Three further cases meet our criteria for emulation but involve task-specific devices of narrower application and lesser everyday impact.

Barbed Wire. Keeping livestock pinned within hedgerows of thorny plants is an old practice, one especially useful where wood or stone for fencing is in short supply.[37] Settlers of the North American prairies faced an ever-worsening wood shortage as they moved westward. The plant of choice for the Midwest was a shrubby tree native to East Texas and nearby areas—the Osage orange (*Maclura pomifera*)—and a small industry during the 1860s and 1870s supplied its seeds and seedlings for use farther north.[38] This thorny bush, though, had substantial disadvantages. Growing an effective hedge took about three years, the grapefruit-size but inedible fruits were a nuisance, and the hedge was both immovable and a nuisance to maintain. Michael Kelly's patent of 1868[39] for an early form

FIGURE 12.10. *Branch and thorns of an Osage orange, Kelly's "thorny fence" of 1868, and the barbs of a modern version.*

of barbed wire was explicit: "My invention [imparts] to fences of wire a character approximating to that of a thorny hedge. I prefer to designate the fence so produced as a thorny fence." Indeed, the wire was produced by an enterprise called the Thorn Wire Hedge Company, perhaps advertising its utility by drawing attention to a familiar antecedent. Figure 12.10 shows the similarity of plant thorns such as those of the Osage orange to this early form of barbed wire.

Kelly barbed wire was eclipsed by two competing brands of cheaper wire after 1874; as with wings, spinnerets, and telephone transmitters, fidelity to nature guarantees no economic magic. Patents for the new types were held by Joseph Glidden and Jacob Haish. With the usual personification of invention, Joseph Glidden is often listed as the inventor of barbed wire. Haish, almost certainly not coincidentally, had a lumberyard that sold Osage orange seed. As the historian George Basalla puts it, "barbed wire was not created by men who happened to twist and cut wire in a peculiar fashion. It originated in a deliberate attempt to copy an organic form that functioned effectively as a deterrent to livestock." Barbed wire has been an enduring success. Current consumption in the United States runs to well over a hundred thousand tons a year.

Chain Saw Cutters. Small gasoline engines were developed early in this century. A saw blade in the form of an endless chain with teeth on the links was patented in 1858. Chain saws both with attached gasoline engines and with electrically powered ones designed to work with adjacent generators appeared in the 1920s and 1930s. Yet commercial logging still made much use of hand-operated crosscut saws. Motorized crosscuts and circular saws were huge, unwieldy things quite unsuitable for felling trees in a forest. Chain saws didn't work smoothly and required frequent resharpening; even with large, heavy motors, they cut slowly. The arrangement of teeth (shown in Figure 12.11) so effective in the cross-

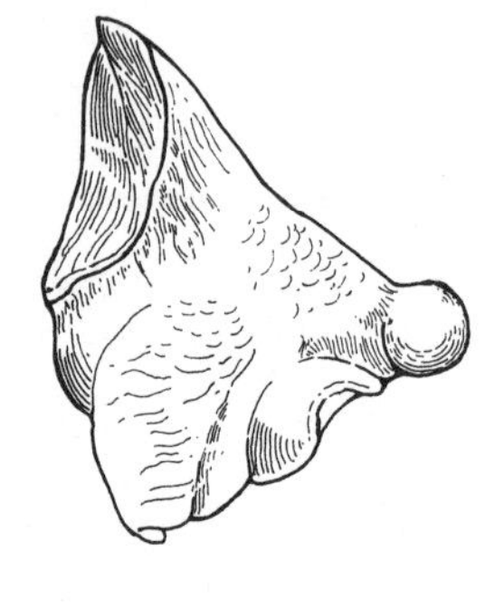

FIGURE 12.11. *The teeth of a crosscut saw, one mandible of a timber beetle, and a section of a saw chain showing cutter and depth feelers.*

cut[40] just doesn't cut the mustard for a saw that cuts a wider kerf (groove) with teeth that move unidirectionally as small, guided links.

During the 1940s a machinist working as a logger took a careful look at the tunneling equipment and technique of the larva of a large wood-boring beetle, *Ergates spiculatus.*[41] These beetle larvae are among the few insects (termites are others) that digest wood—that is, they break the cellulose polymer down into metabolically usable simple sugar. But they start by biting off wood with pairs of mandibles that (in insect fashion) move sideways rather than (as in vertebrate jaws) up and down. Unsurprisingly, these wood borers (the subfamily Prioninae of the family Cerambycidae) have robust mandibles. They splay out from the head so their sharp front edges cut at the walls of the extending tunnel—the equivalent of the bottom and sides of a saw's kerf.

This ex-machinist-logger, Joseph Cox by name, then devised a chain with cutters of the shape and in the equivalent position of those of the beetle larva. They didn't move side to side, and they pointed alternately left and right (adjacent pairs would have jammed immediately), but they were unquestionably beetlelike. These wood-cutting teeth worked so well that millions of portable chain saws now use the design, so well that the company Cox founded in 1947 now dominates the business. If you buy a chain saw, it will probably have an Oregon chain.

Velcro. This flexible and alignment-independent hook-and-loop fastener has found a secure niche in modern life, gradually replacing shoelaces, buttons, zippers, snaps, picture wire, curtain rings, and many other modest but ancient adhesive devices. It does yeoman joinery, socially rehabilitating those disadvantaged by limited manual dexterity. For the more traditional among us, the sound and feel of separating halves still carry a sense of something self-destructing, but we'll eventually adjust. Velcro (from "velvet" and "crochet") gained wide acceptance more slowly than barbed wire or the modern chain saw because it neither satisfied a critical need nor replaced predecessors found wanting. But like them, it capitalizes on a model from the living world.

About 1948 a Swiss engineer and avid walker named Georges de Mestral contemplated the burs that clung to his socks and dog after a hike in the local hills.[42] The burs, variously reported as cocklebur (*Xanthium*) or burdock (*Arctium*), had tiny hooks at their tips that engaged anything fuzzy, as in Figure 12.12. Nylon, then less than ten years old, turned out to be an ideal material to make the hooks—when heated and then cooled, it would retain a curve in the face of all kinds of abuse—but the hooks had to be made from cut loops, which required the invention of some novel machinery. In addition, the hooks were more fastidious in choosing what to grab than were the natural burs, so Velcro ended up as a pair of mating surfaces—hooks and loops. Improvements and derivative products have come along, such as stainless steel Velcro (stronger) and silent Velcro (for military use), but the basic material has changed little.

A person could hang from a patch of standard Velcro less than five inches in diameter, but such a patch can be peeled loose with one hand.

FIGURE 12.12. *The hooked burs of a plant,* Arctium minus *(courtesy of the Duke University Herbarium), and those of Velcro.*

To hold strongly but release with ease, Velcro (like adhesive tape) takes clever advantage of what's called a dimensional reduction. It holds against a force extending over a broad area, while it peels in response to a force concentrated along a line; it attaches in two dimensions but releases in one. A pair of scissors or a zipper does the same thing, except that instead of reducing an area to a line, it reduces a one-dimensional line to a nondimensional point. To bring matters full circle, dimensional reduction is important in nature as well. As they get food, lots of animals peel and tear—that is, they reduce an area to a line or a line to a point; that's how a cat uses its claws and a caterpillar cuts through a leaf. Conversely, some limpets have broad skirts that make them hard to peel off rocks. Also, grasses have lengthwise veins that help them resist crosswise tearing.

WHEN COPYING WORKS, WHY DOES IT?

These cases don't add up to a statistically significant set of data. But we can dimly discern some common threads that, if not definitive, are at least suggestive.

First, mere imitation isn't likely to be productive. George Cayley made a leap of abstraction by carving not a trout but an axisymmetrical body derivative of trout girths. Horatio Phillips's best airfoils gave greater lift-to-drag ratios than his rook's wing. Otto Lilienthal's airfoils were arcs of circles in section, not cross sections of actual bird wings. And while wing warping may be what birds do, Wilbur and Orville Wright did it in a thoroughly unavian way. Besides, wing warping was quickly eclipsed by ailerons, a still less avian device. Barbed wire is a tensile structure, a wire, closer to vines than to branches. The barbs were originally attached rather than integral, with their radial orientation fixed in nonbiological fashion. The way the cutters of a chain saw blade move scarcely resembles the motion of larval mandibles; what's common are the shape of the cutters, their flexible connection to the saw or head, and where in the tunnel or kerf they cut. Velcro is a paired adhesive system like zippers or snaps, while burs are a single element more like the reversible adhesive of Post-It note papers. Only the crucial hooks are a common element. The idea, the inspiration, or the strategy—whatever one chooses to call it—not the details or tactics that humans use, is what nature has provided. Practicality seems to lie somewhere between general inspiration and exact emulation.

Second, success depends inversely on how well we understand the underlying science. Where our science is strong, copying produces at best narrowly targeted items, such as barbed wire, saw chains, and Velcro. But where our science is weak, copying can generate devices of broad utility. Fluid mechanics was a murky business before the present century, so streamlined bodies, cambered airfoils, and ailerons could not easily be deduced from first principles. As Cayley took pains to emphasize, copying was the best one could do under the circumstances. Electrical signals with complex wave forms were almost unknown, so appreciation of the eardrum allowed technology to leap ahead of theory. Papermaking and fiber extrusion involve complex combinations of solid mechanics, fluid mechanics, and chemistry, but such complexity doesn't bother natural selection, so nature provided useful hints both of what's possible and of how to proceed.

A third point, the most important, turns on the difference between the two technologies. Nature's is typically tiny, wet, nonmetallic, nonwheeled, and flexible; human technology is mainly the opposite: large, dry, metallic, wheeled, and stiff. Where one technology operates in what is normally the domain of the other, emulation holds promise. As natural structures go, thorns and beetle mandibles are especially stiff, so they're closer to what we do with our materials. Among devices made by humans, Velcro is relatively flexible, so with it we brush against a world whose possibilities have been more thoroughly explored by nature.

HUMAN FLIGHT AS A CASE HISTORY

To look at successes is to look with perfect hindsight. To counteract such post hoc bias, we might focus on a specific human endeavor. The history of flight, which has given us several examples of successful emulation, contains still more of failure. Even ignoring the completely bizarre, such as men jumping off barns with improvised wings,[43] a nicely multidimensional story emerges. One side of the story was put well by Hiram Maxim, in 1909. He had just spent a lot of money—money made from his machine guns—on a spectacularly large and unsuccessful flying machine: "Man is essentially a land animal, and it is quite possible that if Nature had not placed before him numerous examples of birds and insects that are able to fly, he would never have thought of attempting it himself."[44] The other side of the story is that in most respects the flight of organisms proved to be a notably bad source of guidance.[45] As Maxim also put it, "The successful locomotive was not based upon an imitation of an elephant."[46] Quite beyond their nonflapping wings, the Wrights' aircraft were thoroughly unbirdlike: propeller-pushed biplanes with horizontal stabilizers in front and large vertical control surfaces behind. To sharpen the argument that nature speaks with forked tongue, consider several specific aspects of aircraft design.

Using a single set of structures to produce both upward and forward force is a reasonable scheme only for a flier of low wing loading, which is to say a small one, an argument made in Chapter 10 and not unknown at the end of the nineteenth century. Furthermore, nature's reciprocating motion—up and down, up and down—is a fairly cumbersome way for our motors to apply a large amount of power. So a single pair of flapping wings is an inauspicious route to human flight. No successful human-powered aircraft, where power is most severely limited, works that way. The glider that killed Lilienthal was about to get an engine that would flap the wings with a pair of hydraulic or pneumatic cylinders. It's extremely doubtful that Lilienthal would have flown the first fully powered aircraft but for the accident that killed him.

Most of the early attempts at flying machines had very bird- or bat-like wings; if anything, they were even shorter and broader than those typical of flying animals. Figure 12.13 gives an example. But other things being equal, the best ratio of lift to drag is obtained with the longest wing, although that might not have been completely obvious a century ago. For a given wing area, longer and skinnier are better. Bird and bat wings are compromises; flapping is probably easier for shorter wings, and shorter wings give greater maneuverability and better ability to deal with erratic air currents. A little less attention to animals and a little more experimentation with isolated wings would have been helpful.

Probably the animals misled us worst by giving the impression that control of flight was no big problem. In locomotory machines in general, and with especial severity in aircraft, maneuverability and stability tend to be antithetical characteristics. Each of the three extant lineages of flying animals is of considerable antiquity, so, as pointed out by the evolutionary biologist John Maynard Smith,[47] they've had time to become unstable, time to evolve neural systems that could manage the instability that goes with good maneuverability. After all, maneuverability will almost always be advantageous for animals since they're either aerial predators or targets of aerial predators, or they fly among obstacles, or they soar on atmospheric irregularities.

Lilienthal was especially single-minded about imitating birds. His book on bird flight spells out his operational philosophy: ". . . natural

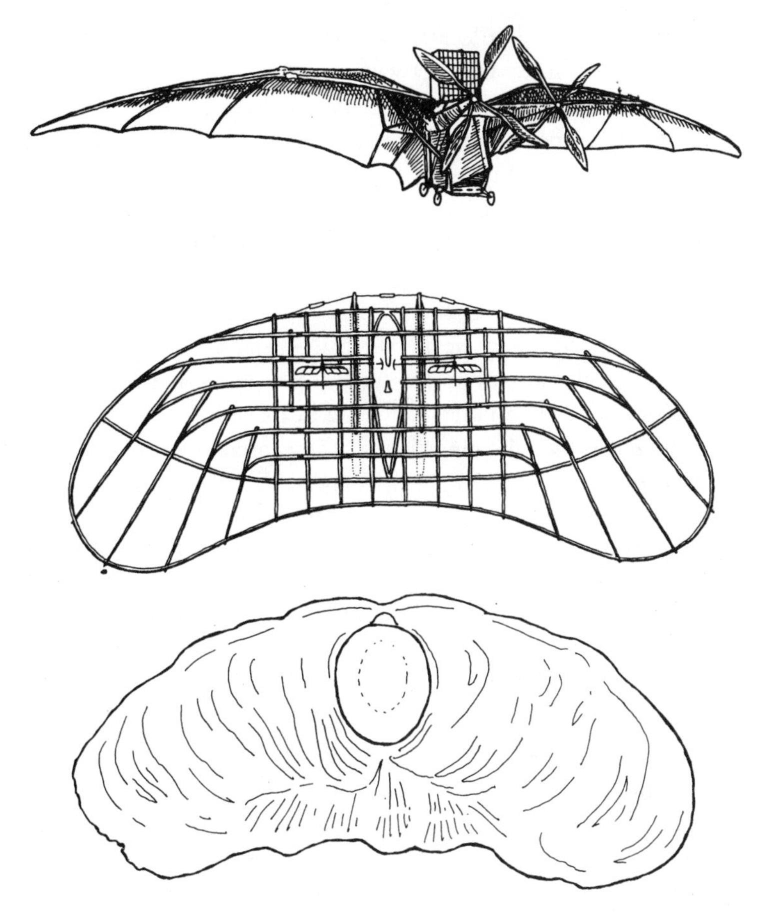

FIGURE 12.13. *Clément Ader's batlike Éole of 1890, which made a partially powered hop; the wing (top view) of Igo Etrich's plane of 1906, which copied the gliding fruit of the Javanese cucumber, and the gliding fruit itself.*

birdflight utilizes the properties of air in such perfect manner, and contains such valuable mechanical features, that any departure from these advantages is equivalent to giving up every practical method of flight." Thus he was fatally trapped in an excessively birdlike and unstable glider. The instability of Lilienthal's craft worried Octave Chanute, an older man who had experimented with gliders earlier. Birds, for instance, have horizontal tail surfaces but no vertical rudders, since they get adequate control of sideways turning with their wings alone, so vertical tail surfaces were small or absent in many hopeful aircraft built between about 1880 and 1905.

Some aircraft, just a little later, went overboard in the other direction, with such great stability that they were almost unmaneuverable. At least one copied another natural flier. Animals can do wonderful things with delicate sense organs and fast-acting feedback loops. Plants obviously can't, so gliding plant parts must be highly stable. The samara (seed in fruit) of a maple, mentioned as an alternative to a parachute, glides helically (it "spirals") downward. The glide slows its earthward movement so it can be carried farther by the wind that blew it off the tree. A few fruits glide along a straight path, gaining distance from the parent tree without the need for wind. Among these is the fruit of the Javanese cucumber, a flying wing that glides earthward in the still air beneath the forest canopy in Southeast Asia (Figure 12.13). After the death of Lilienthal, Ignaz and Igo Etrich bought his remaining gliders, whose instability converted them (scared them?) into advocates of extremely stable craft. Igo obtained flying fruits from the Hamburg Botanical Museum and copied them in a series of craft—unmanned, manned, and then powered. But as gliders they were almost unmanageably stable and lacking in maneuverability. Worse yet, they combined the worst of both worlds since too much stability was lost when motors were installed.[48]

Eventually it became clear that an aircraft, unlike a bird, ought to be so inherently stable that it didn't require continuous activity by the pilot. At the same time it ought not be too stable to be controllable, thus unlike a gliding fruit. Modern aircraft manage the compromise quite well. As a former colleague, Molly Bernheim, describes learning to fly, "So a pilot, before he flies safely, must learn a very difficult lesson, one which is contrary to all his natural instincts. A headlong descent toward the ground? Pulling back won't help you! You must let go of the stick, so that it moves forward. A wing that won't come up? Let go! The airplane can look after itself better, now, than you can do! Turn it loose! Then, and only then, you may guide it gently where you want it to go."[49]

One domain, though, differs. Small, high-performance military aircraft are made deliberately unstable to achieve great maneuverability. But they feel stable to their pilots because they've adopted a crucial feature of animal flight. Modern control technology, which means sensors and actuators in rapidly acting feedback loops, gives them the best of both worlds.[50] The pilot makes the strategic decisions while the tactical details are handled by servomechanisms, just as a mammal or bird divides control between the cerebral executive and the middle management of the rest of the central nervous system.

Sticking close to nature's methods has proved no boon in at least one other form of locomotion. We're most often told that Robert Fulton invented the steamboat in 1807. In fact, steamboats existed, and Fulton knew about them. He had a better engine (from Boulton and Watt, in England) and better financial backing, but the main advantage of his boat lay in that very unbiological device the rotating paddle wheel. James Rumsey's boat of 1787 used a steam-driven piston to make a squidlike pulse jet, taking water in beneath the bow and expelling it forcefully at the stern. A piston engine naturally reciprocates, so a few valves suffice to make it power a pulsating squirter. But Rumsey's boat suffered from the squid's problem: the low-propulsion efficiency that comes with a high-speed, low-mass output. John Fitch's boat of 1790 used a set of reciprocating rear paddles that alternately dipped into the water and pushed rearward like a person doing a crawl stroke with duck feet as hands. Paddle wheels, first in the rear and then at the sides of boats, were simpler, more effective, and more efficient.[51]

This has been a skeptical and polemical chapter. But its message of disbelief and irritation at excessive romanticism and self-deception is hard to paint in brighter colors. Historians are fated to look backward, whereas scientists usually look forward. In all the debunking, however, a positive message should not be lost: The record of copying may be sparse, but good cases do exist. Blind emulation may have nothing to recommend it, but we've sometimes been much smarter than that. Indeed, a good argument can be made that as our technology takes on more of the characteristics of nature's—more flexible materials and structures, increased miniaturization, greater use of nonmetallic materials, and so forth—nature will be an increasingly useful teacher.

Chapter 13

COPYING, PRESENT AND PROSPECTIVE

History isn't destiny. If it were, then we'd have known in 1875 that cities were doomed, that by 1925 the ever-increasing volume of horse manure would produce total urban inundation. We've looked at the past; its messages about what's to come in the business of copying nature are both mixed and unreliable. The best we can claim is that an informed guess is better than none. Bioemulation makes the news—or at least a lot of corporate advertising copy—but balance and perspective aren't the hallmarks of day-to-day journalism.

A MIXED BAG OF MAYBES

Problems: Copying may work but hold insufficient advantage over an alternative. Or copying may be assumed—mistakenly—to have worked. Or else the "nature-copied" device works but turns out not to have been copied. Here are examples of each, intentional fuel for the fires of skepticism. The concentration on fluid mechanical cases should be blamed on the author's experience and not taken as defining the current mainstream of work in bioemulation.

Guts and Peristaltic Pumps. Our intestines act as their own pumps, pushing along slurries of digesting food. Waves of muscle contraction move lengthwise down our twenty feet of small intestine and eight feet of large intestine in the process known as peristalsis—much the way we squeeze toothpaste from the bottom of a nearly empty tube. The digestive systems of animals most often use this kind of pump; even the circulatory systems of a lot of worms pump peristaltically. Peristaltic pumps—the name borrowed from physiology—find occasional small-scale use in human technology, especially in biomedical applications; an example is shown in Figure 13.1. The various versions have in common an arrangement for flattening successive elements of a tube with elastic walls, pushing the contents along. Peristaltic pumps have two selling points. The viscosity of the material being pumped can vary from that of water to that of a sloppy slurry without affecting the pumping rate, as you know from everyday personal experience. And the material doesn't have to leave the elastic tube while it passes through the pump, which avoids contamination from contact with the rest of the pump and simplifies sterilization. Otherwise, peristaltic pumps are a nuisance. Elastic tubing must be replaced often, and as pumps go, they're inefficient; deforming the elastic tube takes a lot of energy, little of which gets recovered.

Was the peristaltic pump copied from nature's peristalsis? The name suggests emulation, but the evidence argues otherwise. In 1894 a British inventor, John G. A. Kitchen, patented a pump much like one of our

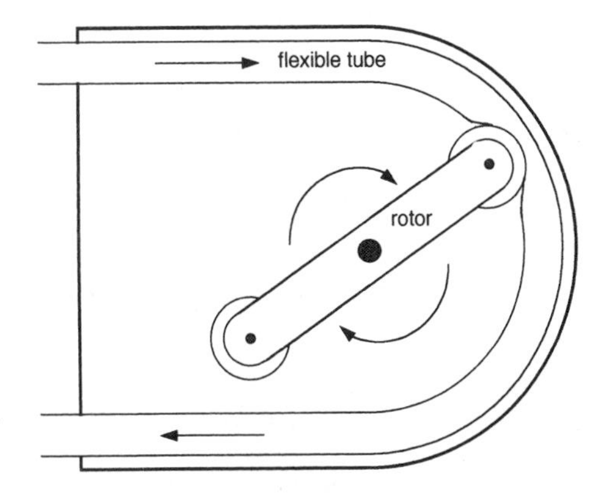

FIGURE 13.1. *The major components of a modern peristaltic pump; it's belt-driven through a pulley immediately behind the plane of the drawing. John Kitchen's version differed mainly in having one rather than two peripheral rollers and a hand crank on the central shaft.*

contemporary ones but without using the name. He intended the pump as a hand-operated, portable device for reinflating bicycle tires, a low-speed, occasional use in which rubber tubes wouldn't quickly fail from the repeated squashing.[1] The present name most likely comes (perhaps as advertising) from its use by a biomedical community.

Dolphins and Low-Drag Submarines. A lovely story that tells how a dolphin's almost dragless swimming was discovered and then mimicked. In the mid-1930s Sir James Gray, one of the founders of modern biomechanics, got a curious result when he calculated how hard a dolphin had to work at its top swimming speed. Flow around a body can be either laminar or turbulent, depending on such factors as the body's size, speed, smoothness, and shape. Gray calculated that only if flow around a dolphin was laminar could a dolphin's muscles produce the power needed for the speeds it attains; turbulent flow would cause too much drag. But a body as big and fast as a dolphin's normally experiences turbulent flow. The result became known as Gray's paradox.

About twenty years later Max Kramer claimed that dolphins maintain laminarity with a soft, compliant skin that damps incipient turbulence and that his Lamiflo coating system for submarines imparted the same impressive effect. Unfortunately the story doesn't hold water. First, we now know that Gray underestimated how much aerobic power can come from mammalian muscle and overestimated how fast a dolphin could swim continuously. A dolphin should manage well enough even if flow is turbulent. Second, no one has demonstrated that dolphins work as Kramer suggested, although they have soft skin and do appear to disturb the water remarkably little as they swim. Finally, the Lamiflo coating never gave the promised reduction in drag.[2]

Fish Slime and Polymers for Drag Reduction. If you add to water substances made of big molecules, the water, now more viscous, flows less readily. Under some circumstances, though, the opposite happens. Adding small amounts of a long, linear, soluble polymer to turbulent flows can reduce friction and speed up flow through a pipe or across a body. Some fish seem to use such polymers on some occasions, shedding fish slime into the water as they swim. Fish slimes and other biological polymers can lower resistance to flow on fish, on flow through pipes, or on solid test bodies. But practical use of the trick requires continuous shedding, since the flow carries the polymer away from the surface. Even fish probably secrete polymers only during brief predator-prey chases. Furthermore, while a lot of effort has focused on fish, since they apparently use the stuff, the original discovery and investigation of friction reduction by adding polymers didn't involve biological systems. It turned up in investigations of so-called non-Newtonian liquids, complex liquid systems in which viscosity isn't constant from time to time and place to place.[3]

Shark Scales and Drag Reduction. Not all fish are slimy. Sharkskin, for instance, feels like medium-grit sandpaper. Examined closely, the scales of sharks turn out to have ridges running atop them, and the ridges look as if they were aligned with the local direction of flow. These ridges are small, less than one-tenth of a millimeter apart and still less in height, but they appear to have evolved independently in several lineages of especially speedy sharks. Experiments with artificial versions have worked well enough to spawn a commercial coating (riblets) for use on high-performance racing yachts. But this coating reduces skin friction at most by 10 percent and usually much less, and skin friction, in any case, isn't as important as wave drag for surface ships. Moreover, whether sharks use their own ridging in the same way remains a bit uncertain.[4]

One final note: We learn things by word of mouth more than we ordinarily admit to students. Recently word of mouth has gained a whole new dimension as we've plugged into another informal avenue of inquiry, the Internet. I explained on several Internet news groups[5] that I was looking for successful emulations of nature. I gave the cases I already knew about and the guidelines I was following. A number of the stories here originated in the replies, and I got a fine introduction to work in progress. I also became more confident that I'd left no major stone unturned.

THE PROMISE OF CURRENT WORK

Never have more people been chasing biomimetics. Moreover, we're now seeing careful and systematic investigation of living systems in concert with efforts at emulation. An increasing number of people are now familiar with both sides of the street: biomechanics and engineering. The past history of copying nature may be unportentous, but its scattershot character is no longer representative. In short, it's a cautionary tale, not a trajectory. Here, then, are a few of the endeavors to which biomimetic thinking has recently taken on relevance.[6]

Nanotechnology. Nature builds upward from molecules and cells, while we usually don't. Some things, though, are more easily made that way. For instance, you hook together in specific sequences the different amino acids that form your proteins as directed by your genetic code, not by brewing up a batch of identical little molecules and making them stick to each other in random order. You also make composites like bone by synthesizing and positioning the various components just where they're supposed to be, not by mixing the components and letting the product solidify. One way for human technology to build upward is to use microorganisms as host factories with DNA inserts of our choosing. We already do this when we make small amounts of human insulin and other highly valuable biological molecules. Nothing stands in the way of our making either normal or modified structural proteins—nothing but horrible economics. Even with the most devoted microorganisms, the cost of making such products by the ton instead of by the gram would require that they have enormous advantages over ordinary polymers. Where might they be worthwhile? Perhaps in medical applications, where we're willing to spend extravagantly. For instance, we might want to make something that a human body, with its immune system, accepted as a normal part of itself. Bacterially synthesized human insulin has that advantage. Or imagine solid, prosthetic materials that not only would be accepted but that would join in the body's growth and replacement processes. Another way to build upward is to design large molecules that self-assemble into microscopic structures like microtubules and other cellular components.[7]

Muscle Analogs. Engines that work by contraction are rare in human technology. So are room-temperature engines made of soft material that can be made arbitrarily small. But nothing prohibits us from making musclelike devices that convert chemical to mechanical energy; such devices just haven't yet reached the stage of practical applications. Anything really similar to muscle, though, would need a well-developed nanotechnology for production. Just how similar to muscle we'd want our engine to be is uncertain. While muscle is fine stuff, it's an engine that can operate only when cool and wet. Except, again, for direct replacement for body parts, we don't have to be limited to low-temperature operation and shouldn't have to do everything underwater.[8]

Composite Materials. Bone, enamel, coral, wood—nature is so good at making hierarchical and composite materials that we surely should be able to learn from her products. In fact, this area is especially active at present. Seashells seem to hold lessons for making cermets—ceramic and metal composites. Beetle body wall—cuticle—seems to have a superior system of bonding fibers and matrix. The hard parts of sea urchins and other echinoderms are apparently single calcite crystals, but they incorporate a bit of protein that makes them interestingly fracture-resistant. And so on, from materials through large-scale structures derived from bamboo stems and molluscan teeth. Even if close emulation awaits effective nanotechnology, products that combine presynthesized components in nature's ways are clearly feasible in the short term. One wood-derived product, for instance, has been patented and can be produced with the equipment now used to make corrugated paperboard.[9]

Smart Materials. Muscle, bone, wood, skin, and some other natural materials adjust their composition or quantity in response to changes in load. This capacity for load-dependent reconstruction depends on sensing loads—on the sensory physiology of bones, tree trunks, blood vessels, and other structures that we don't usually think of as capable of sensation. Sensory systems have classically been the province of neurobiology, but in these structural materials well-understood neural feedback loops play little or no role. Loading some living materials such as bone and wood sets up electrical changes within them. The phenomenon, called piezoelectricity, has long been known in crystals and synthetic ceramics; press on such a material, and electrical charges appear on its surface. We use piezoelectricity to convert mechanical changes to electrical signals in devices such as phonograph pickups and microphones. Thus something unusual for the biologists is standard fare for the engineers, and piezoelectric sensation should be more readily emulated than neural sensation. Still, sensation is only one element of a feedback system that must also be capable of processing and acting on information. The self-awareness of hunger is of no value without mechanisms to obtain food. The other elements—how electrical charges trigger growth, for instance—remain mysterious. So making a responsive material is no trivial matter. Still, the ability to put material where and only where it's needed and to compensate for wear and for changes in loading would be no small advantage.[10]

Robotic Manipulators. Robots, industrial robots in particular, are of special interest these days. No one outside the film industry gives serious thought to visually anthropomorphic robots, but humans do have enviable dexterity and tactile ability. One tricky task that a good robot should master is picking up and positioning delicate, nonrigid objects. Humans do this well, using jointed appendages and a combination of visual, tactile, and proprioceptive (telling where your various parts are at any instant) senses. Various multijointed handlike manipulators have been built, with nature's versions (as with aircraft, composites, and smart materials) a steady reminder of what's possible. One interesting approach uses flexible, pneumatically inflated structures rather than jointed appendages, as in Figure 13.2, copying the muscular hydrostats of tentacles, tongues, and elephant trunks in order to handle fragile objects. We mammals mainly use jointed appendages, but they have no intrinsic superiority over muscular hydrostats for biological uses, much less as models for robotic technology. Indeed, close emulation of a tentacle may be a better bet than copying a hand since the hand requires many more discrete and independently controlled elements. The tentacle, though, pulls us into the unfamiliar world of structures that lack hard parts.[11]

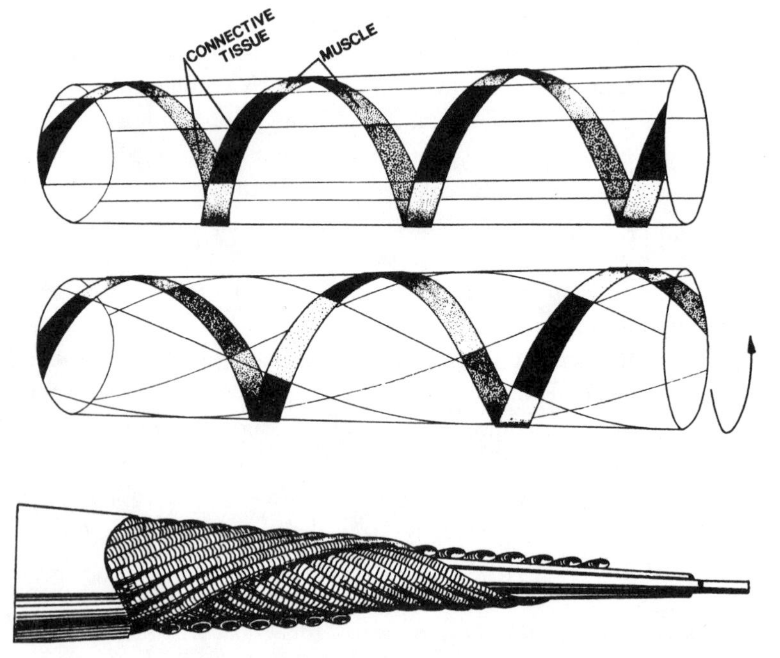

FIGURE 13.2. *William Kier's diagram of how a squid tentacle twists and (from U.S. Patent 4,792,173 of 1988) a diagram of one of James F. Wilson's pneumatic manipulators.*

Walking Vehicles. Wheels are better than legs in hard, flat contexts. The messier the terrain, the better legs become. On a soft substratum—say, sand—that a foot can still push against effectively, a wheel must climb out of its own rut as it moves, doing something analogous to a ship sailing above hull speed. The military worry about getting people and things around without first making roads, so they're attracted to legged vehicles and have paid for building some rather fancy ones. Both the motor arrangements and the sensors and feedback systems of these are enormously complex. One six-thousand-pound walker goes only five miles per hour and carries only five hundred pounds; on the other hand, it can climb a 60 percent gradient. A strong consensus favors six-legged machines, so the walkers (such as the one in Figure 13.3) look a bit like giant insects. Balance comes harder to four-legged machines; if one leg is lifted, the thing may tip. But a hexapod can lift every other leg while retaining a stable tripedal stance, or it can lift almost any two without loss of stability, as you can demonstrate with clay (or a brownie) and toothpicks. Current work on walking vehicles pays serious attention to insect biomechanics.[12] Nonetheless, the logic of hexapedalism doesn't mean that good walkers will look like insects; such things as the difference between muscles and motors and the difference in scale will inevitably drive divergence.

Swimming by Bending. The fish most familiar to us swim with obvious ease and agility in a manner unlike either surface ships or submarines.

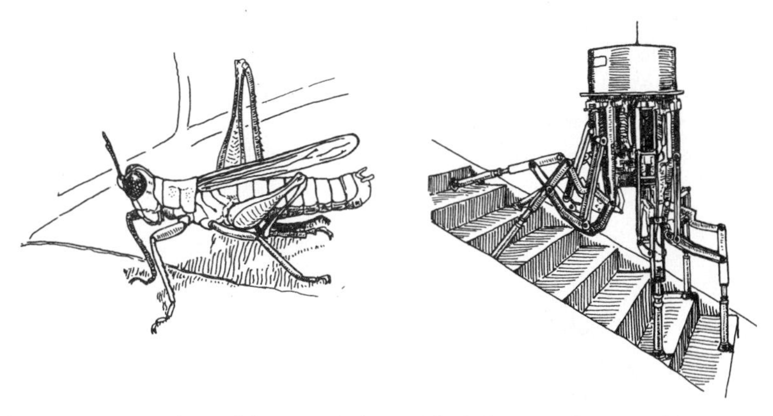

FIGURE 13.3. *A hexapedal insect and a hexapedal vehicle, not at the same scale!*

FIGURE 13.4. *A pair of twiddlefish invented by Stephen Wainwright and Charles Pell (courtesy of Twidco, P.O. Box 542, Durham, NC 27702).*

These fish—trout, bass, and the like—accelerate impressively when they start and turn sharply at any speed. In their mode of swimming, waves of bending pass from head to tail. The waves increase in amplitude as they move aft, so the whole body is an integrated propulsive unit powered by a large mass of edible muscle. For our machines, that's an even more awkward arrangement than flapping wings. Even so, the waving motion has been simulated using a complexly interconnected set of solid segments called the robotuna.[13] It can also be simulated—and much more cheaply—by a simple fish-shaped casting of soft plastic. A twiddlefish, as in Figure 13.4 is persuaded to oscillate side to side by twisting, first one way and then the other, by a vertical shaft that emerges upward just behind its head.[14] While the twiddlefish is now sold as a toy, it's a serious proposition. A version a foot or two long can propel a small person-powered boat; a person sits on the boat and twists a shaft that goes down through the hull to the twiddlefish. Larger ones, various geometries, combinations of propulsive units, and different degrees of stiffness are currently being constructed and tested.

Will such devices supplant marine propellers? Doubtful: Rotating propellers are probably more efficient than any possible twiddlefish. Really fast marine swimmers—tunas and whales—swim by swinging wide tails back and forth behind stiff bodies, not by passing back waves of bending. That strongly suggests (as does other evidence) that a trout, for instance, trades off propulsive efficiency for maneuverability and acceleration.[15] But just as legged vehicles have specific uses that their intrinsic complexity and inefficiency don't rule out, so perhaps do soft, oscillating propulsive units. Imagine, for instance, a trolling motor that moves a fishing boat through a tangle of submerged vegetation. An oscillating propulsor ought to get fouled a lot less easily than a rotating propeller.

QUO VADIMUS?

If you want to make very small things or to fabricate materials that are heterogeneous at a microscopic level, nature may hold both lessons and assistance. The smaller the scale, the better the prospects for emulation. We're relatively better on large scales, where our metals and wheels come into their own. This suggests that materials science is a fertile place for biomimesis, while structures and mechanical systems are less so. Since the early cases of copying were all large, we glimpse the possibility of a new path.

If you develop new materials that are more like nature's, her designs should show how to use those materials to best advantage. Using new materials is a nontrivial business. The earliest iron bridges looked like those made of wood or stone, while the earliest steel bridges looked like the bridges that had been made of iron. Many of us remember that early plastic buckets, yard carts, and such looked more like their metallic antecedents than do current designs. As composites, including ones particularly like those of nature, become more competitive in price with metals and homogeneous plastics, natural designs may take on new attractiveness for us.

If you intend to make things of pliant rather than stiff materials, recognize that nature has been there first. We need just point again to tentaclelike robotic handlers and to twiddlefish, both cases in which flexible devices similar to their natural analogs work well. Designing with pliancy in mind puts strength, not stiffness, uppermost, and that holds promise of material economy: doing more with less.

If you want to make prostheses, some advantages may come from using materials and structures as close as possible in properties and operation to nature's originals. Here one has to fit an element of one technology into the other. And to be crassly commercial, a prosthesis can command a spectacular price relative to the amount of material in it.

Even though put as admonitions, each of these items points to a future considerably brighter than implied by a simple historical account.

We're moving toward ever smaller components in our various contrivances, in effect getting closer to nature's miniature world; remember that an average animal is only a millimeter or so in length. We're developing a great array of flexible materials to supplement or supplant stiff metals and brittle ceramics. We're exploring the use of composite materials, composites far better than our old standby fiberglass, and thus entering a world in which nature (perhaps for lack of metals) is an experienced and versatile player. With improvements in small actuators and complex controls, devices made of muscles, tendons, bones, and nerves are increasingly attractive as models.

But once again the caveat. Nature may show what's possible, but she's a poor guide to what's worth doing. Flashy corporate advertising repeatedly pops up heralding the imminent industrial synthesis of spider silk. Spider silks (there are many variants) do have unusual properties. They're strong, extensible, and tough (although they have low stiffness and resilience). Even so, they're not, as the ads imply, leagues better than existing polymers such as Kevlar, which happens to be stronger than any known silk. Making a spider silk would be an impressive biotechnological accomplishment. The task extends well beyond getting the amino acids linked together in the right order; it has to be extruded with the right organization of crystalline and noncrystalline regions.

How badly do we want to make a spider silk? Getting manufacturing costs down far enough is hard to imagine, since almost any application would need a large amount of the stuff. For that matter, thinking of applications isn't all that easy. The properties of spider silk and indeed of structural proteins in general are extremely dependent on their water content—level of hydration, strictly—and on temperature. To complicate any uses even farther, changes that happen when hydration and temperature are altered often aren't reversed when hydration and temperature come back to normal. What applications do we have that are sufficiently permissive to put up with such trouble? Yet another trouble spot is the low resilience of silk—perhaps its most unusual and thus potentially desirable feature. The combination of high toughness and low resilience means that lots of energy can be absorbed in a stretch and that the energy doesn't come out again as elastic rebound. In other words, spider silk is stretchy, but it doesn't behave at all like a rubber band; it absorbs and then keeps the energy. But energy, according to the first law of thermodynamics, can't be destroyed, and the energy appears as heat.[16] Now that may not pose a problem for a skinny thread. None of its inside is at all far from its surface, so heat loss to the surrounding air or water keeps it from getting hot. But imagine using a rope of spider silk to stop a falling body or a moving aircraft. If the rope is thick enough to do such a big job, the silk will be immediately ruined by the resulting increase in its temperature.

So having easy access to large ropes of spider silk might hold little advantage. They might be like some other biological items that are widely available but not especially useful. The big and prosperous stems (properly "culms") of bamboo in my yard are a good example; they produce a substantial annual harvest for which I've never found an application. What might be more useful would be to understand how spider silk comes by its curious combination of properties. That could help us devise materials that have its unusual and desirable properties without its disabilities, materials that can be made cheaply by our kind of manufacturing.

Poor prospects get weeded out quickly enough either because the people doing the work get discouraged or because the sources of funding get discouraged. Even skeptic that I am, I view the present avenues being explored as highly promising. A kind of natural selection will operate among technological projects, one driven by experience, wisdom, and financial support.

History may teach another lesson: that ignoring nature can be as much a peril as mindless copying. Recasting history with hindsight may not be the fairest of games, but consider the following. Screw propulsion for ships was invented in the first half of the nineteenth century, mainly by Francis Pettit Smith and John Ericsson. Their model was a pump long used in human technology, the Archimedean screw (Figure 13.5). Contained in a close-fitting pipe, the device will do service as a simple, if not especially efficient, pump. For propulsion of a vessel such a screw holds some advantage over a paddle wheel: It's more compact and works entirely underwater. Smith's first commercial vessel (launched in 1838) even bore the name *Archimedes.* But as a thrust producer the Archimedean screw isn't much better than it is as a pump. An accident is what established screw propulsion as the way to go. The outer half of a wooden screw broke off, and the shortened screw worked better. A set of flat blades, none extending for more than a fraction of a turn around the shaft turned out to work best of all.[17] (Recognition of the advantage of curving—cambering—the blades awaited the twentieth century and the development of aircraft propellers.)[18]

Smith and Ericsson ignored a better model for an underwater propul-

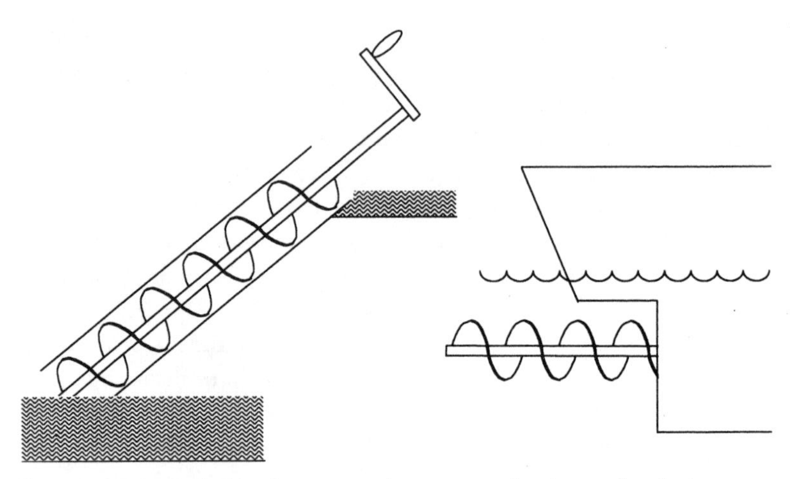

FIGURE 13.5. *An Archimedean screw used as a pump and as the propeller of a ship.*

sor. A ship's propeller employs the same fluid mechanical scheme as do the flukes of a whale or the wide tails of fish such as tuna. The only real difference is what we've repeatedly noted: rotational instead of reciprocating motion. If these developers of screw propulsion had understood that they were making rotating analogs of whale flukes or tuna tail fins, they might have done a lot better a lot sooner.

In viewing on one hand biomechanics and on the other bionics, biomimetics, and robotics, we're looking at the difference between pure science and applied science. The old Dewey decimal system for book classification makes precisely this distinction. Books on science, the 500s, are found in one part of the library; books on applied science, the 600s, are found in another. And the people who care about each are all too often in different places as well. Bringing them together has to be mutually beneficial.

Chapter 14

CONTRASTS, CONVERGENCES, AND CONSEQUENCES

The more closely we look at the technologies of natural selection and of human contrivance, the less similar they appear. We might well have guessed otherwise in the light of their common situation. Life has proliferated on our planet for several billion years, and we've been making things for a million or so—ample time for underlying imperatives to make themselves felt. Yet those basic differences persist:

- Nature uses fewer flat and more curved surfaces than we do.
- Ours is a far more rectilinear world while nature shows little bias in favor of right angles.
- Corners in our technology are abrupt; nature's are more often rounded.
- Numerous mechanically separate but individually homogeneous components make up our devices; nature uses fewer components whose properties vary internally.
- Nature's designs take advantage of diffusion, surface tension, and

laminar flow; gravity, thermal conductivity, and turbulence matter more for ours.

- We most often design to a criterion of adequate stiffness, while nature seems more commonly concerned with ample strength.
- Partly as a consequence, our artifacts tend to be more brittle while nature's are tougher.
- As another consequence, our things move on sliding contacts between stiff objects whereas nature's objects bend, twist, or stretch at predetermined places.
- As an additional result, we minimize drag with streamlined bodies of fixed shape, but nature often does so with nonrigid bodies that reconfigure in flows.
- Human technology makes enormous use of metals, while metallic materials (as opposed to materials containing metal atoms) are totally absent in nature.
- As a result, we use the ductility of metals to prevent crack propagation; nature does as well, but with foams and composites instead.
- We more commonly load materials in compression while nature more often loads in tension.
- Concomitantly, we make greater use of shear preventatives such as nails and mortar to keep stacked objects aligned.
- Structures with tensile sheaths outside and pressurized fluid inside are both more common and more diverse in natural designs than in ours.
- For such hydrostatic and aerostatic systems, nature's predominant fluid is water while our structures mostly contain air or some other gas.
- We make profuse and diverse use of rolling devices based on the wheel and axle; but things rarely roll in nature, and only one true wheel and axle is known.
- Our prime movers—engines—are based on rotation or expansion; most of nature's are based on sliding or contracting.
- Many of our engines extract mechanical energy from temperature differences, whereas all natural engines are isothermal.
- Levers in human technology most often amplify force at the expense of distance, while nature's commonest levers amplify distance at the expense of force.
- Our devices store mechanical work as electrical, kinetic, gravitational, or elastic energy; nature mainly uses the last two and most often the last one.

- Our fluid transport devices often interchange pressure drop and volume flow, but equivalent transformers are rare in nature.
- Surface ships have long played an important role in human technology, but nature overwhelmingly prefers submarines.
- Our factories dwarf the items they produce; nature's factories make products far larger than themselves.
- We judge our devices best when they need only minimal maintenance, but nature's devices get continuously rebuilt.
- Our technology is as dry as nature's is wet.

Listing differences hints at interrelationships among them. If gravity dominates, then stiff materials to resist it find special utility. Stacking becomes a reasonable way to build things, with shear preventatives like nails to prevent sliding. Permitting such sliding here and there then defines joints. And so on. Using metals makes their special properties available (for instance, high thermal and electrical conductivities) and allows using otherwise impractical devices (such as wires) and construction methods (like pressing and forging). Composites drop from crucial to simply useful. And so forth.

Each domain, nature's and ours, thus develops its distinctive coherence, consistency, and rationality, each a well-integrated entity in its particular context. Might we mix and match among the features of the two technologies, generating a vast number of further ones?[1] All but a few would surely lack that degree of coherence, consistency, and rationality; the combinations of features that mark each of our two transcend historical accident. What determines the kinds of devices that a technology finds effective? For one thing, its physical situation: the size of a technology's artifacts, whether its basic medium is air or water, whether it works at a surface or suspended in a gas or liquid, and so forth. For another, how it goes about doing things: production methods, degrees of resistance to revolutionary change, relative ease of technological diffusion, and, once again, so forth.

Even social interactions matter to how a technology goes about its business. Nature faces severe limits when organizing and coordinating the efforts of individuals; one might say "institutional limits" to sharpen the comparison with human efforts. Casual verbiage to the contrary, organisms don't naturally work "for the good of the species." Except (perhaps) for humans, neither the individual organism nor its genes know anything about its species or feels any obligation to it. Only to a limited extent do organisms work for the good of some community, ecosystem, or biosphere.[2] In nature, coordination of effort happens with any regularity only under two specific conditions. First, an organism may do things that increase a close relative's chance of reproduction while decreasing its own. Relatives have many genes in common, so a self-sacrificing action can still enhance an individual's genetic contribution to the next generation.[3] Second, if by helping another, an organism helps itself, then public-spirited action is acceptably selfish in evolutionary terms.[4] These stringent conditions must preclude all kinds of cooperative activities.

Human technology, for instance (since we're now including hypothetical ones), operates in a much different social context. Our large enterprises depend on cooperation among people with no close kinship, although a paycheck does represent respectably selfish reciprocity. Governmental and religious institutions—the military being the most ancient and extreme—coordinate the activities of people of the most remote familial relationships located in far-distant places. Tin from mines in British Cornwall converted Mediterranean copper into bronze well before a single political entity bound the areas. Our remarkable facility for inducing people to pool their efforts permits great task specialization; think how many individuals had some hand in bringing you a car or computer. Our facility for generating transportation systems also means that we can transcend nature's need to make things out of locally available materials.

SIMILARITIES

A list of similarities turns out to be such a scattershot mix of major matters and minor details that we gain little from the itemization. Most similarities between the technologies emerge from inescapable physical rules and environmental circumstances, both matters we've already dwelled upon. The best gravity-resisting vertical column will be circular in cross section, whether it's evolved or manufactured, grown or cast, of wood or of concrete, of bone or of steel. Jet engines find use by either technology only when some other special advantage (such as high speed or mechanical simplicity) outweighs their inefficiency. Low-pressure pumps capitalize on fluid-dynamic phenomena while high-pressure pumps employ fluid statics. And so forth.

At this point more subtle and abstract similarities hold more interest: similarities of process and historical trajectory rather than of product, similarities for which common physical context provides no adequate explanation. Addressing them returns us to issues quiescent since the second chapter.

- Major innovation is no easy thing for either technology, but for different reasons. We persistently believe that the progress of human technology depends on an adequate scientific base. Probably that's only occasionally true. George Basalla argues (and I don't disagree) that we too easily and too often exaggerate the contribution of science to technology.[5] When the two interact, technology more often drives science than the other way around; steam engines stimulated thermodynamics, and aviation provided impetus for aerodynamics. More often, I suspect, the main difficulty of innovation in human technology comes from the intrinsic complexity of introducing something truly novel. A good idea isn't enough. Steam turbines were mentioned around 1800 by James Watt, but despite obvious advantages over piston engines, they didn't power ships for another century. Turbines made unmanageable demands on metallurgy, techniques for precision fabrication, and lubrication. A sail is a great device to spare human labor, but sailing in all but the calmest of waters requires drastic redesign of oar-driven ships. A computer as good as Babbage designed awaited electronics, a century later; doing the job mechanically cost too much to hold sufficient appeal to launch the technology. A developed technology has a great deal of momentum, more often than not barring from the market competing schemes, even intrinsically superior ones. About this last argument, more later.

  Evolutionary innovation faces even greater difficulty. Natural selection requires advantage all too immediately. That mandates continuity of design and must rule out lots of life-forms that might ultimately have proved superior. A system of inheritance that includes sexual recombination and recessive genes does allow changes in several features at once. But it doesn't encourage large-scale change in any particular way. Small changes encounter different troubles. Competing with established designs must be especially hard when a new design differs only marginally. In both technologies refuges from the full burden of competition have obviously been important—for instance, military programs for our products and geographical isolation for nature's.

- The time course of change or progress, depending on one's outlook, contains several curious parallels. When we look at relative rather than absolute durations, both technologies seem to change slowly and steadily. Nature achieved terrestrial life long after accomplishing large-scale organization and mastered flight a long time later. Step by step, humans elaborated devices to apply human muscle in ever more diverse ways; we then added, with increasing effectiveness, the muscles of other animals; we further added, again step-wise, nonliving energy sources. The same incremental increase in potency marked food acquisition, construction of tools and dwellings, and destructive social interactions.

  Viewed more closely, slow and steady characterizes neither. In each culture one can recognize some takeoff point that follows a very slow prepaleontology (for visibly large fossils) or prehistory (for record-keeping cultures). Cells of sorts persisted for several billion years, both singly and in small chains and aggregates, before they managed the specialization and coordination that led to large creatures. But the latter appeared suddenly—over only ten or so million years—and have been around for a paltry six hundred million years. Humans occupied most of the earth long before the modern era—less than ten thousand years—of specialized occupations and elaborate coordination of individuals. Most of human history preceded all the complex trade, transportation, and political organization that we think of as definitively human. Whether change in nature or human societies continues to accelerate beyond the period of takeoff is much less clear. The fossils from the Cambrian, from more than five hundred million years ago, are impressively complex, and a look at the artifacts, both large and small, of ancient Egypt gives the same impression for things that humans make. Has our technological progress in the twentieth century been greater than that during the nineteenth? The nineteenth saw the spread of self-powered locomotory systems, instantaneous communication with electrical devices, cheap metals, and mass production. To these the twentieth has added mainly electronics, aviation, modern medicine, and the development of polymeric materials. The culture shock of moving back a hundred years would probably be a lot less for one of us than for someone living in the 1890s.

- What about changes on a finer scale, changes in specific devices and characters? In both technologies they're at least as episodic as broad organizational changes. A trigger gets the ball rolling, to mix metaphors, initiating a period of rapid change. Some critical environmental circumstance may alter: A seed lands on an island; a ship reaches a new continent. Some novel material or general component becomes available: a growing skeleton or edible grass in nature, a stronger metal or ship's propeller for us. Or some key constraint gets removed: Teeth appear that can tolerate the abrasives in grass; dynamos supplant batteries as primary electrical sources. For either, the range of possible triggers is wide.

Whether evolution is predominantly steady or episodic has engendered considerable controversy, but we needn't worry here about the debate between "gradualists" and "punctuationalists." At the molecular level, evolution proceeds gradually, with the genetic material changing at a relatively constant rate. At the level of the functioning organism, things change more erratically; a given amount of genetic change simply doesn't cause a fixed amount of organismic alteration. No paradox attaches to this difference; many complex and interactive processes intervene between a sequence of bases in DNA and a functioning, multicellular structure. At the molecular level gradualism reigns. At the organismic level a considerable degree of unsteadiness has long been at least tacitly recognized; the present debate just concerns the degree to which a punctuational model should be considered the default. Evolutionary biologists care about evolutionary change, so they're jarred a little by the idea that most species most of the time are well adjusted and do very little overt evolving.[6]

The episodic character of change in human technology—or in human institutions in general—isn't often questioned. Or at least not recently, however persuasive was the steady "onward and upward" view of human progress a generation or two ago. "Revolution" may be a buzzword, with an agricultural revolution, an industrial revolution, and even a postindustrial revolution, but it's not inappropriate. Any history of technology provides lots of examples of rapid change on the heels of some key innovation, whether early bronze or cheap steel, steam engines or internal-combustion engines, or the fast electronic switches of tubes and transistors. Not that "rapid" means instantaneous. Applications lead to improvements in cost and efficacy of the basic innovation, which then lead to other applications, but even such a positive feedback process takes time to operate. A century elapsed between Newcomen's bulky steam-driven pump and engines sufficiently light and efficient for self-propelled land vehicles—for Trevithick's steam carriage and Stephenson's locomotive. But that much delay may be unusual, the special result of engines worked by pressure differences in a technology still unable to handle large pressures. Telegraph, telephone, and electric motors followed closely on the easy availability of batteries and wire. Horse-drawn urban transport completely disappeared within a few decades of the development of practical internal-combustion engines.

The technologies share another factor that reduces the steadiness of change. Improvements in new designs come more easily than improvements in well-established designs. For nature, the chance (always low, of course) that a mutation will be favorable and ultimately fixed in the population will therefore be greater if it occurs in an organism that has recently undergone other change or recently invaded a new habitat. A larger fraction of mutations will be neutral or detrimental when they happen in creatures that have been doing the same thing in the same place for eons. In our domain, improvement moves from the province of the casual amateur to the experienced professional—whether one considers farm implements, automobile engines, or computers.

• If revolutions are easy to recognize, cases of stasis are no more obscure. In nature we needn't invoke famous living fossils, such as the lobe-finned coelacanth fish, the ginkgo, or the horseshoe crab. On the evidence of either very similar fossils in beds of different ages or members of stay-at-home species on different and distant islands (where accidental introductions must be rare), change must be neither continuous nor inevitable.

Even amid the ostensible rush of contemporary change, many things endure. Once stripped of a lot of adventitious machinery, the basic four-cylinder engine and manual transmission of my car wouldn't much puzzle a mechanic who'd been asleep since 1930. A clothes dryer in our house replaced one that served for twenty-five years; most of its parts are interchangeable with those of its predecessor. Typewriters developed rapidly after their introduction but then changed little for the next half century. Jet aircraft have changed only slightly in the last thirty or so years. Stasis is evident in the basic types of electric motor, in almost all our pumps and industrial power transmission devices, and in the techniques of manufacturing most of our basic materials. Information handling—the nervous system's analog—may have changed dramatically, but the final effectors—the industrial equivalents of muscles and bones—have been much more stable. Sensors, computers, and robots may replace human operators, but they mostly run the same tools for cutting, shaping, and assembly.

- In both technologies a reduction in diversity can follow an episode of innovation and initial diversification. Temporarily reduced competition permits more experimentation than the norm, creating a climate amenable to creativity and proliferation. Restoration of fully competitive selection by nature or marketplace then prunes the progeny.

The first phase of the natural phenomenon is called ecological release: A species expands and diversifies to take advantage of multiple habitats. Particularly famous examples include the proliferation of cichlid fishes in the great lakes of East Africa and of finches in the Galápagos Islands off Peru. Cases of the second phase—ecological displacement—emerge when one compares the wide variability of particular kinds of organisms on islands where they face no competition with their more limited variability where they do.[7] On a vastly larger scale, Stephen Jay Gould views the earliest macroscopic fauna, the Ediacara of the late Precambrian, something more than half a billion years ago, as a diversification of which only a small element remained as ancestors for later life.[8] But his view of the Ediacara remains controversial. Other cases in which a few better designs displaced many less effective ones are probable but problematic. Still, while evidence of overall diversity reduction may be lacking, a decent consensus of paleontologists agrees that the past half billion years haven't seen a steady increase in the diversity of life-forms.

The consolidation more clearly characterizes human technology, with a concatenation of forces driving the shakedown. Industrywide, national, or universal standards get established, from systems of screw threads and drill gauges to computer codes. Superior supply and repair networks make some designs more attractive to consumers. The familiar design is the comfortable one—our fascination with novelty can easily be exaggerated—increasing the dominance of whatever version gains initial preeminence in the marketplace. And true technological superiority does certainly matter, despite all the factors that dilute its impact. So the "mature" technology often ends up less diverse than its "revolutionary" antecedent.[9] (Sacred literature, curiously, may undergo an analogous proliferation and then standardization; the early Gospels, for instance, got pruned as they got canonized.)

- Another factor homogenizes both technologies. We do the same things in the same ways over the entire earth, the result both of fad and fashion and of our long-distance transportation and communication. The same global transportation has spread many kinds of organisms beyond their original geographic ranges. Some have been spread accidentally,[10] and some deliberately, most of the latter through movement of our domesticated plants and animals. Whatever the impetus, homogenization decreases diversity in both technologies; indigenous crafts (along with many other aspects of indigenous culture) are supplanted, and local species are driven to extinction.

DRAWING ANALOGIES

Similarity of product need not imply similarity of process. I'm persuaded that comparing the products of the two technologies lends breadth to our thinking and gives insights not otherwise evident. About processes I'm more equivocal. Natural selection is a most peculiar process, and its limitations are inadequately appreciated. One often encounters analogies between the processes by which human technology changes and evolution by natural selection. I think these badly need the scrutiny of a biologist.[11] To start with, a clear distinction ought to be made between mechanism and history—the distinction we biologists make between natural selection and natural history. The more basic problem is that an analogy doesn't explain. One judges an analogy not by whether it's true, trivial, or untrue but by whether it's useful, nonuseful, or misleading.

The evolutionary analogy counteracts the personified, heroic view of technological change that afflicts our early education. With that I have no quarrel and will say little more. The analogy helps little, if at all, in understanding the role of personal creativity in technological change. Too bad—the most crucial and most mysterious element of our technology is the origin and nature of human creativity, something without parallel in nonhuman affairs.[12] Where change hinges on inheritance and the latter cares nothing about personal effort, no great inventor or political leader can play a role. Nature can only show that a system need not be creative in intent in order to be innovative in result. While the point isn't intuitively obvious, it's also not all that important. Who can doubt the central role of creativity in the advance of human technology? For our technology, increasing the role of marketplace selection relative to that of creative intention would just add irrationality, inefficiency, and danger.

An analogy or point of similarity may either provide a good goad or a poor substitute for further analysis, but in itself it's nonanalytic. For instance, Robert Heilbroner makes the statement that "advances, particularly in retrospect, appear essentially incremental, evolutionary. If nature makes no sudden leaps, neither, it would appear, does technology."[13] Nice words, but where does one goes next? Incidentally, note his use of "advances," with its implication of progress. That presumes some equivalent type of progress in nature, something not as obvious as we once thought. Perhaps evolution was once progressive but long ago reached a state in which complexity and diversity change only randomly. Perhaps, by contrast, human technology has yet to reach such a plateau and remains progressive. If so, then we mislead ourselves by comparing their recent alterations.

Technological determinism is another notion for which natural selection supplies an analogy of uncertain value. Does (or, more realistically, to what extent does) technology drive history? Figures who debate the issue run the gamut from Karl Marx to recent social critics of technology, such as Lewis Mumford. Inventions suggested as historically critical include the stirrups, breast straps, horse collars, and heavy plows that allowed efficient application of the power of large animals in the Middle Ages.[14] Equivalent determinism in natural history is so obvious that the analogy helps very little. For organisms, little matters beyond natural selection—that is to say, reproductive success. No social consciousness, political expediency, informed judgment, or mass hysteria can confuse the story.

A Darwinian or selectionist analogy attracts us because natural selection is more straightforward and better understood than less dependably irrational systems; the latter resist reduction to syllogisms, such as that of Chapter 2. But no matter how attractive a selectionist model might be, it at least oversimplifies and may do even worse. For instance, humans intentionally minimize selection among competing designs by analyzing before fabricating and by field testing before marketing; we do our best to anticipate the effectiveness of our devices.[15] As Joel Mokyr puts the crux of the difference, "new ideas and mutations are inherently different in that mutations are copying errors, while ideas are deliberate attempts to make a change." Furthermore, we learn from failure, a factor well argued and illustrated in several books by my colleague Henry Petroski. In addition, we can ask, "Wouldn't it be nice if we had something that would . . . ?" Human technology emerges from some complex combination of the rational and the irrational, and we shouldn't let familiarity disguise its resistance to tidy analysis. At the same time we shouldn't let its analytical complexity disguise its practical effectiveness.

The selectionist model is dangerous in another way, one more of application than applicability. What's done by nature is by definition natural. But the word "natural" carries a strong connotation of rightness, even of sanctity. We must reject any general notion that nature's ways indicate proper procedures for people. Sometimes nature provides a model; in other instances veneration of the natural is just snake oil for our contemporary troubles. Not that we don't face unprecedented and possibly insuperable problems, and not that some of these aren't unintended consequences of human technology, but we can't indulge in the simplistic confidence that these problems will be solved by going back (whatever that may mean) to nature, to what's natural, or to natural selection.

What nature does show is that a complex technological system can appear, function, and persist without moment-to-moment micromanagement by some prescient guiding hand. That, though, the continued existence of complex, technological, capitalistic economies demonstrates as well. Furthermore, the latter should be judged on their own merits, in particular on how they contribute to our social goals, and not on any analogy with nature. Nature may be nicer than the "red in tooth and claw" monster (Tennyson's phrase) of a century ago, but nothing should impel us to hold it up as model for emulation. Darwin followed Adam Smith, not the other way around.

I don't know a biologist who's a nature worshiper in this quasi-theological sense. We're not more rational, merely battle-scarred by repellent social doctrines that have used natural processes for justification, doctrines perpetually reemerging with new names and new sponsorships. One is social Darwinism (mentioned in Chapter 11), the turn-of-the-century idea that since nature indulged in unfettered competition, it provided a proper operating principle for human society. Another is biological determinism, in its extreme form the denial of any possible amelioration of the deficiencies of one's heritage, whether personal or racial. But we're wandering from the present argument.

Not only is natural selection attractively logical, but similarities between the histories of life and of human technologies tempt wondering about common mechanisms. One parallel is what economic historians call lock-in and some paleontologists have termed the privilege of incumbency.[16] Each observes that an established device or life-form isn't readily displaced, even by something economically or selectively superior. In economics this strikes yet another blow to the notion of the ideally competitive market. In biology it denies full applicability of what's called the principle of competitive exclusion—the idea that if two groups (at the species level or above, so they can't mix genetically) are in complete competition, one will displace the other.

The privilege of incumbency for a well-established group of organisms is the resistance to extinction conferred on it by a large geographic range, a large number of species, or a large number of individuals. Its reality rests on less than fully satisfying indirect evidence; because of the time spans necessarily involved, we see only outcomes and not competitive battles. But the privilege of incumbency rationalizes a lot of otherwise paradoxical features of the fossil record. In Chapter 6, the possibility was raised that the nonmetallic character of natural technology might be a case of incumbent privilege, but as noted then, how can we ever know for sure?

For human technology, the evidence for lock-in is better but still imperfect, even if the general idea is too logical to deny. The persistence of the QWERTY typing keyboard provides the usual example of lock-in. The QWERTY arrangement of keys was initially chosen to minimize jamming rather than to maximize typing speed. After technical improvements fixed the jamming problem, one might expect that an ergonomically superior keyboard would have replaced QWERTY. But retraining costs time and money, and facility with two keyboards may be counterproductive to performance with either. A faster keyboard, the Dvorak, has never gained acceptance.[17] But the case may not be as good as it looks. Comparative tests of the two keyboards were badly flawed and almost certainly biased. The QWERTY arrangement turns out to be better than an alphabetical or a random one; forcing the hands to alternate to avoid jamming enhances typing speed as well.[18]

Another flawed example is the persistence of the VHS videotape system in competition with the purportedly superior Beta system. Any slight advantage of Beta was offset by the disadvantage of a shorter recording time—initially an hour, less than the normal length of a movie.[19] The argument that the internal-combustion engine got locked in, that we'd have good electric cars if we'd standardized on them a century ago suffers from similar flaws. The argument presumes that any technology can be improved without limit given sufficient incentive or investment. But only battery weight limits electric cars, and military interest in battery weight for submarine use (before the advent of nuclear power in the 1950s) as well as similar imperatives of the space program must have put maximum muscle (and a nice competitive shelter) into the search for weight-efficient storage of electricity.

Nonetheless, lock-in—market failure because of persistence of initial choices—must be common. Is the present computer screen the best shape or merely a persistent version of a television set? Is the thirty-five-millimeter film format for still cameras, with all that space given to perforations, an accident traceable to the initial adoption of thirty-five-millimeter movie film? And lock-in may represent no strict market failure. After all, a more complex item may be cheaper to produce if it can be done in larger quantities since the cost of designing and tooling up gets spread more widely.[20] We have to ask, though, whether an equivalent phenomenon in nature says anything about the existence and role of lock-in. Market dominance, irrational behavior of consumers, existing infrastructure to support initial choices—none has a precise analog in the factors underlying nature's privilege of incumbency.

CONES AND SPIRALS—ONE FINAL TALE

We've seen the deep pitfalls of arguments based on similarities between the processes of natural and human technologies. Arguments based on similarities between their products incur less risk of self-deception. But

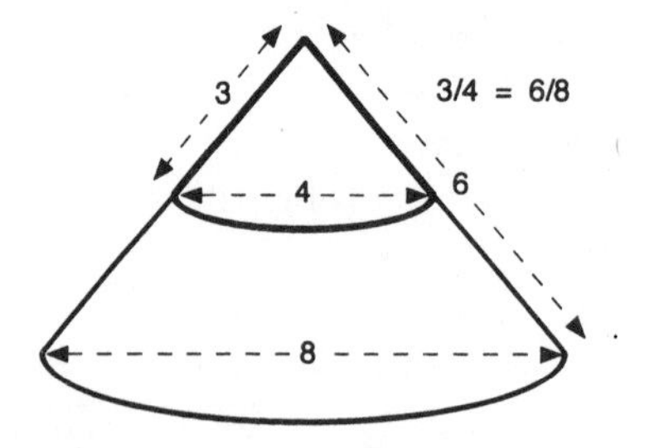

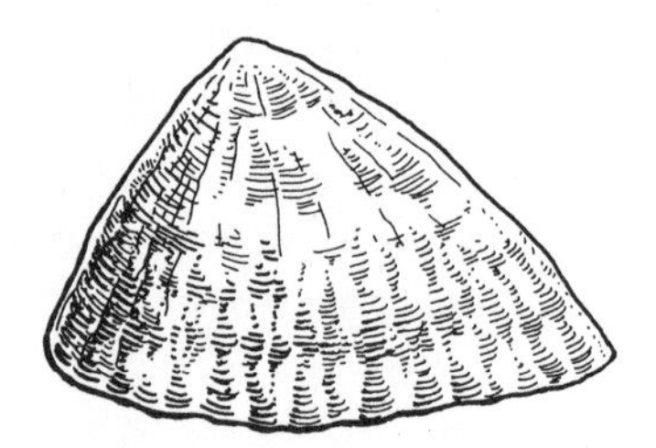

FIGURE 14.1. *Increasing the size of a cone without changing its shape, along with a limpet. Limpets are essentially low, uncoiled snails.*

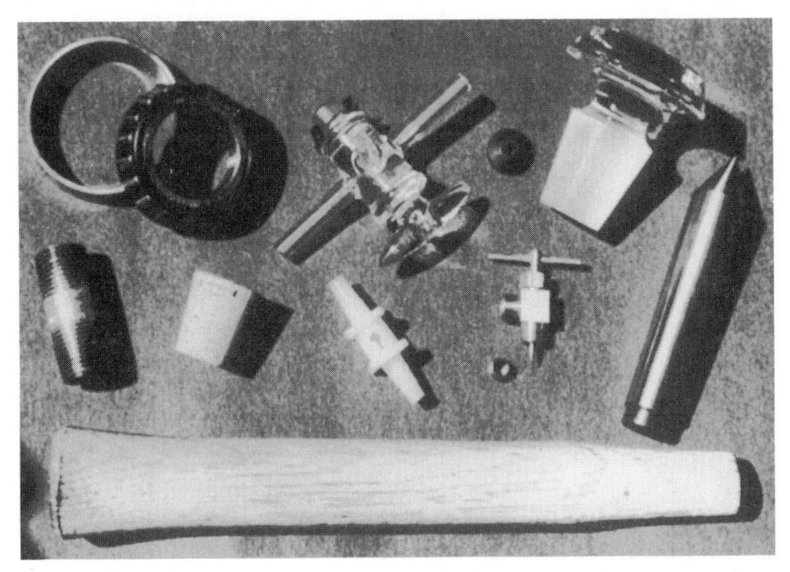

FIGURE 14.2. *Common cones of human technology: (top row) conical wheel bearings and their race from an automobile, ground glass stopcock, rubber faucet washer, ground glass stopper; (middle row) threaded pipe, cork, tubing connector, the needle valve of the water supply for an ice-making refrigerator, the centering device of a metal lathe; (bottom) hammer stock with conical top.*

the risk is reduced, not eliminated. A second look at a subject mentioned in Chapter 2 will do better than any admonition.

Consider a simple cone. A cone can be characterized by two measures, its diameter and its edge to apex distance, as in Figure 14.1. Enlarging the cone by extending its lower edge changes its size but not its shape; doubling the diameter doubles the edge to apex distance. By contrast, extending the end of a cylinder makes it skinnier and skinnier.

Cones have another interesting property. If their inside and outside tapers are the same, cones can be nested together as tightly as one wishes and in whatever number one wishes—the way we ship and store cones for ice cream and conical paper cups. Pressing them together snugs the connection. Cones of identical taper are easy to manufacture, whether by cutting on a lathe, by casting in a mold, or by any of several other techniques. Again, the contrast with cylinders is sharp. Nesting cylinders find use for such things as telescoping radio antennas. But their precise sizes fix once and for all the tightness with which they fit together. Nor will identical cylinders nest together.

While both technologies make great use of cones, they do so for different reasons. For nature, the essential advantage must be the first one noted here, the ability to grow by adding at the edge without change of shape. That makes cones good for things like mollusk shells, where growth takes place only by such addition. Since we don't make artifacts that grow, that advantage means little to us. For us, the advantages that matter are their ability to nest and the way they can be press-fitted with controllable snugness; Figure 14.2 shows several such applications.

(We're looking at severe biases, not absolute dichotomies. Some identical cones do nest in nature. Many jellyfish in their sessile stage produce a stack of what are called strobili, each of which will break off to form a recognizable swimming jellyfish. The individual muscular units of the bodies of fish form nested cones, as you can see by picking apart cooked specimens. And you make use of the way cones permit ungrowth, if not growth, without change of shape, every time you sharpen a pencil.)

The axial cutting tools on a lathe (drills, reamers, and so forth) fit into the hole in the tailstock (the opposite end from the motorized headstock) as the male halves of a pair of nesting cones. Pressing inward, as happens when the tool is used, tightens the fit, but if the tool suddenly jams in the work, it comes loose from the tailstock and spins nondestructively. Not only are such nesting cones easy to manufacture, but the angle of taper isn't sensitive to the changes in size that accompany heating and cooling. Old-fashioned ground-glass stoppers were always conical for the same reasons: ease of achieving the necessary fit and dependable behavior when tightened or loosened. By contrast, the pistons of glass hypodermic syringes have to be individually fitted to their respective cylinders. They're often numbered, so they can be remated after communal washing or sterilization. Corks for single use, as in wine bottles, are cylindrical; corks for repetitive use (as in laboratories or in an earlier generation of Thermos bottles) are conical so they will fit a range of bores and fit as well after long use.

Few of us notice how modern technology prefers cones over cylinders. The threads on the end of a metal pipe differ from the threads of ordinary screws; they're not uniformly deep grooves on a cylinder but get deeper toward the pipe's end, forming part of a long cone. As a result, pipes can be screwed to the same tightness even if their threading depth varies a bit. Automobiles have conical rather than cylindrical roller bearings within

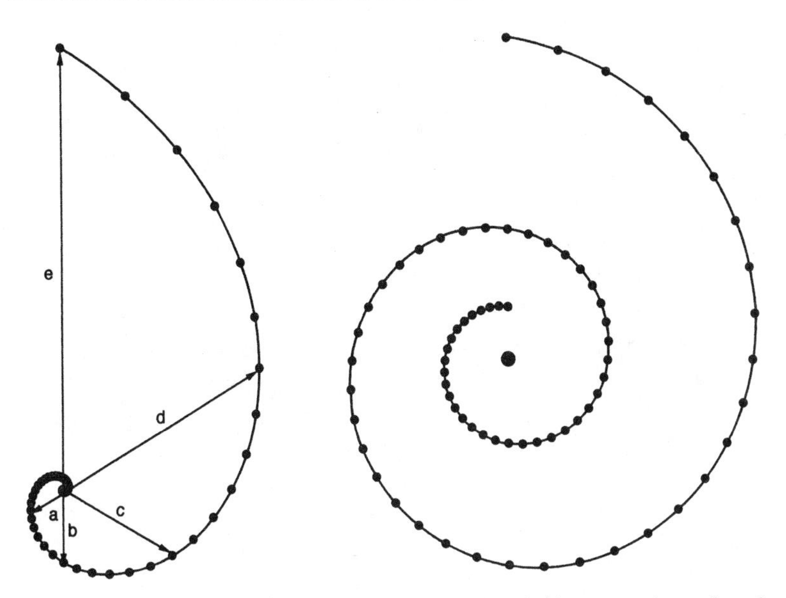

FIGURE 14.3. *Two logarithmic spirals. On the left, each radial line is sixty degrees from the previous one and is twice as long.*

their wheels, so tightening an outer bolt to a standard torque can counteract wear or variation in manufacture.[21] But—again—the advantages of these cones differ from anything gained by the shell of a mollusk.

Nature likes not just ordinary cones but a derived kind of cone as well. Horns and tusks, clamshells and snail shells, even the tiny protozoa that formed the white cliffs of Dover in southern England—all are conic derivatives. Recall how height and diameter are proportional in cones. That same relationship holds for a more general class of bodies, logarithmic or equiangular spirals; using curved instead of straight lines produces these particular spirals instead of ordinary cones. A line spirals outward from a starting point in such a way that the channel between turns gets wider in proportion to how far outward the spiral has grown.[22] For most of us a drawing, as in Figure 14.3, works better than words. Like cones, logarithmic spirals can grow without change of shape by the simple addition of material. Moreover, they're consistent with biologically typical logarithmic growth—growth in which how fast material is added is proportional to how much material is already present. In fact, what were loosely described a few paragraphs back as cones, things like limpet shells, are

FIGURE 14.4. *Logarithmic spirals in nature: a pinecone, horns of a sheep, a section of a chambered nautilus, and a microscopic foraminiferan (a shelled protozoan).*

really logarithmic spirals of very low curvature.

Early in this century three biologists became fascinated with the way these spirals appeared again and again in nature; Figure 14.4 gives a few examples. For James Bell Pettigrew, they revealed a curious bias of the Creator, the centerpiece of a huge antievolutionary tome. For D'Arcy Thompson, they provided the best example of the geometric idealism or at least the mathematical tidiness of living forms. Thompson's house at 44 South Street in St. Andrews, Scotland, now wears a commemorative plaque well adorned with logarithmic spirals. For Theodore Andrea Cook, the issue turned on aesthetics, without such cosmic overtones.[23] We're still attracted to them, but by their informational simplicity and by their handiness for the way organisms most often grow. The same mathematics that determines the growth rate of a fresh bacterial culture (and the yield of an investment with steady and reinvested interest) generates a logarithmic spiral.

Should our designs pay more attention to these logarithmic spirals because natural designs demonstrate their desirability? No. Their prevalence in nature reflects their compatibility with the way she does things. They share that compatibility with cones, in essence unturned spirals.

Our interest in cones doesn't carry over to spirals; spirals don't nest. So for the way we build things, the logarithmic spiral is irrelevant.

CONVERGENCES

Biologists since Darwin have known that similar organisms don't always come from close common ancestors; similar structural arrangements often evolve again and again. The image-forming eye of fish or mammal bears a remarkable resemblance to that of squid or octopus. That they differ in certain fundamental but functionally irrelevant aspects confirms our conviction that no common ancestor had such an image-forming eye. We know of innumerable cases of this phenomenon (mentioned in the second chapter) of convergence—that between marsupial and placental mammals, for instance. For biologists (from Darwin, incidentally), it makes trouble—not for the theory of evolution, which happily accepts convergence, but for the practical business of classifying organisms. Since similarity of structure doesn't indicate relatedness, convergence makes it harder to decipher lineages. For the biologist interested in function, by contrast, convergence is a handy tool. Functional significance (plus a few other things, admittedly) ought to drive convergence as a direct consequence of natural selection: What's useful will be selected. Therefore, cases of convergence ought to tell us what's functionally significant—that is, what aspects of the design of an organism matters to its reproductive success. What's selected must have been useful.

In addition, convergence ought to point to what's easy for the evolutionary process to generate. If something turns up repeatedly, it must be achievable without heroic alteration of hereditary information. Biologists who reconstruct evolutionary lineages look for continuous scenarios and descent schemes that have the fewest innovative steps. So "easy" isn't a trivial consideration.

Perhaps the same approach might prove productive when we look at how human technology has developed. All our communicative channels deprive our contemporary cultures of the independence of nature's separate lineages. But that hasn't always been so. Humans may have originated from a small number of ancestral hominids, but we spread over the planet before we acquired much in the way of technology. Perhaps convergence signifies the same things for both technologies: that cases of convergence are fingers pointing to what matters and what's easy. The idea got proper notice from Joseph Needham in his monumental work *Science and*

Civilisation in China, but Needham, probably not coincidentally, was an excellent biologist as well as a spectacular sinologist.[24]

Consider the early histories of our use of metals. Without a persuasive alternative, one should prefer the most incremental scenario, the "easy" scenario that requires the least genius. From evidence from the Middle East, a likely sequence runs as follows. Copper, almost alone among the mechanically useful metals, occurs naturally in metallic form. Copper ores commonly contain pieces of metallic copper. Melting copper with some associated ore (perhaps to separate the two) will yield more product than will a pure melt, so the transition from melting to smelting presents no great difficulty. Ores with certain impurities will yield better metal; antimony and arsenic bronzes find more uses than pure copper. Tin bronze, less likely to poison the artisan, is still better, but it's also less likely to be fortuitously made. In most ways iron is a still-better metal, but producing it demands higher temperatures.

How reasonable or likely is that sequence from metallic copper through smelted ore, arsenical bronze, and tin bronze to iron? We gain confidence in that reconstruction when we learn that the same sequence from copper to tin bronze occurred in South America as well as in the Middle East, if a few thousand years later in the New World.[25] One can alternatively invoke what the anthropologists and archaeologists call diffusion as an alternative explanation, implying that the technology spread through occasional contacts between separate civilizations. But I view diffusion as (to pun slightly) a farfetched scenario for how the Americans went metallic—to the limited extent that they did.[26] (Assuming that they didn't get there on their own also happens to be insulting, patronizing, or worse.)

The development of woven textiles in the Old World and New World provides a similar story of convergence. While some elements may have crossed the Bering land bridge with early invaders of North America, much of the technology must have been developed independently. Both Old World and New World used looms, but features consistent within either but differing between them point to independent innovation. The sequence in the two hemispheres looks similar, running from untwisted to twisted cordage, from there to basketry and knotted netting, and on to weaving, with improvements in spinning technique and the labor efficiency of looms. In both instances, long fibers, such as flax and sisal, seem to have been used before short ones, such as cotton and wool; the short ones can be produced in greater quantity but are harder to make into adequately strong thread. Both cotton and wool were developed in both hemispheres, the cotton from different species of the genus *Gossypium* and the wool from different animals—sheep and goats in the Old World, llamas and alpacas in the New.[27] A careful comparative analysis might reveal some general principles of early textile evolution.

The courses of development of both metals and textiles point to the importance of a sequence with an easy continuity, one that requires a minimum of large-scale innovation and one the evolutionary biologist would find familiar. Nor are metals and textiles the extent of possible subjects for comparisons that focus on convergence. Fired pottery was widely used, but with different techniques for shaping clay in the two hemispheres. Other candidates include techniques for trapping, fishing, plowing, seeding, tanning, and boat making, as well as weaponry for both hunting and warfare.[28]

Convergences might also serve in a way that's almost the opposite, showing not what's easy but what's hard for one or the other and forcing us to ask why something is hard. To do this, one should look for elements that are ordinary in one technology but rare or unknown in the other. Consider the use of tools. While we know many examples of minor tool use by mammals and birds, one is struck not by their existence but by their scarcity and casual character. The limited use of tools contrasts sharply with the way animals use materials from the environment to construct domiciles: caddisfly larval tubes; worm tubes; hermit crab shells; nests of fish, reptiles, birds, and mammals; beaver dams; and so forth. The limited use of tools contrasts quite as sharply with the way tools proliferate in human societies. Why the difference? My only suggestion, a casual and uninformed one, is that elaborate tool use may happen only after some critical takeoff point. That takeoff was reached only once by any animal—by humans at some stage in our prehistory. Perhaps it came with the acquisition of learned, symbolic language and the fabulous improvement in cultural information transmission that such language permitted. But would one ask the question in this way except in the context of a comparison between the technologies?

MESSAGES

This view of nature as a technology has provided an unusual perspective on the world around us. Recognizing that nature deals with the same variables as do human designers—in the present case mostly mechanical engineers—led us to compare both products and processes. But the main test of the comparison is utility. In short, what do we learn?

Identifying specific devices that we might profitably emulate constitutes the least of what we gain. What's the chance that in searching through the pieces of an electric motor you'll recognize a part that will improve the performance of an internal-combustion engine? How much more remote the chance when searching through the proteins of a muscle! By contrast, one may well expect insight on design where natural technology is preeminent, but enough noise about that has already been made. That each technology is a coherent entity, remarkably distinct from the other, can be either advantage or disadvantage. Perhaps the best encapsulation, if a trifle trite, is that nature shows what's possible.

Disparaging things were said about analogies, but real utility balances the risk analogies pose of short-circuiting proper analysis and explanation. A useful tool emerges from recognition that the technologies involve analogous time courses of development and ways of operation. One can test the logic and credibility of hypotheses about how one system operates by examining what happens in the other.

We've seen places where the products of the two technologies coincided. And we've seen places where the products proved surprisingly different. Each may carry a prescient message. Coincidence between these vastly different technological contexts directs attention to constraints that neither can escape, constraints that we must try to identify. Different solutions to the same problems or different devices for the same task imply something equally interesting: the possibility of a third or fourth solution or device. Thus the comparison can suggest where major innovation might be possible and where it's unlikely. Since major innovation isn't easy, a little help might go a long way.

The histories of both science and technology are replete with cases in which someone with an unusual background or outlook—an outsider—solved a long-standing problem.[29] Can one deliberately become an outsider? Or, better, how might one get the fresh perspective of an outsider without sacrificing the insider's expertise? For the human designer, a perceptive look at nature's technology can do just that: It can provide the wide-angle view that reveals possibilities that would otherwise escape consideration.

I began with statements extolling the superiority of nature. In a sense this entire book is my skeptical response. Sure, nature is wonderful. But bear in mind what we do that she doesn't: However they came about, the unique achievements of our species deserve our full appreciation. Metallic materials, ropes of short fibers, woven fabrics of crossed threads, the wheel and axle, thermal expansion engines, fast surface ships, electrochemical energy storage, lighter-than-air aircraft—humans, our engineers in particular, have made extraordinary things. How we use them is another matter altogether.[30]

In these pages the reader has been introduced to my immediate scientific interest, biomechanics. As thoroughly comparative as any area of science, it looks at the mechanical problems of existence, in particular at how solutions to these problems vary among organisms of different sizes, different lineages, and different ways of managing growth, reproduction, and dispersal. It has always borrowed its basic concepts from engineering, and it has done so to its great advantage. We've made unfettered and hugely productive use of the existing technology of humans in our efforts to understand what nature does. Our books and articles make no attempt to disguise the process; indeed, we're continually asked by biology students encountering biomechanics, "What is this, engineering?" The importance to us of human technology as a critical reference is best put by reiterating the admission made at the start. We've only rarely recognized any mechanical device in an organism with which we weren't already familiar from engineering. Perhaps we're insufficiently creative and imaginative, or perhaps the engineers enjoy a long head start and numerical superiority. Or perhaps the situation indicates the value of an external reference for any attempt at understanding.

NOTES

CHAPTER 1: NONCOINCIDENT WORLDS

1. The real counterpart of biology isn't engineering but a tiny field concerned with the nature and history of human technology. A few recent books (to which further reference will be made) are those of Basalla (1988), Vincenti (1990), Adams (1991), Petroski (1992), and Cardwell (1995). A key journal is *Technology and Culture*; a central organization is the Society for the History of Technology (SHOT).
2. General books on the subject are those of Wainwright et al. (1976), Alexander (1983, 1992), Vogel (1988), and Niklas (1992).
3. *Nicomachean Ethics* 1099B, 23. Quoted in Mackay (1991).
4. Quoted in Schneider (1994).
5. T. Ewbank (1842), p. 514. The author has a fine (and enthusiastic) appreciation of physiology and natural history.
6. Papanek (1971).
7. The antitechnological literature is large and diffuse. Examples are

books by Ellul (1964) and Mumford (1967) and pieces by Commoner, Mumford, and Ellul in the general collection of Burke and Eakin (1979). Florman (1975) gives a good look at it as well as providing a sober rejoinder.

CHAPTER 2: TWO SCHOOLS OF DESIGN

1. Thus *The Evolution of Useful Things, The Evolution of Technology*, and *The Evolution of Design*—books by Petroski (1992), Basalla (1988), and Steadman (1979). See also Mokyr (1991), Dasgupta (1996), and the final chapter here.
2. This logical scheme isn't original here; its origin, though, is obscure. See, for instance, Endler (1986) or Futuyma (1986) for exposure to the full richness of the subject.
3. Wainwright et al. (1976).
4. Raup (1966). A good recent book on shells from an evolutionary perspective is that of Vermeij (1993). More recent computer-generated "virtual shells" are given by Meinhardt (1995).
5. Our special attachment to soft-shell crabs is a case, perhaps trivial, of a predator's preference for a newly molted arthropod.
6. Pierce (1961).
7. Crane (1950).
8. But evolution hasn't scorned the possibility. Having two nonidentical sets of genetic information in the cells of most organisms (diploidy) and engaging in recombination (sex) permit sheltering variation and testing it in different combinations.
9. Dawkins (1986).
10. These are clearly separate evolutionary developments. As a biologist would put it, the nearest common ancestor of birds and bats, for instance, lacked wings and didn't fly—according to paleontological, anatomical, and molecular evidence.
11. Termites, close relatives of cockroaches, constitute the only case of hypersociality in insects outside of the Hymenoptera, the group consisting of ants, bees, and wasps. Repeated evolution of hypersociality among Hymenoptera is a consequence of a predilection peculiar to their particular (and peculiar) genetic system; some other explanation (such as infectious symbiosis) is needed for the termites. See, for instance, Wilson (1980).
12. Dyer and Obar (1994) give a little further information and a useful

context; the basic source is Cleveland and Grimstone (1964). One hopes not to be too familiar with spirochetes, one of which is the pathogen in syphilis.

13. Tamm (1982).
14. See, for instance, Thompson and Bennett (1969), for a case in which nematocysts of sufficient potency to harm human swimmers were expropriated. The basic phenomenon has been known for almost a century.
15. The case is described by Fisher and Hinde (1949) and Hinde and Fisher (1951). Griffin (1984) summarizes it in prose, and Burns (1975) does the same in verse. My colleague Peter Klopfer tells me that the birds could do even better when challenged. The types of milk were distinguished only by the color of the bottle caps (presumably so returned bottles needed no sorting). Not only did the birds choose the bottles with cream beneath the caps, but they rapidly altered their color preference when caps were deliberately switched.
16. Galbraith (1967).
17. Eldredge and Gould (1972). What must be emphasized is that sudden bursts of change are claimed to be sudden only in a paleontological sense—that is, when one is looking over very long periods of time. No serious challenge to the underlying Darwinian model is implied, merely the recognition that the relative gain in fitness by change in form is enormously circumstance-dependent. For that matter, Darwin himself recognized (in *The Origin of Species*) that most species remain largely unchanged most of the time.
18. Grasses and grassland are among the earth's more recent acquisitions. Grasses are especially tolerant of fire and herbivory, and many grasslands revert to forest if these factors are removed. But the nutritive value of grass is low, and it's especially mean stuff to masticate in quantity.
19. The portable electric fan was in fact invented a lot earlier—by Nikola Tesla around 1890.
20. See, for instance, Basalla (1988) and Cardwell (1995). The former stresses continuity while the latter has a more saltatory or heroic bent but, even so, is more incrementalist than are low-level textbooks. I take no stand on the controversy, which isn't central here anyway. But I do believe that any analogy with biological evolution is almost entirely irrelevant. Vincenti (1990) and Adams (1991) flesh out the issue with specific cases; in essence the former is about what engineers know (as in its title) while the latter is about what they do. One

might also look at basic textbooks of mechanical design as used in engineering courses.

21. The discovery and biology of the coelacanth are a fine story, well told by Thomson (1991).
22. Houston et al. (1989). Some other flies have colonized the water-filled traps under our sinks.
23. Some biologists make a distinction between "parallelism" and "convergence." But it's a hairsplitting one (see Roth, 1996) that we'll ignore.
24. See, for instance, Stanley (1987) and Raup (1992).

CHAPTER 3: THE MATTER OF MAGNITUDE

1. Good general accounts of the biology of size are given in several books, especially McMahon and Bonner (1983) and Schmidt-Nielsen (1984). For brief and engaging introductions, see Haldane (1928); Chapter 2 of Thompson (1942); or Went (1968).
2. Bennet-Clark (1977).
3. Descartes's *Principia,* quoted by Kearney (1971), p. 156. Size mattered to Horace, much earlier: "mountains will be in labor, the birth will be a laughable mouse." I've thus, as James Thurber once wrote, put Descartes before Horace.
4. Thompson (1942), p. 68.
5. The matter is called Cope's law by evolutionary biologists, after the paleontologist E. D. Cope (1840–1897).
6. Robinson and Frederick (1989). We've occasionally done this as a class exercise, with people tracing the feet and recording the heights of each other and of their particularly large and small acquaintances. Of course that may not be the sole reason why big people have bigger feet.
7. Haldane (1928). Galileo's rule about all bodies falling at the same rate works only in the absence of air or as an approximation for large, dense bodies.
8. Orwell (1937), p. 26.
9. I've taken most of the material on fluid mechanics—lift, drag, gliding, types of flow, diffusion, surface tension—from my earlier book (Vogel 1994a), which contains copious references to the basic literature. Tennekes (1996) gives an admirably accessible account of size and flight.

10. Cope's law is mentioned in note 5. The three lineages of extant flying animals—birds, bats, and insects—appear to be significant exceptions to the rule. That's probably because flapping flight evolved from gliding, and gliders work better if large. The tiniest insects, the microchiropteran bats, the hummingbirds—all are members of relatively new groups within their general kinds.
11. The skeptical reader will ask whence comes the power source for staying aloft with fixed wings; after all, propellers, but not wings, are attached to engines. Nothing ethereal is happening; a fixed wing incurs more drag when it's producing lift, and the propeller and engine have to work harder to offset that drag.
12. These pores are in the feltwork of fibers that make up the walls of cells within leaves. They're not the stomata on the surface of leaves through which gases pass in and out, which are around a hundred times larger.
13. The efficacy of diffusion depends substantially on the size of the diffusing molecules, and good perfumes are complex mixtures. Were diffusion the main agency of spread, the character of a perfume would depend on one's distance from the source! Writers of science fiction might take note. My curmudgeonish diatribe on the matter is in Vogel (1994d).
14. We touch here on a general argument for why cell size is fairly constant and why organisms use a cellular scheme of organization, one elaborated in Vogel (1988, 1992).
15. We do use the scheme of the maple seed occasionally. Some obscure aircraft (called autogiros, gyroplanes, or gyrocopters) use the principle, and a helicopter whose engine has quit slows its descent with the maple seed's trick. I've provided a more extensive explanation of these various ways to fall more slowly elsewhere (Vogel, 1994a). Current interest centers on using such single winged autogyrating aircraft to deliver payloads from orbit to the surface of Mars; see Stephen Morris's work at http://aero.stanford.edu/MapleSeed.html.

CHAPTER 4: SURFACES, ANGLES, AND CORNERS

1. We commonly refer to the component of the load coming from the weight of the structure as dead load and that from useful activity as live load.
2. $36^3/30^3 = 1.73$, a 73 percent increase. $36^4/30^4 = 2.07$, a 107 percent

increase. I did a quick introduction to the relevant part of beam theory in a book called *Life's Devices.* A better source (this isn't false modesty) is Gordon (1978).

3. One more consequence: If sag increased with length and decreased with thickness to the same power, then bookshelves could be built with the two in simple proportion—twice as long being twice as thick. But since sag increases with length to the fourth and decreases only with thickness cubed, the longer shelf has to be disproportionately thick; twice as long needs to be four times as thick. That's a major reason why larger structures must have a disproportionate amount of supportive material, an argument made with a slightly different example in the last chapter.
4. The reason why camber is aerodynamically important is fairly complex, more so than implied by the usual prevarication. I give a fairly equation-free account in *Life in Moving Fluids*; other accessible sources are Sutton (1949) and von Kármán (1954). The phenomenon will come up again in Chapter 12.
5. A. Roland Ennos (1988) shows that camber and lengthwise twist were largely reversed passively—that is, by aerodynamic forces—rather than, as had previously been assumed, by highly coordinated muscular action.
6. The rule often goes nameless; in any case it's not clear that the French mathematician Pierre-Simon, marquis de Laplace (1749–1827) is its originator. ("Laplace's equation" is something else again.) For a cylinder the rule is similar, omitting only the one half; cylinders are curved in one dimension, spheres in two. Since the pressure is resisted by tension running in only one direction, twice as much tension is generated. A cylinder with hemispherical ends (such as a balloon or boiled sausage), pressurized until it explodes, will usually split its sides before blowing off its ends.
7. Shortses are marvelously roomy by comparison with the flying cigars into which we're more often inserted. They also seem to rattle a lot. Is that real and related or just my imaginative anticipation of logical consequences?
8. This was the Israel Museum in Jerusalem; it has the special virtue of dealing with an unusually long and continuous history of human occupation of a place exposed to input from diverse cultures.
9. Walsby (1980).
10. Put another way, is drawing graphs in Cartesian coordinates—that is,

with mutually orthogonal *x*, *y*, and sometimes *z* coordinates—in some sense natural for us?

11. The key reference is Flannery (1972); see also Hodges (1972) and Saidel (1993).
12. The original paper is by Allport and Pettigrew (1957). The illusion was reported by Ames (1951) and is described well by Hochberg (1978). The work with the Zulu is discussed by Owen (1978) along with work of Annis and Frost (1973) on Cree Indians.
13. Frazzetta (1966); Gans (1974).
14. Westneat and Wainwright (1989).
15. Parkes (1965), for instance. Neither three- nor four-strutted devices exhaust the world of mechanisms and statically determined structures; we've the tip of an iceberg here.
16. Gould (1980).
17. Well-evolved walkers in fact aren't consumed by some search for static stability. After all, when running, you have no legs at all on the ground a fair fraction of the time.
18. Which the U.S. military has tried, very expensively, to do for many years. One might ask about using more than six legs. Aside from the obvious increase in mechanical complexity and potential interactions and stride-length limitations, another factor enters. Recall from the last chapter that skinny columns buckle under lower loads than fat columns of the same length. The material investment in using a large number of thin columns for support turns out to be greater than using a small number of fatter columns, as nicely explained by Gordon (1988).
19. Between different languages and systems of mathematical notation, it seems unlikely that knowledge of the theorem spread after some single genius discovered it. See Bose (1971), Anderson (1972), Swetz and Kao (1977), and Prakash (1987).
20. Grant (1968), Buckley and Buckley (1977), and Barlow (1974), respectively.
21. D'Arcy Thompson (1942), mentioned earlier in connection with size, has a nice discussion of the relevant geometry and biology. His book, still available in its entirety or in an abridgment by J. T. Bonner (Thompson, 1961), is at once unique in content, imaginative in approach, and magnificent in the elegance of its prose.
22. Hulbary (1944) faces the complications of our less than ideal world.
23. A splendid treatment of the phenomena associated with cracking—

fracture mechanics—is given in any of three books written by James E. Gordon (1976, 1978, and 1988). All three are simply wonderful works in general. The 1978 one, *Structures*, gets my vote for the best book ever written on a technical subject for a nontechnical readership.

24. An engaging and accessible treatment of such aspects of the construction of trees is given by Mattheck (1991).

CHAPTER 5: THE STIFF AND THE SOFT

1. In part, of course, because we can measure very small changes in length. Parts per million are no trick at all for modern technology applied to very stiff samples.
2. From Gordon (1978).
3. Note that "elasticity" is being used in a most peculiar sense. First, the more easily stretched a material, the lower is its Young's modulus, and second, whether and how forcefully the material returns to its original length are immaterial.
4. "Toughness" sometimes is used for the energy involved in the actual process of breakage as opposed to that en route to the breaking point. In breaking, new surface is created, and that takes an additional input of energy, an amount related to the resistance of the material to making surface and to the amount of surface that's made in a break. Real materials, after all, don't all break evenly across their diameters. While work of extension is an energy per unit volume because a whole specimen gets stretched, "toughness" in this latter sense is an energy per unit area because the energy is invested in making surface. The overall work of breakage of an object that's initially unstressed involves of course both.
5. That's one of many consequences of one of the most important of all rules in science, the second law of thermodynamics. Stated anecdotally, it declares that in any real process you get out something less than you put in, with the difference appearing as heat. The comments of C. P. Snow (1959) on it are more relevant than ever as we become increasingly afflicted with people who view science strictly as social construct.
6. Such as hardness, based on who scratches whom; properties such as ductility, important in manufacturing; a host of time-dependent properties relating to differences in response depending on how fast loads are applied and recovery after loading for various times; and various thermal properties.

7. One might argue that rubber is (or originally was) entirely a natural product. But in the rubber tree it's a liquid, used by humans mainly as a coating for waterproofing. So rubber (like rayon) is something merely derived from a natural product; its mechanical utility began with Charles Goodyear's discovery of the vulcanization process in 1839.
8. Spider silk isn't actually a single material. Different kinds of spiders produce silks of somewhat different composition and properties, and those that spin nice orbs use several kinds in a single web. Silk moth silk is still another group of related materials. On the mechanics of spider silk, see Gosline et al. (1986) and Vincent (1990).
9. Harrar and Harrar (1962).
10. The pejorative "crackpot" must be a tacit cultural recognition of the lack of toughness of ceramics.
11. In particular, I like "A Chapter of Accidents" in Gordon (1978) and books by Levy and Salvadori (1992) and my colleague Henry Petroski (1985, 1994).
12. A version of this section has been previously published as Vogel (1995).
13. See Koehl et al. (1991).
14. See Grace (1977) or Miller et al. (1987).
15. See Vogel (1984a, 1989, 1993).
16. O'Neill (1990).
17. Koehl (1977). We sometimes demonstrate the relationship between the diameter of a cylinder and the ease with which it bends by using cylindrical pasta—spaghetti—of various diameters but of the same brand. A strand is supported between two bricks, fishing sinkers are added, and sag is measured with a ruler. Both the fourth power rule for weight versus sag and the third power rule for distance between supports versus sag emerge reasonably well, at least for small sags—Vogel (1988), p. 337.
18. Shadwick et al. (1990); Shadwick (1994). Squid and octopus are cephalopod mollusks. Vertebrates split from the ancestors of mollusks and arthropods (insects, spiders, crustaceans, etc.) at a simple, wormy stage of animal evolution roughly half a billion years ago.
19. Alexander (1984a, 1988).

CHAPTER 6: TWO ROUTES TO RIGIDITY

1. Tungsten, for instance, has recently been shown (see Chan et al., 1995) to be a crucial component of certain very unusual bacteria that grow best at temperatures up to the boiling point of water.

A general source on metals in organisms is Kendrick et al. (1992).

2. All these data are from Schmidt-Nielsen (1990), a textbook that really explains things rather than just naming and mentioning them.
3. The main source here is Vincent (1990). Lowenstam (1967) points out that the teeth of chitons contained enough magnetite so that substantial amounts of the magnetite in the sediments beneath shallow seas might be of biological origin.
4. Blakemore (1975).
5. Magnetite is, for the chemically minded, Fe_3O_4 or Fe_2O_3FeO, distinguished from ferrous oxide, FeO, and ferric oxide, Fe_2O_3. (*Ferrum* is the Latin word for "iron.") Accessible references include J. L. Gould et al. (1978) and Walcott (1979).
6. Photosynthesis, in case you've forgotten, splits water as a source of hydrogen and combines the latter with carbon dioxide, liberating oxygen. Before life invented the process, the atmosphere had little or no oxygen in it; thus oxygen is the first and greatest pollutant.
7. Association of exposure to aluminum with subsequent development of Alzheimer's disease looks less likely than it did a few years ago. The *Merck Manual* (Berkow and Fletcher, eds., 1987) mentions only "dialysis dementia," a side effect of administration of aluminum to counteract hyperphosphatemia in chronic renal failure. Aluminum dissolved in natural waters is distinctly bad for fish, though.
8. Mainly materials such as muscle and collagen in animals, the last a major component of tendon, skin, bone, and cartilage; also, cellulose in plants, the main constituent of wood.
9. See Wayman (1989a, b), Tylecote (1992), and Schmidt (1996).
10. In the trade, materials that give straight plots of stress versus strain are spoken of as Hookean, after Robert Hooke (1635–1703) who stated what's now known as Hooke's law, that the amount of deformation follows the deforming force, that strain is proportional to stress. Calling it a law smacks of hyperbole, since it does little more than describe how some, but not all, materials behave.
11. Currey (1984). This is a lovely integration of biology and mechanics, sophisticated without being formidably technical.
12. It's worth reminding ourselves that what I'm calling micromaintenance is unusual even in living systems for large, stiff, structural elements. Bone is intracellular, by contrast with hair, arthropod exoskeletal material, mollusk shell, and wood. The latter are built once and for all, with only very limited reabsorption and replacement.

13. Hodges (1970) has a short but insightful account of how the use of metals may have begun. A more elaborate analysis that focuses on the archaeological evidence is that of Singer et al. (1954).
14. Gordon (1976).
15. Note for the purists: Plexiglas is a trade name for a certain plastic mainly of polymethylmethacrylate that has no common generic name. Perspex and Lucite are the same plastic from other manufacturers. By contrast, "fiberglass" is a generic, with a double *s* and no capitalization.
16. Singer et al. (1954).
17. I'm following the account given by Gordon (1976), who was a major player in the crack game.
18. Wainwright et al. (1976). Vincent (1990) is a newer source, which, while less general, is especially good on biological foams and composites.
19. "Conductivity" refers to a property of a material, while "conductance" refers to the property or behavior of a particular object.
20. The advantage of metallic leaves has been inadvertently demonstrated in several studies of the thermal behavior of leaves that used models inappropriately cut from sheets of metal.
21. I did some work a few years ago on thermal behavior and thermal conductivity of leaves. The upshot of a paper on convective cooling at very low wind speeds (Vogel, 1970) and another on conductivity (Vogel, 1984b) is that broad leaves either wouldn't get so hot during lulls in the wind or could be less constrained in shape and size—if leaves had the thermal properties of, say, copper.
22. But even that isn't instantaneous. The largest and fastest computers are arranged to minimize the lengths of wires connecting their components since conduction times constitute significant delays in their operation.

CHAPTER 7: PULLING VERSUS PUSHING

1. A recent book (Barber, 1994) gives a fine introduction to both the technology and the social context of spinning as well as to the etymology of the words, such as "life-span," that we derive from "spin."
2. Salvadori (1980) makes much of this point in a most enlightening book.
3. But an analog of a crack does occur in compression. Wooden beams such as diving boards accumulate compression creases on their undersides. As Vincent (1990) cautions, subjecting such creases to tensile loads is an exceedingly bad idea. Don't turn an old wooden

diving board upside down to expose an unworn surface.

4. Several good sources on the mechanics of cathedrals are available—for instance, Mark (1978 and 1982).
5. Bronowski (1973) makes quite a point of this empiricism in cathedral construction.
6. The data for these comparisons are from Wainwright et al. (1976), Currey (1984), Vincent (1990), and Niklas (1992). See note 3, above.
7. French (1994) has a good discussion of the matter. For considerations of size and scale, going back to the second chapter of D'Arcy Thompson's classic work (1942) is certainly worthwhile. For bridges in particular, I enthusiastically recommend Billington (1983).
8. Isaacs et al. (1966) suggest the possibility.
9. Niklas (1994).
10. I give more data and a more detailed account in *Life's Devices* (1988); the best source on scaling in mammals is Schmidt-Nielsen (1984). For aquatic mammals such as whales (and for fish), skeletal mass is very nearly proportional to body mass, as it ought to be where the most severe loading is provided by muscle contraction rather than gravity. If size is doubled, then both resistance to buckling and muscle cross-sectional area (which muscle force follows) increase fourfold.
11. See, on brachiation, Hallgrimsson and Swartz (1995) and the references in that paper.
12. Salvadori (1980) gives a good account of Brunelleschi's achievement.
13. More complete accounts and further references are given by Petroski (1985, 1991, and 1995).
14. Tendons usually connect muscles to bones; ligaments usually connect bones to bones.
15. Laithwaite (1984) makes quite a point of nature's invention of the underslung suspension.
16. The wishbone of a bird bends when the bird flaps its wings up and down, according to Jenkins et al. (1988). Springs can work by twisting as well, as in torsion bar automobile suspensions. I know of no examples in nature unless one includes such structures as the trunks and branches of trees and the wing feathers of birds that are loaded torsionally and have reasonable resilience. No relatively massive components, though, are attached to their outboard ends, so it's arguable whether they're springs in the present sense.
17. Vines of course are normally loaded in tension. But trees aren't ordinarily held upright by the vines they harbor. The large but thin but-

tresses of many tropical trees do seem to be tensile guys; they're thus the exact opposite of the buttresses (flying and otherwise) of Gothic cathedrals. On the buttresses of trees, see Mattheck (1991) or Ennos (1993). Of course plenty of material is loaded in tension wherever a trunk forks or a branch diverges; on this matter, see Mattheck (1991).

18. Bath sponges are unusual in that they lack spicules and have only the tension-resisting stuff spongin. They're not very large, and a meshwork of spongin does have some compressive stiffness.
19. See, in particular, U.S. number 3,063,521, November 13, 1962 (Fuller, R. B.; Tensile-Integrity Structures).
20. The best paper I know on spicular skeletons in general is Koehl (1982). At this point good mechanical analyses of the supportive system of sponges don't seem to exist.
21. Liquids care about shear, but what matters isn't how far they're sheared but how fast; they resist only the rate of shear, and the coffee adjusts its shape to fit any cup fairly quickly. The resistance of liquids to tension will come up again; it's important in getting the sap up every tall tree, but our technology finds the property impossible to employ. Another way to put the distinction, incidentally, is to recognize that solids have shape and size, liquids have only size, and gases have neither shape nor size.
22. Notice how balloons keep arising here. My colleague Stephen Wainwright (with hydrostatic tongue only slightly in membranous cheek) says that if you can't demonstrate it with a balloon, it's probably not important anyway.
23. French (1994) gives these examples and a short treatment of the technology involved.
24. General descriptions of hydrostats in nature are given by Wainwright et al. (1976), Alexander (1983), and Vogel (1988). Bending in hydrostats is analyzed by Alexander (1988).
25. See Smith and Kier (1989). Note the photograph of our late cat Fred with his tongue extended. To make Fred perform it took two Vogels, one Kier, and an entire can of tuna.

CHAPTER 8: ENGINES FOR THE MECHANICAL WORLDS

1. Various sources (perhaps copying one another) suggest the origin of windmills in seventh-century Persia, although Tokaty (1971) men-

tions earlier cases. Derry and Williams (1960) note that more than six thousand water mills for grinding grain were reported in the Domesday Book of England in 1086. As Cardwell (1995) points out, technological innovation in Europe during the Middle Ages was much more impressive than during the Roman Empire.

2. The proportion is largely independent of the size of the mammal, which means that the edible fraction doesn't depend on how big an animal is. See Schmidt-Nielsen (1984).
3. Data for cilia, muscle, electric motor, and automobile engine from Nicklas (1984); for motorcycle, from McMahon and Bonner (1983); for aircraft, from French (1994). The figure for the Newcomen engine is my guess based on its 5.5-horsepower output (Derry and Williams, 1960) and an assumed weight of half a ton. The output, at least, is reasonably reliable since the engine was employed lifting water—in energetic terms a well-defined task.
4. A very fine exhibit of working models of early steam engines may be seen at the Science Museum in South Kensington, London.
5. A good source on heat engines is Atkins (1984). This limitation on efficiency emerges from what's called the second law of thermodynamics.
6. $100 + 273 = 373$; $1000 + 273 = 1273$; $100(1 - 373/1273) = 71$.
7. This isn't a straw man or rhetorical device. See, for instance, Paturi (1976).
8. These are an electric catfish, the so-called electric eel, and electric rays; the trick has clearly evolved several times. Some other freshwater fish generate low-strength electric fields that serve sensory functions; see Denny (1993).
9. See Basalla (1988) for a description, a figure, and a good argument for the real rarity of technological discontinuity. Laithwaite (1989) traces the development of linear electric motors.
10. Datum calculated from Edsall and Wyman (1958).
11. The ability to use transformers for easy and efficient conversion was the original reason for switching to alternating current (60 Hz in North America and 50 Hz in most other places) from the direct current used by Edison and others. Transformers don't work with direct current.
12. Intermediate storage is one place where nature really eclipses anything in common use in our electrical technology. We use storage of a form of carbohydrate, glycogen, in the liver for short-term use (min-

utes to days) and storage as fat elsewhere for longer-term use. Analogous arrangements (such as roots and tubers in plants) are nearly universal. Demand for electricity and for muscle power vary similarly during the day; the power companies mainly accommodate the variation by adjusting the output level of their fossil-fueled generators. (Nuclear generators, by contrast, are more economically run at high and constant output since the plants are relatively more costly and the fuel is cheaper.) Intermediate-term storage takes such awkward devices as pumped storage plants, in which water is pumped to an uphill reservoir when power demand is low and released through generators when demand increases.

13. Usher (1954) gives a good introduction to the subject. An experimental analysis from the days when waterwheels were major engines (and a paper still worth reading) is that of John Smeaton (1759).
14. An exception, one using rotating cylinders and the same Magnus effect that makes spinning balls take curved paths, was built by Anton Flettner in the 1920s and is described by Tokaty (1971).
15. See Tokaty (1971) for an account of various versions of Sigurd Savonius's design, along with some performance data.
16. I've given proper attention to this distinction between drag-based and lift-based thrust production in connection with different modes of swimming in animals (Vogel, 1994a, pp. 154–55, and 283–87).
17. I'm avoiding proper explanation of why flow over the higher opening is more rapid as well as mention of a secondary physical mechanism for driving the flow, one based on the viscous stickiness of real fluids.
18. Dynamic soaring, it's called. See, for instance, Vogel (1994a).
19. Quite a lot more on both mechanisms and cases can be found in Vogel (1978, 1994a). The original suggestion was made by Vogel and Bretz (1972).
20. The flagella of bacteria, different things altogether, will get proper attention in the next chapter.
21. A good introduction to the mechanics of motility in animals is given by Schmidt-Nielsen (1990); McMahon (1984) provides more information on muscle in particular.
22. The physical scientist will quickly protest that "lift" by suction to greater than ten meters isn't possible since what's really doing the work is the atmosphere pushing from the bottom, and the maximum pressure difference of one atmosphere can raise water only that high.

The rejoinder to this objection is that in tall plants the suction is real and that when properly contained in thin hydrophilic pipes, water can withstand enormous tension. Zimmermann (1983) has a good account of our present understanding of the process.

23. Nijhout and Sheffield (1979).
24. The figures come from a variety of sources. For human-built engines, I used Salisbury (1950), Croft (1981), and Atkins (1984); for muscle, Heglund and Cavagna (1985).

CHAPTER 9: PUTTING ENGINES TO WORK

1. I'm avoiding the conventional classification into first-, second-, and third-class levers to which the reader may have been subjected. It's a structural rather than a functional scheme and proved confusing and irrelevant to readers of the first version of this chapter. First-class levers, in case you care, have their fulcra in the middle; second and third classes have fulcra at the ends. First-class levers may be either force or distance amplifiers; second-class ones are always force amplifiers; third-class ones are always distance or speed amplifiers. A first-class lever in which effort and load are equidistant from the fulcrum is the exception; it changes only the direction of the force.
2. On rare occasions a crank is geared up to spin something rapidly—hand-operated centrifuges and grinding wheels, for instance. Less drastically but more commonly, bicycle tires move faster than the pedals except in the lowest gears.
3. The lower end of the humerus, no joke, is the "funny bone," with no etymological pun whatsoever. Distance advantages are 5.4 and 22, respectively (Currey 1984).
4. Alexander (1983) does a short and effective job on jaw mechanics and the topics that immediately follow here.
5. The fibers are quite as easy to see after a few minutes' immersion in boiling water to kill the creature humanely. Leave the final, clawed segment intact, and work on the penultimate one. After looking at and then consuming the muscle, you may be able to expose a pair of flat straps (properly called apodemes rather than tendons since these are arthropods). The larger (pull on it) closes the claw while the smaller opens it.
6. Smith and Kier (1989).

7. Vogel (1992) puts the calculation in context.
8. Berg and Anderson (1973) and Berg (1974). The figures in the next paragraph are from the latter paper, to which my attention was drawn by R. Bruce Nicklas.
9. Mitochondria, the main sites of aerobic energy production in most cells, and chloroplasts, the sites of the photosynthetic machinery of plant cells, appear to have originated as symbiotic bacteria. Another case, the use of bacteria as locomotory organelles, is described in Chapter 2. For both, see Margulis (1993) or Dyer and Obar (1994).
10. Gould (1981). "Wheels are not flawed as modes of transport; I'm sure that many animals would be far better with them."
11. LaBarbera (1983) and Basalla (1988). I'm not going to sort out their separate arguments.
12. Ekholm (1946). I don't take seriously suggestions that these toys must have been copies of real devices, now lost, since the wheels are on the ends of animal legs rather than on wagons.
13. On this last point, see Bulliet (1975).
14. These figures are all from a source I'd not be without, *Machinery's Handbook* (Olberg et al., eds., 1984).
15. On bearings in human technology, see French (1994); on those between bones, see Currey (1984).
16. Watt's device incorporated an orbital gear largely as a way to evade infringing on a recent patent, so it seems to us excessively complicated.
17. Sleeswyk (1981) talks about the general problem of coupling humans to machines in the context of Egyptian antiquity and is my source for guided and misguided cranks.
18. A good source for this topic as well as for a lot of other matters relevant to this book is a two-volume set by Amerongen (1967).
19. These transmissions use a version of a fluid coupling called a torque converter, which adds a little complexity to this basic fluid coupling to get a trade-off of speed and force. And to improve fuel economy, the input and output shafts may actually clamp together at high speeds to prevent any slip at all. In addition, gearing is provided (hydraulically shifted!) to increase the car's usable range of speeds.
20. Dunwell (1991) recounts the controversy, with the appropriate maps and pictures.
21. See Alexander (1984b or 1988) or (secondhand) Vogel (1988) on this topic and those that follow. If you try the calculation, remember to

keep your units consistent—for instance, height in feet, acceleration in feet per second squared, and speed in feet per second.

22. Sotavalta (1953).

CHAPTER 10: ABOUT PUMPS, JETS, AND SHIPS

1. A more elaborate version of the present section appears in Vogel (1994b).
2. Miller (1945).
3. But some doubt. Martin Canny (1995) has cast an informed and skeptical eye on the evidence for the extreme negative pressures in trees and suggests an alternative explanation for the rise of sap.
4. Zimmermann (1983) gives a good review of the system. The record for pull, 120 atmospheres, was measured by a colleague, William Schlesinger (1982).
5. For example, no less than 22.4 atmospheres are needed to offset the propensity of a one-molar solution of a nonelectrolyte to absorb pure water.
6. The underlying distinction between propellers and paddles or flukes and cilia, elaborated in Vogel (1994a), is between lift-based thrust producers and drag-based thrust producers.
7. LaBarbera and Vogel (1982).
8. We do have an exception. A field called exobiology considers extraterrestrial life; on occasion one or another of its devotees has proudly proclaimed this distinction and the uniqueness of a field that doesn't know if it has a subject matter.
9. Some of the material in this section appeared as Vogel (1994c).
10. A crude steam rocket made a fine toy for bathtub or pond back when small metal containers were ubiquitous as 35 mm film canisters, tooth powder cans, and the like. One mounted a candle in the hull of an open boat, with a partly filled can of water above. A pinhole in the rearward-facing top of the can let steam escape to propel the boat forward. One shudders when thinking of its efficiency, much less of its safety. A device distantly related to Hero's engine did see service in the late nineteenth century: the De Laval steam turbine, used to run high-speed electrical generators.
11. We're assuming that the working fluid of the jet is of about the same density as the surrounding fluid. A rocket in space isn't concerned about the speed at which it passes through nothing at all.

12. Except, perhaps, for very tiny insects for which covering distance is more a matter of ambient wind than their own navigation.
13. The life and times of both of these are the subjects of engaging popular accounts, Crouch (1989) and Grosser (1981) respectively.
14. But oversimplifying is all too easy. The inboard part of a flapping wing goes up and down very little, so it does work very nearly as a fixed wing.
15. The motion that matters isn't really the lateral progress of the wave—very little actual wave-wise transport of water happens—but the orbital motion of the water immediately beneath the wave. See Denny (1993), Vogel (1994a), or any introduction to oceanography.
16. Full support by surface tension is quite another subject from the speed limit set by waves. Whirligig beetles are supported in part by surface tension and in part by their buoyancy, while water striders are almost entirely supported by surface tension. Whirligigs are what matter here; water striders move in a specialized way, one in which surface waves are less directly involved.

CHAPTER 11: MAKING WIDGETS

1. See, for instance, Schopf's own contribution in Schopf (1992).
2. Vincent (1990). Of course I compound the sin by repetition.
3. Such multicellularity was in fact achieved in several separate lineages, and traces of these separate origins exist in the different chemical and physical devices that present multicellular lineages use to keep themselves operative.
4. See Chapter 3, note 10.
5. These are all cases of *negative* feedback: Action is taken to *decrease* the difference between the actual state and some desired state. Positive feedback has another role altogether and isn't immediately relevant to the present discussion.
6. Mayr (1970). He raises such fascinating questions as why float valves were not used between the twelfth and eighteenth centuries.
7. My colleague Fred Nijhout (1990) argues, in an elegant and persuasive essay, that we've been poorly served by the metaphor of gene as unique and complete causative agent. Part of his argument is the basis of the next paragraph.
8. Electronic equipment that operates with very high precision is ordinarily made from far less precise components by using a variety of

self-correcting schemes, most of which fit comfortably within the present invocation of feedback.

9. Morowitz (1968) gives figures for the percentage of an average protein that will spontaneously spoil (denature) per day as a function of temperature. In Celsius degrees, they are 1.1 percent at thirty-seven degrees; 4.4 percent at forty degrees; 13.8 percent at forty-three degrees; 46.2 percent at forty-six degrees; and 161 percent at forty-nine degrees. It does look as if mammals and birds, with body temperatures around thirty-seven degrees to forty degrees, hug an upper practical limit for use of ordinary protein.
10. The relative diameters of all the blood vessels in a vertebrate are constructed and, when necessary, readjusted to minimize the operating cost of its circulatory system. Astonishingly, no overall coordination is required or involved; each of the cells lining the vessels needs only to respond to changes in the shear stress of the blood flowing past it. For details, see LaBarbera (1990) or Vogel (1994a).
11. A reasonable introduction to a fairly diffuse literature on how trees respond to environmental forces is a recent volume edited by Coutts and Grace (1995).
12. The doctrine was at its zenith in Britain around the turn of the century; it used the slogan "survival of the fittest" to justify all manner of social conservatism, unrestrained capitalism, class stratification, racial discrimination, and imperialism. Considerable opposition to evolution by natural selection has been generated by this combination of an inappropriate analogy and the notion that a social system based on natural selection has some natural and intrinsic merit. So in parallel with fundamentalist opposition to evolution, we have left-wing opposition based on rejection of the predestination and social Darwinism often thought to be part and parcel of the biological world view.
13. The famous paper here, into which more is often read than a literal look suggests was intended, is Gould and Lewontin (1979).
14. See the literature on symmorphosis, a term coined by Taylor and Weibel (1981); typical criticisms will be found in Garland and Huey (1987) and Dudley and Gans (1991).
15. This isn't to deny that sometimes single substitutions make dramatic differences. The sensitivity of the product to such alterations isn't at all uniform.
16. Two good references on safety factors in organisms are Alexander (1981) and Currey (1984). I sum up their arguments in Vogel (1988).

17. Niklas (1992) sums up what information is available.
18. As with most historical accounts, complexities lurk beneath the surface. Thus Whitney may have been one of the first with the basic idea of interchangeable parts, but full achievement was decades away (see Woodbury, 1960). Interesting general (if short) sources are Derry and Williams (1960), Boorstin (1965), and Reynolds (1991). Fridenson (1978) describes the transfer of this American system to Europe.
19. During World War II Liberty ships, cheap and expendable freighters, were truly mass-produced, not just made in quantity. But the case is exceptional.
20. About twenty-five years ago I took the standard tour of the local cigarette factory. The tour guide halved a cigarette and tossed it back into the machinery; a few seconds later a pack was ejected some distance away that contained 19.5 rather than 20.0 items.
21. Norbert Wiener's now rather ancient book (1950) on the interrelationship between humans and feedback devices, the field he named cybernetics, is still worth reading.
22. J. E. Gordon (1978) regards them as essentially factors of ignorance forming part of the theology of design.

CHAPTER 12: COPYING, IN RETROSPECT

1. I'm indebted to Francis Newton, a neighbor and colleague in classical studies, for bringing in Ovid. The quotations are from Book VIII, lines 183–259. Note, in the first, that the wings were curved; as we'll see, the crucial role of wing camber was not otherwise appreciated until the 1880s. The ancients, especially the Greeks, took a dim view of human creativity; everything was given by the gods, stolen from the gods, or copied from nature. Thus, according to Democritus of Abdera (ca. 420 B.C.E.), "We are pupils of the animals in the most important things: the spider for spinning and mending, the swallow for building, and the songsters, swan and nightingale, for singing, by way of imitation" (Freeman, 1948). Poseidonius (135–51 B.C.E.) viewed millstones as copied from teeth (Cole, 1967).
2. E. O. Wilson (1984). He defines it as "the innate tendency to focus on life and lifelike processes."
3. Quoted in Gérardin (1968) and elsewhere. Other books specifically about bionics include Bernard and Kare (1962), Halacy (1965), and Marteka (1965). Winfield et al. (1991) have done an annotated bibli-

ography covering a broad sweep of bionics and biomimetics; more recent material is available by searching under these headings in the relevant engineering and biological databases (NTIS, BIOSIS, etc.)

4. "Biomimesis" was defined by Warren McCulloch in 1962 as all areas in which one organism copies another; thus mimicry of distasteful insects by innocuous insects would be included. But that usage seems not to have caught on.
5. Several disclaimers: Potentially profitable bits of technology are rarely reported in peer-reviewed journals; more commonly they form the bases for patents, with laws and lawyers deliberately compromising any semblance of historical verisimilitude. A patent tries to impress upon the world just how original is the invention, giving as little benefit as possible to any notion of continuity; it fully intends to be heroic literature. That nature might have done the trick first is not usually something to be forthrightly admitted, and the idea that the invention just copied nature is certainly best kept private. As Basalla (1988) puts it (p. 60), "All of patent law is based on the assumption that an invention is a discrete, novel entity that can be assigned to the individual who is determined by the courts to be its legitimate creator." For ancient and prehistoric possibilities, the chance of documentation is even less.
6. From Smeaton's *A Narrative of the Building and a Description of the Construction of the Edystone Lighthouse with Stone*, abridged by T. Williams (1882).
7. In *Lighthouses, A Rudimentary Treatise*, quoted by Majdalany (1960).
8. A wealth of history of the tunnel is provided by two biographies of the more famous Isambard Kingdom Brunel, who as a young man supervised the work for his father. See Rolt (1959) and Vaughan (1991).
9. Beamish (1862), p. 207. But the story must be older, since Ewbank (1842), p. 258, alludes to it, as he does to Smeaton's tree trunk.
10. Miller (1924).
11. Paxton and the Crystal Palace have not suffered obscurity; see Chadwick (1961), Beaver (1970), or Kihlstadt (1984).
12. "[A] gardener as well known as his patron, the Duke of Devonshire," according to Hix (1974), p. 50.
13. No. 13,186, dated July 22, 1850, enrolled January 22, 1851; in "Alphabetical Index of Patentees of Inventions; Building Materials 1850–51, Vol. 184."
14. Hertel (1963), Paturi (1976), Tributsch (1982), and Laithwaite (1994) all comment on the copying.

15. *Illustrated London News, Supplement 17*: 317–24 (October 19, 1850).
16. Forget about low-drag cars, whose shape is severely compromised by the requirement that they work near the ground and not produce lift. And as noted just ahead, forget surface ships, whose drag comes from surface waves.
17. Quoted by von Kármán (1954), in part, and more fully by Pritchard (1961).
18. Quoted by Gibbs-Smith (1962).
19. Phillips (1885); also reported by Chanute (1893).
20. The book was first published in 1889 as *Der Vogelflug als Grundlage der Fliegekunst*; a second edition was prepared by his brother Gustav and published in 1910. That second edition was translated a year later into English, under the title in the text. A short account, translated, is given as an appendix in Chanute (1893).
21. A brief account of the difficult birth of airfoil theory is given by Giacomelli and Pistolesi (1934). They also provide a glossary that's needed to read Lanchester's work; the latter coined and used a host of wonderful Greek- and Latin-based terms that never came into common usage. Von Kármán (1954), Sutton (1949), and Vogel (1994a) describe in nonmathematical terms how lift originates without relying on the usual polite fiction. In the latter, bits of fluid diverge in front of the wing and then rejoin at the rear; since the path across the top is longer, those half bits have to go faster; by Bernoulli (that at least is okay), faster means lower pressure, so the wing is forced upward. The trouble with the explanation is that no rule requires that the bits of fluid reassemble with their same partners at the rear, something crucial to its logic.
22. At least that's my best guess as to the trouble. My first attempt to calibrate an electronic anemometer I'd built foundered on this particular rock. Von Kármán (1954) notes that the problem corrupted a lot of nineteenth-century data.
23. Wilbur Wright to Octave Chanute, May 13, 1900, in McFarland (1953). The famous first powered flight came in December 1903.
24. Recent biographies of the Wrights are those of Howard (1987) and Crouch (1989). See also Wright (1953).
25. I'm ignoring old Chinese and Japanese paper craft as out of the historical sequence that led to present mass-production technology, whatever its intrinsic interest or claims of priority.
26. Quoted in translation by Hunter (1947). Hunter is the best overall source I've seen, but Ainsworth (1959) and Schlosser (1980) are also

useful. Flatow (1992) gives a quick account without documentation. About the wasps, see Hansell (1989).

27. Koops (1800). I was able to examine an original of Koops's book; two centuries have treated it kindly, with no deterioration obvious to my admittedly untutored eye. The slightly yellow-beige paper has a rough surface, but the print is clear and dark, with no capillary "wicking" of the ink.
28. Hooke (1665), with modern spelling.
29. He demonstrated the process for the annual meeting of the British Association for the Advancement of Science at its Manchester meeting in 1842; the event is noted in the report of the meeting.
30. Ozanam (1862). The occasional attribution of the invention of the spinneret to Ozanam on the basis of these few vague and hopeful words (without the name) must be taken as an exaggeration. Schwabe's claim is both earlier and better.
31. Sources here are Avram (1927) and Leeming (1949). The quotation is from Chardonnet's entry in the *Dictionary of Scientific Biography.*
32. The *Oxford English Dictionary* traces it back to Kirby and Spence's *Entomology* of 1825.
33. From Worden (1911).
34. Schober (1930).
35. From Bell's lecture of October 31, 1877, to the Society of Telegraph Engineers, in London; quoted in Prescott (1884). A good account is that of Bruce (1973); M. Gorman and W. B. Carlson, of the University of Virginia, are making available Bell's notebooks, currently as http://jefferson.village.virginia.edu/~meg3c/id/AGB/index.html. John Wourms, of Clemson University, called this case to my attention.
36. No. 174,465; March 7, 1876.
37. For instance, the prickly-pear cactus, a New World plant (as are almost all cacti), was imported into the Middle East for hedgerow use about three hundred years ago. It's now a nuisance.
38. A very nice encapsulation of this story is given by Basalla (1988); a more elaborate recounting is that of the McCallums (1965). According to the former, in 1860 alone enough seed was sent north to produce sixty thousand miles of hedge.
39. The Kelly patent is no. 74,379 (1868); the Glidden patent (commercially the most important one) is no. 157,124 (1874).
40. A crosscut timber saw with well-set and sharp teeth cuts impressively. "Cutters" splay outward very slightly and make slits that define the outside of the kerf, while "rakers" work like chisels and shear off the

wood between those slits. With strong and determined sawyers, each pull of the saw may deepen the kerf by more than half an inch.

41. The account is mainly that of Lucia (1975 and 1981 contain essentially the same material) together with entomological information from Craighead (1915, 1923). I'm indebted to Professor George Pearsall for calling my attention to the case.
42. A short history of Velcro is given by Flatow (1992) in an engaging, if poorly documented, book on inventions. See also Budde (1995).
43. A good history of misguided people and misbegotten craft is that of Hart (1985).
44. Maxim (1909). His huge steam-engined biplane self-destructed after an expenditure of twenty thousand pounds.
45. See, on this point, J. S. Harris (1989).
46. Quoted by von Kármán (1954).
47. Maynard Smith (1952)—who did aerodynamics before being lured into biology.
48. The basic account here is from Hertel (1963) and Paturi (1976), with the aerodynamics largely confirmed by Bishop (1961). The plant was known to the Etrichs as *Zanonia macrocarpa* but is now in the genus *Alsomitra* and hence *A. macrocarpa.* Good aerodynamic data on the fruit are now available, in Azuma and Okuno (1987). In fact, it's not an especially good wing, descending with a lift-to-drag ratio of 3.5. For comparison, the hind wing of a locust, which operates under comparable conditions, manages a ratio of 8.2.
49. Bernheim (1959).
50. A particularly good introduction to the history of stability in aircraft is that of Vincenti (1988). Von Kármán (1954) gives one a good feel for the various controls themselves.
51. See Boyd (1935) and C. M. Harris (1989). One wonders why neither Rumsey nor Fitch used a paddle wheel; the device is nothing more than the then familiar undershot waterwheel used in reverse. Perhaps Watt's crankshaft for converting linear to rotary motion was insufficiently appreciated.

CHAPTER 13: COPYING, PRESENT AND PROSPECTIVE

1. British patent 3587/1894. From Wilson (1975).
2. The relevant references are Gray (1936), Kramer (1965), and Riley et al. (1988).

3. The earliest report seems to be Toms (1948). See also Davies and Porter (1966), Wells (1969), and Bushnell and Moore (1991).
4. See Reif (1985), Choi (1990), and Bushnell and Moore (1991).
5. sci.aeronautics, sci.engr, sci.mech.engr, sci.engr.marine.hydrodynamics, sci.mech.fluids, and soc.history.science. I deliberately biased the selection toward news groups frequented by engineers, and I received a number of very thoughtful and interesting responses. I thank all the people who responded and hope that I managed to reply to each.
6. Two general sources are Winfield et al. (1991) and Sarikaya and Aksay (1995). Quite a lot of information emerges if one searches databases such as that of the National Technical Information Service (NTIS) or Biological Abstracts (BIOSIS) using "bionics" or "biomimetics" as keywords.
7. See Douglas and Clark (1990), Lombardi et al. (1990), Mann (1990, 1995), and Fournier et al. (1995).
8. See De Rossi et al. (1985), Pool (1989), Caldwell and Taylor (1990), Urry (1993), and Salehpoor (1996).
9. See Gunderson and Schiavone (1989), Berman (1990), Mann (1990), and Sarikaya et al. (1994). The patent (United States, 1983) is to C. R. Chaplin, J. E. Gordon, and G. Jeronimidis, no. 4,409,274, entitled "Composite Material."
10. This is somewhat further from practical realization, but it has been the subject of much speculation and some experimentation. See Winfield et al. (1991).
11. See Mason and Salisbury (1985), Hayward (1993), Wilson et al. (1993), and Wainwright (1995).
12. See Raibert and Sutherland (1983), Song and Waldron (1989), and Manko (1992).
13. Stix (1994). The principals are Robert Barrett and Michael Triantafyllou. The principles are prefigured in Triantafyllou et al. (1993).
14. The twiddlefish is the invention of Charles Pell and Stephen Wainwright, working at the Biodesign Studio, at Duke. See McHenry et al. (1995).
15. Blake et al. (1995) have shown that tuna can't turn especially sharply and suggest that their efficiency in steady swimming involves some loss of maneuverability.
16. My main sources on spider silks are Gosline et al. (1986 and 1995).

17. Chatterton (1910) and Rouse and Ince (1957).
18. Maybe naval design is naturally conservative. Needham (1965) notes the cambered surfaces of ancient Chinese kites. The matter is confused by the way sails automatically camber in a wind. See also Smeaton (1759).

CHAPTER 14: CONTRASTS, CONVERGENCES, AND CONSEQUENCES

1. It's now late enough in the game to admit something that has been kept sub rosa. Other mechanical technologies do exist—in particular those of a variety of social organisms, such as bees, ants, termites, and beaver. Each seems (adjusting for such things as scale) largely a compromise between the two main ones under scrutiny here, so they have only minimal relevance to the present discussion.
2. Stable systems are more persistent than unstable systems. A system in which destructive interactions can arise is more likely to disappear than one with self-improving interactions. This argument for large-scale integration beyond what natural selection per se can produce was made by Pantin (1964). It's crucial to the Gaia hypothesis popularized by Lovelock and Margulis (1974).
3. The phenomenon is most often known as kin selection. It rationalizes things ranging from parent-child altruism to the repeated evolution of complex societies in the genetically peculiar ants, bees, and wasps (Hymenoptera). See, for instance, Dawkins (1976) or Hölldobler and Wilson (1994).
4. We use the term "mutualism" to make a needed distinction from symbioses (living together) in which mutual and reciprocal benefit is absent—parasitism, for instance.
5. Basalla (1988). See especially pp. 91–92. Derek deSolla Price has noted the same thing on various occasions.
6. The classic statement of the notion of "punctuated equilibria" in evolution—the extreme of episodic change—is that of Eldredge and Gould (1972). The idea pervades many of the essays of Gould (1980, and other collections from *Natural History* magazine). Less protagonistic statements can be found in contemporary evolution textbooks such as Futuyma (1986).
7. For a more extensive but admirably accessible treatment, see Wilson (1992).

8. Gould (1989). A somewhat less opinionated view of the Ediacara is given by Runnegar (1992).
9. Petroski (1992) gives lots of examples involving some of the simplest of our industrial artifacts, such as paper clips, forks, and hand tools.
10. See, for instance, Crosby (1986) or Vitousek et al. (1996).
11. See Chapter 2, note 1.
12. Dasgupta (1996) gives a lot of attention to this issue. Unless of course one includes the few odd cases of acquired tool use and other minor elements of acquired behavior in animals.
13. Heilbroner (1967), an important and influential essay, reprinted in Smith and Marx (1995).
14. The main reference here is White (1962).
15. Dasgupta (1996) emphasizes this point.
16. Gilinsky and Rambach (1987); the idea was followed up by Rosenzweig and McCord (1991).
17. The usual citation for the argument is David (1985). It's put in a larger context in Rogers (1983).
18. Liebowitz and Margolis (1990).
19. Liebowitz and Margolis (1995).
20. See, for a fuller argument, Arthur (1990). I'm not sure that his analogy of positive feedback in economics with punctuated equilibrium in evolution is close enough to be of use.
21. One can get a good look at the importance of tapers—that is, conical surfaces—in contemporary technology by glancing at handbooks such as Oberg et al. (1984).
22. A lovely recent book (Maor, 1994) puts the logarithmic spiral into both mathematical and historical context.
23. See Pettigrew (1908), Cook (1914), and Thompson (1942, originally 1917). A general and realistic view of alternative geometries is given by Pearce (1978).
24. Needham (1954), vol. 1, p. 229. At every possible occasion he compares Western and Chinese technological achievements.
25. Tylecote (1992) is my main reference here; further information on Andean metallurgy is given by Lechtman (1988) and on the Old World by Hodges (1970). Native copper, incidentally, was mined for about five thousand years in the Lake Superior/Isle Royale area of North America (Wayman, 1989a), and meteoritic iron was hammered out into teeth for blades by prehistoric people in the Cape

York area of northern Greenland (Wayman, 1989b). Metallic technologies have arisen on quite a few occasions!

26. An extreme example of diffusionist argument, one relevant to the present case, is that of Needham and Lu (1985).
27. My main sources here are Bird (1979), King (1979), Anton (1984), and Barber (1994).
28. The approach I'm suggesting isn't really all that different from the analysis of weapons technology by Churchill (1993) and probably other anthropological work.
29. The example of Bell, Elisha Gray, and the invention of the telephone is analyzed by Hounshell (1975) and Gorman and Carlson (1990). Many other cases can be cited. Molecular biology, for instance, was in large measure a creation of recruits from physics—part of the reason for using a name other than the already established and familiar biochemistry.
30. See, for some recent musings, Tenner (1996).

REFERENCES

Adams, J. L. (1991) *Flying Buttresses, Entropy, and O-Rings: The World of an Engineer.* Cambridge, MA: Harvard University Press.

Ainsworth, J. H. (1959) *Paper: The Fifth Wonder.* Kaukauna, WI: Thomas Printing and Publishing Co.

Alexander, R. M. (1981) Factors of safety in the structure of animals. *Sci. Prog.*, Oxford 67: 109–30.

———. (1983) *Animal Mechanics,* 2d ed. Oxford, UK: Blackwell Scientific Publications.

———. (1984a) Elastic energy stores in running vertebrates. *Amer. Zool.* 24: 85–94.

———. (1984b) Walking and running. *Amer. Sci.* 72: 348–54.

———. (1988) *Elastic Mechanisms in Animal Movement.* Cambridge, UK: Cambridge University Press.

———. (1992) *Exploring Biomechanics: Animals in Motion.* New York: Scientific American Library.

Allport, G. W., and T. F. Pettigrew (1957) Cultural influence on the perception of movement: the trapezoidal illusion among Zulus. *J. Abnormal Soc. Psychol.* 55: 104–13.

Amerongen, C. van (1967) *The Way Things Work*, vols. 1 and 2. New York: Simon and Schuster. (Originally *Wie Funktioniert Das*, 1963.)

Ames, A. (1951) Visual perception and the rotating trapezoidal window. *Psychol. Monogr.* 65: 324.

Anderson, C. N. (1972) *My Fertile Crescent; Travels in the Footsteps of Ancient Science.* Fort Lauderdale, FL: Sylvester Press.

Annis, R. C., and B. Frost (1973) Human visual ecology and orientation anisotropies in acuity. *Science* 182: 729–31.

Anton, F. (1984) *Ancient Peruvian Textiles.* New York: Thames and Hudson.

Arthur, W. B. (1990) Positive feedbacks in the economy. *Sci. Amer.* 262 (2): 92–99.

Atkins, P. W. (1984) *The Second Law.* New York: Scientific American Library.

Avram, M. H. (1927) *The Rayon Industry.* New York: D. Van Nostrand Co.

Azuma, A., and Y. Okuno (1987) Flight of a samara, *Alsomitramacrocarpa. J. Theor. Biol.* 129: 263–74.

Barber, E. W. (1994) *Women's Work: The First 20,000 Years.* New York: W. W. Norton.

Barlow, G. W. (1974) Hexagonal territories. *Anim. Behav.* 22: 876–78.

Basalla, G. (1988) *The Evolution of Technology.* Cambridge, UK: Cambridge University Press.

Beamish, R. (1862) *Memoir of the Life of Sir Marc Isambard Brunel.* London: Longman, Green, Longman, and Roberts.

Beaver, P. (1970) *The Crystal Palace.* London: Hugh Evelyn, Ltd.

Bennet-Clark, H. C. (1977) Scale effects in jumping animals. Pp. 185–201 in T. J. Pedley, ed., *Scale Effects in Animal Locomotion.* London: Academic Press.

Berg, H. C. (1974) Dynamic properties of bacterial flagellar motors. *Nature* 249: 77–79.

———, and R. A. Anderson (1973) Bacteria swim by rotating their flagellar filaments. *Nature* 245: 380–82.

Berkow, R., and A. J. Fletcher (1987) *The Merck Manual of Diagnosis and Therapy.* Rahway, NJ: Merck and Co.

Berman, A., L. Addadi, Å. Kvick, L. Leiserowitz, M. Nelson, and S. Weiner (1990) Intercalation of sea urchin proteins in calcite: study of a crystalline composite material. *Science* 250: 664–67.

Bernard, E. E., and M. R. Kare (1962) *Biological Prototypes and Synthetic Systems.* New York: Plenum Press.

Bernheim, M. (1959) *A Sky of My Own.* New York: Rinehart and Co.

Bijker, W. E. (1995) *Of Bicycles, Bakelites, and Bulbs: Toward a Theory of Sociotechnological Change.* Cambridge, MA: MIT Press.

Billington, D. P. (1983) *The Tower and the Bridge: The New Art of Structural Engineering.* Princeton, NJ: Princeton University Press.

Bird, J. B. (1979) Fibers and spinning procedures in the Andean area. Pp. 13–17 in A. P. Rowe, E. P. Brown, and A. L. Schaffer, eds., *Junius B. Bird Pre-Columbian Textile Conference.* Washington, DC: Textile Museum.

Bishop, W. (1961) The development of tailless aircraft and flying wings. *J. Roy. Aero. Soc.* 65: 799–806.

Blake, R. W., L. M. Chatters, and P. Domenici (1995) Turning radius of yellowfin tuna (*Thunnus albacares*) in unsteady swimming maneuvers. *J. Fish Biol.* 46: 536–38.

Blakemore, R. (1975) Magnetotactic bacteria. *Science* 190: 377–79.

Boorstin, D. J. (1965) *The Americans: The National Experience.* New York: Random House.

Bose, D. M. (1971) *A Concise History of Science in India.* New Delhi: Indian National Science Academy.

Boyd, T. (1935) *Poor John Fitch, Inventor of the Steamboat.* New York: Putnam.

British Association for the Advancement of Science (1842) Notes and Abstracts of Communications to the B.A.A.S. at the Manchester Meeting, June 1842. P. 114 in *Rept. of Annu. Meet.*

Bronowski, J. (1973) *The Ascent of Man.* Boston: Little, Brown and Co.

Bruce, R. V. (1973) *Bell: Alexander Graham Bell and the Conquest of Solitude.* Boston: Little, Brown and Co.

Buckley, P. H., and F. G. Buckley (1977) Hexagonal packing of royal tern nests. *Auk* 94: 36–43.

Budde, R. (1995) The story of Velcro. *Physics World* 8(1): 22.

Bulliet, R. W. (1975) *The Camel and the Wheel.* Cambridge, MA: Harvard University Press.

Burke, J. G., and M. C. Eakin (1979) *Technology and Change*. San Francisco: Boyd and Fraser.

Burns, J. M. (1975) *BioGraffiti: A Natural Selection*. New York: Quadrangle/New York Times Book Co.

Bushnell, D. M., and K. J. Moore (1991) Drag reduction in nature. *Annu. Rev. Fluid Mech.* 23: 65–79.

Caldwell, D. G., and P. M. Taylor (1990) Chemically stimulated pseudo-muscular actuation. *Int. J. Engineering Sci.* 28: 797–808.

Canny, M. J. (1995) A new theory for the ascent of sap: cohesion supported by tissue pressure. *Ann. Bot.* 75: 343–57.

Cardwell, D. (1995) *The Norton History of Technology*. New York: W. W. Norton.

Chadwick, G. F. (1961) *The Works of Joseph Paxton 1803–1865*. London: Architectural Press.

Chan, M. K., S. Mukund, A. Kletzin, M. W. W. Adams, and D. C. Rees (1995) Structure of a hyperthermophilic tungstopterin enzyme, aldehyde ferredoxin oxidoreductase. *Science* 267: 1463–69.

Channell, D. F. (1991) *The Vital Machine*. New York: Oxford University Press.

Chanute, O. (1893) Progress in flying machines. *Amer. Engineer and Railroad J.* 67 (3): 135.

Chatterton, E. K. (1910) *Steamships and Their Story*. London: Cassell and Co.

Choi, K. S. (1990) Drag-reduction test of riblets using ARE's high speed buoyancy propelled vehicle - MOBY-D. *Aeronaut. J.* 94: 79–85.

Churchill, S. E. (1993) Weapon technology, prey size selection, and hunting methods in modern hunter-gatherers: implications for hunting in the Palaeolithic and Mesolithic. In G. L. Peterkin, H. M. Bricker, and P. Mellars, eds., *Hunting and Animal Exploitation in the Later Palaeolithic and Mesolithic of Eurasia. Archeological Papers of the American Anthropological Association* 4: 11–24.

Cleveland, L. R., and A. V. Grimstone (1964) The fine structure of the flagellate *Mixotricha* and its associated microorganisms. *Proc. Roy. Soc. Lond.* 159B: 668–86.

Cole, A. T. (1967) *Democritus and the Sources of Greek Anthropology*. Cleveland, OH: American Philological Association/Western Reserve University Press.

Cook, T. A. (1914) *The Curves of Life*. London: Constable and Co.

Coutts, M. P., and J. Grace, eds. (1995) *Wind and Trees.* Cambridge, UK: Cambridge University Press.

Craighead, F. C. (1915) Larvae of the Prioninae. *U.S. Dept. Agr. Rept. 107.*

———. (1923) North American Cerambycid larvae. *Bull. Dept. Agr. Can.* 27 (NS): 1–237.

Crane, H. R. (1950) Principles and problems of biological growth. *Sci. Monthly* 70: 376–89.

Croft, T. (1981) *American Electrician's Handbook,* 10th ed., W. I. Summers, ed. New York: McGraw-Hill.

Crosby, A. W. (1986) *Ecological Imperialism: The Biological Expansion of Europe, 900–1900.* Cambridge, UK: Cambridge University Press.

Crouch, T. D. (1989) *The Bishop's Boys: A Life of Wilbur and Orville Wright.* New York: W. W. Norton.

Currey, J. (1984) *The Mechanical Adaptations of Bones.* Princeton, NJ: Princeton University Press.

Dasgupta, S. (1996) *Technology and Creativity.* New York: Oxford University Press.

David, P. A. (1985) Clio and the economics of QWERTY. *Amer. Econ. Rev.* 75: 332–37.

Davies, G. A., and A. B. Porter (1966) Turbulent flow properties of dilute polymer solutions. *Nature* 212: 66.

Dawkins, R. (1976) *The Selfish Gene.* Oxford, UK: Oxford University Press.

———. (1986) *The Blind Watchmaker.* Harlow, UK: Longmans Scientific and Technical Press.

Denny, M. W. (1993) *Air and Water: The Biology and Physics of Life's Media.* Princeton, NJ: Princeton University Press.

De Rossi, D., P. Parrini, P. Chiarelli, and G. Buzzigoli (1985) Electrically induced contractile phenomena in charged polymer networks: preliminary study on the feasibility of muscle-like structures. *Trans. Amer. Soc. Artificial Internal Organs* 31: 60–65.

Derry, T. K., and T. I. Williams (1960) *A Short History of Technology: From the Earliest Times to A.D. 1900.* New York: Dover Publications.

Douglas, K., and N. A. Clark (1990) Biomolecular/solid-state nanoheterostructures. *Appl. Phys. Lett.* (USA) 56: 692–94.

Dudley, R., and C. Gans (1991) A critique of symmorphosis and optimality models in physiology. *Physiol. Zool.* 64: 627–37.

Dunwell, F. F. (1991) *The Hudson River Highlands.* New York: Columbia University Press.

Dyer, B. D., and R. A. Obar (1994) *Tracing the History of Eukaryotic Cells.* New York: Columbia University Press.

Edsall, J. T., and J. Wyman (1958) *Biophysical Chemistry.* New York: Academic Press.

Ekholm, G. F. (1946) Wheeled toys in Mexico. *Amer. Antiquity* 11: 222–28.

Eldredge, N. and S. J. Gould (1972) Punctuated equilibrium: an alternative to phyletic gradualism. Pp. 82–115 in T. J. M. Schopf, ed., *Models in Paleobiology.* San Francisco: Freeman, Cooper, and Co.

Ellul, J. (1964) *The Technological Society.* New York: Alfred A. Knopf.

Endler, J. A. (1986) *Natural Selection in the Wild.* Princeton, NJ: Princeton University Press.

Ennos, A. R. (1988) The importance of torsion in the design of insect wings. *J. Exp. Biol.* 140: 137–60.

———. (1993) The function and formation of buttresses. *TREE* 10: 350–51.

Ewbank, T. (1842) *Hydraulic and Other Machines for Raising Water Including the Progressive Development of the Steam Engine.* London: Tilt and Bogue.

Fisher, J., and R. A. Hinde (1949) The opening of milk-bottles by birds. *Brit. Birds* 42: 347–57.

Flannery, K. V. (1972) The origins of the village as a settlement type in Mesoamerica and the Near East: a comparative study. Pp. 23–53 in P. J. Ucko, R. Tringham, and G. W. Dimbleby, eds., *Man, Settlement, and Urbanism.* Cambridge, MA: Schenkman Publishing Co.

Flatow, I. (1992) *They All Laughed.* New York: HarperCollins.

Florman, S. C. (1975) *The Existential Pleasures of Engineering.* New York: St. Martin's Press.

Fournier, M. J., T. L. Mason, and D. A. Tirrell (1995) Role of molecular genetics in polymer materials science. Pp. 263–75 in M. Sarikaya and I. A. Aksay, eds.: *Biomimetics: Design and Processing of Materials.* Woodbury, NY: American Institute of Physics.

Frazzetta, T. H. (1966) Studies on the morphology and function of the skull in the Boidae (Serpentes). Part II. Morphology and function of the jaw apparatus in *Python sebae and Python molurus. J. Morph.* 118: 217–96.

Freeman, K. (1948) *Ancilla to the Pre-Socratic Philosophers.* Cambridge, MA: Harvard University Press.

French, M. (1994) *Invention and Evolution: Design in Nature and Engineering.* Cambridge, UK: Cambridge University Press.

Fridenson, P. (1978) The coming of the assembly line to Europe. Pp. 159–75 in W. Krohn, E. T. Layton, Jr., and P. Weingart, eds., *The Dynamics of Science and Technology.* Boston: Dordrecht.

Futuyma, D. J. (1986) *Evolutionary Biology,* 2d. ed. Sunderland, MA: Sinauer Associates.

Galbraith, J. K. (1967) *The New Industrial State.* Boston: Houghton Mifflin Co.

Gans, C. (1974) *Biomechanics: An Approach to Vertebrate Biology.* Philadelphia: Lippincott.

Garland, T., and R. B. Huey (1987) Testing symmorphosis: does structure match functional requirements? *Evolution* 41: 1404–09.

Gérardin, L. (1968) *Bionics.* New York: McGraw-Hill.

Giacomelli, R., and E. Pistolesi (1934) Historical Sketch. Pp. 305–94 in W. F. Durand, ed., *Aerodynamic Theory.* New York: Dover Publications (1963 reprint).

Gibbs-Smith, C. H. (1962) *Sir George Cayley's Aeronautics, 1796–1855.* London: H. M. Stationery Office.

Gilinsky, N. L., and R. K. Bambach (1987) Asymmetrical patterns of origination and extinction in higher taxa. *Paleobiol.* 13: 427–45.

Gordon, J. E. (1976) *The New Science of Strong Materials.* Harmondsworth, UK: Penguin Books. (Reprint, Princeton, NJ: Princeton University Press, 1984).

———. (1978) *Structures; or, Why Things Don't Fall Down.* New York: Plenum Press.

———. (1988) *The Science of Structures and Materials.* New York: Scientific American Books.

Gorman, M. E., and W. B. Carlson (1990) Interpreting invention as a cognitive process: Alexander Graham Bell, Thomas Edison, and the telephone, 1876–1878. *Science, Technology, and Human Values* 15: 131–64.

Gosline, J. M., M. E. DeMont, and M. W. Denny (1986) The structures and properties of spider silk. *Endeavour* 10 (1): 37–43.

———, C. Nichols, P. Guerette, A. Cheng, and S. Katz (1995) The macromolecular design of spiders' silks. Pp. 237–61 in M. Sarikaya and I. A. Aksay, eds., *Biomimetics: Design and Processing of Materials.* Woodbury, NY: American Institute of Physics.

Gould, J. L., J. L. Kirschvink, and K. S. Deffeyes (1978) Bees have magnetic remanence. *Science* 201: 1026.

Gould, S. J. (1980) *The Panda's Thumb.* New York: W. W. Norton.

———. (1981) Kingdoms without wheels. *Nat. Hist.* 90 (4): 42–48.

———. (1989) *Wonderful Life: The Burgess Shale and the Nature of History.* New York: W. W. Norton.

———, and Lewontin, R. C. (1979) The spandrels of San Marco and the panglossian paradigm: a critique of the adaptationist programme. *Proc. Roy. Soc.* London B205: 581–98.

Grace, J. (1977) *Plant Response to Wind.* London: Academic Press.

Grant, P. R. (1968) Polyhedral territories of animals. *Amer. Nat.* 102: 75–80.

Gray, J. (1936) Studies on animal locomotion. VI. The propulsive powers of the dolphin. *J. Exp. Biol.* 13: 192–99.

Griffin, D. R. (1984) *Animal Thinking.* Cambridge, MA: Harvard University Press.

Grosser, M. (1981) *Gossamer Odyssey: The Triumph of Human-Powered Flight.* Boston: Houghton Mifflin.

Gunderson, S., and R. Schiavone (1989) The insect exoskeleton: a natural structural composite. *J. Minerals, Metals, and Materials Soc.* 41: 60–62.

Halacy, D. S. (1965) *Bionics: The Science of Living Machines.* New York: Holiday House.

Haldane, J. W. S. (1928) On being the right size. Pp. 20–28 in *Possible Worlds.* New York: Harper.

Hallgrimsson, B., and S. Swartz (1995) Biomechanical adaptation of ulnar cross-sectional morphology in brachiating primates. *J. Morph.* 224: 111–23.

Hansell, M. H. (1989) Wasp papier-mâché. *Nat. Hist.* 98 (8): 52–61.

Harrar, E. S., and J. G. Harrar (1962) *Guide to Southern Trees.* New York: Dover Publications.

Harris, C. M. (1989) The improbable success of John Fitch. *Amer. Heritage of Invention and Technology* 4(3): 24–31.

Harris, J. S. (1989) An airplane is not a bird. *Amer. Heritage of Invention and Technology* 5(2): 19–22.

Hart, C. (1985) *The Prehistory of Flight.* Berkeley: University of California Press.

Hayward, V. (1993) Borrowing some ideas from biological manipulators to design an artificial one. Pp. 139–52 in P. Dario, G. Sandini, P. Aebischer, eds., *Robots and Biological Systems: Toward a New Bionics.* NATO ASI Series F vol. 102.

Heglund, N. C., and G. A. Cavagna (1985) Efficiency of vertebrate locomotory muscles. *J. Exp. Biol.* 115: 283–92.

Heilbroner, R. L. (1967) Do machines make history? *Technology and Culture* 8: 335–45.

Hertel, H. (1963) *Structure, Form, and Movement.* New York: Reinhold Publishing Co. (translation).

Hinde, R. A. and J. Fisher (1951) Further observations on the opening of milk bottles by birds. *Brit. Birds* 44: 393–96.

Hix, J. (1974) *The Glass House.* Cambridge, MA: MIT Press.

Hochberg, J. E. (1978) *Perception.* Englewood Cliffs, NJ: Prentice-Hall.

Hodges, H. (1970) *Technology in the Ancient World.* New York: Alfred A. Knopf.

Hodges, H. W. M. (1972) Domestic building materials and ancient settlements. Pp. 523–30 in P. J. Ucko, R. Tringham, and G. W. Dimbleby, eds., *Man, Settlement, and Urbanism.* Cambridge, MA: Schenkman Publishing Co.

Hölldobler, B., and E. O. Wilson (1994) *Journey to the Ants.* Cambridge, MA: Harvard University Press.

Hooke, R. (1665) *Micrographia,* facsimile ed., 1961. New York: Dover Publications.

Hounshell, D. A. (1975) Elisha Gray and the telephone: on the disadvantages of being an expert. *Technology and Culture* 16: 133–61.

Houston, J., B. N. Dancer, and M. A. Learner (1989) Control of sewer flies using *Bacillus thuringiensis var. israelensis.* I. Acute toxicity tests and pilot scale trial. *Water Res.* 23: 369–78.

Howard, F. (1987) *Wilbur and Orville: A Biography of the Wright Brothers.* New York: Alfred A. Knopf.

Hulbary, R. L. (1944) The influence of air spaces on the three-dimensional shapes of cells in *Elodea* stems, and a comparison with pith cells of *Ailanthus. Amer. J. Bot.* 31: 561–80.

Hunter, D. (1947) *Papermaking: The History and Technique of an Ancient Craft.* New York: Alfred A. Knopf.

Isaacs, J. D., A. C. Vine, H. Bradner, and G. E. Bachus (1966) Satellite elongation into a true "sky-hook." *Science* 151: 682–83.

Jenkins, F. A., Jr., K. P. Dial, and G. E. Goslow, Jr. (1988) A cineradiographic analysis of bird flight: the wishbone in starlings is a spring. *Science* 241: 1495–98.

Kearney, H. (1971) *Science and Change 1500–1700.* New York: McGraw-Hill.

Kendrick, M. J., M. T. May, M. J. Plishka, and K. D. Robinson (1992) *Metals in Biological Systems.* New York: Horwood.

Kihlstedt, F. T. (1984) The crystal palace. *Sci. Amer.* 251 (4): 132–43.

King, M. E. (1979) The prehistoric textile industry of Mesoamerica. Pp. 265–78 in A. P. Rowe, E. P. Brown, and A. L. Schaffer, eds., *Junius B. Bird Pre-Columbian Textile Conference.* Washington, DC: Textile Museum.

Koehl, M. A. R. (1977) Mechanical organization of cantilever-like sessile organisms: sea anemones. *J. Exp. Biol.* 69: 127–42.

———. (1982) Mechanical design of spicule-reinforced connnective tissue: stiffness. *J. Exp. Biol.* 98: 239–67.

———, T. Hunter, and J. Jed (1991) How do body flexibility and length affect hydrodynamic forces on sessile organisms in waves versus in currents? *Amer. Zool.* 31: 60A.

Koops, M. (1800) *Historical Account of the Substances Which Have Been Used to Describe Events and to Convey Ideas from the Earliest Date to the Invention of Paper.* London: T. Burton.

Kramer, M. O. (1965) Hydrodynamics of the dolphin. *Adv. Hydrosci.* 2: 111–30.

LaBarbera, M. (1983) Why the wheels won't go. *Amer. Nat.* 121: 395–408.

———. (1990) Principles of design of fluid transport systems in zoology. *Science* 249: 992–1000.

———, and S. Vogel (1982) The design of fluid transport systems in organisms. *Amer. Sci.* 70: 54–60.

Laithwaite, E. (1984) *Invitation to Engineering.* Oxford, UK: Basil Blackwell.

———. (1989) *A History of Linear Electric Motors.* San Francisco: San Francisco Press.

———. (1994) *An Inventor in the Garden of Eden.* Cambridge, UK: Cambridge University Press.

Lechtman, H. (1988) Traditions and styles in central Andean metalworking. Pp. 344–78 in R. Madden, ed., *The Beginning of the Use of Metals and Alloys.* Cambridge, MA: MIT Press.

Leeming, J. (1949) *Rayon, the First Man-Made Fiber.* Brooklyn, NY: Chemical Publishing Co.

Levy, M., and M. Salvadori (1992) *Why Buildings Fall Down: How Structures Fail.* New York: W. W. Norton.

Liebowitz, S. J., and S. E. Margolis (1990) The fable of the keys. *J. Law and Econ.* 33: 1–25.

———, and ——. (1995) Path dependence, lock-in, and history. *J. Law, Economics, and Organization* 11: 205–26.

Lilienthal, O. (1910) *Birdflight as the Basis for Aviation.* London: Longmans, Green.

Lombardi, S. J., S. Fossey, and D. L. Kaplan (1990) Recombinant spider silk proteins for composite fibers. *Proc. Amer. Soc. for Composites, 5th Technical Conference*: 184–87.

Lovelock, J. E., and L. Margulis (1974) Atmospheric homeostasis by and for the biosphere: the Gaia hypothesis. *Tellus* 26: 1–9.

Lowenstam, H. A. (1967) Lepidocrocite, an apatite mineral, and magnetite in teeth of chitons (Polyplacophora). *Science* 156: 1373–75.

Lucia, E. (1975) *The Big Woods: Logging and Lumbering, from Bull Teams to Helicopters, in the Pacific Northwest.* New York: Doubleday.

———. (1981) Joe Cox and his revolutionary chain saw. *J. Forest Hist.* 25: 159–65.

McCallum, H. D., and F. T. McCallum (1965) *The Wire That Fenced the West.* Norman: University of Oklahoma Press.

McCulloch, W. S. (1962) The imitation of one life form by another—biomimesis. Pp. 393–97 in E. E. Bernard and M. R. Kare, eds., *Biological Prototypes and Synthetic Systems.* New York: Plenum Press.

McFarland, M. W., ed. (1953) *The Papers of Wilbur and Orville Wright,* Vol. 1: 1899–1905. New York: McGraw-Hill.

McHenry, M. J., C. A. Pell, and J. H. Long, Jr. (1995) Mechanical control of swimming speed: stiffness and axial wave form in an undulatory fish model. *J. Exp. Biol.* 198: 2293–2305.

McMahon, T. A. (1984) *Muscles, Reflexes, and Locomotion.* Princeton, NJ: Princeton University Press.

———, and J. T. Bonner (1983) *On Size and Life.* New York: Scientific American Books.

Mackay, A. L. (1991) *A Dictionary of Scientific Quotations.* New York: Adam Hilger.

Majdalany, F. (1960) *The Eddystone Light.* Boston: Houghton Mifflin.

Manko, D. J. (1992) *A General Model of Legged Locomotion on Natural Terrain.* Boston: Kluwer Academic Publishers.

Mann, S. (1990) Crystal engineering: the natural way. *New Scientist* 125 (1707): 42–47.

———. (1995) Biomineralization, the inorganic-organic interface, and crystal engineering. Pp. 91–116 in M. Sarikaya and I. A. Aksay, eds.: *Biomimetics: Design and Processing of Materials.* Woodbury, NY: American Institute of Physics.

Maor, E. (1994) e: *The Story of a Number.* Princeton, NJ: Princeton University Press.

Margulis, L. (1993) *Symbiosis in Cell Evolution.* New York: W. H. Freeman.

Mark, R. (1978) Structural experimentation in Gothic architecture. *Amer. Sci.* 66: 542–50.

———. (1982) *Experiments in Gothic Structure.* Cambridge, MA: MIT Press.

Marteka, V. (1965) *Bionics.* Philadelphia: J. B. Lippincott.

Mason, M. T., and J. K. Salisbury, Jr. (1985) *Robot Hands and the Mechanics of Manipulation.* Cambridge, MA: MIT Press.

Mattheck, C. (1991) *Trees: The Mechanical Design.* Berlin: Springer-Verlag.

Maxim, H. S. (1909) *Artificial and Natural Flight.* New York: Whittaker and Co.

Maynard Smith, J. (1952) The importance of the nervous system in the evolution of animal flight. *Evolution* 6: 127–29.

Mayr, O. (1970) *The Origins of Feedback Control.* Cambridge, MA: MIT Press.

Meinhardt, H. (1995) *The Algorithmic Beauty of Sea Shells.* New York: Springer-Verlag.

Miller, K. F., C. P. Quine, and J. Hunt (1987) The assessment of wind exposure for forestry in upland Britain. *Forestry* 60: 179–92.

Miller, M. (1945) *The Far Shore.* New York: McGraw-Hill.

Miller, R. C. (1924) The boring mechanism of *Teredo. Univ. Calif. Publ. Zool.* 26: 41–80.

Mokyr, J. (1991) Evolutionary biology, technological change, and economic history. *Bull. Economic Res.* 43: 127–47.

Morowitz, H. (1968) *Energy Flow in Biology.* New York: Academic Press.

Mumford, L. (1967) *The Myth and the Machine: Technics and Human Development.* New York: Harcourt Brace Jovanovich.

Needham, J. (1954) *Science and Civilisation in China,* vol. 1. Cambridge, UK: Cambridge University Press.

———. (1965) *Science and Civilisation in China,* vol. 4, part 2. Cambridge, UK: Cambridge University Press.

———, and G.-D. Lu (1985) *Transpacific Echoes and Resonances: Listening Once Again.* Philadelphia: World Scientific.

Nicklas, R. B. (1984) A quantitative comparison of cellular motile systems. *Cell Motility* 4: 1–5.

Nijhout, H. F. (1990). Metaphors and the role of genes in development. *BioEssays* 12: 441–46.

———, and H. G. Sheffield (1979) Antennal hair erection in male mosquitoes: a new mechanical effector in insects. *Science* 206: 595–96.

Niklas, K. J. (1992) *Plant Biomechanics.* Chicago: University of Chicago Press.

———. (1994) *Plant Allometry.* Chicago: University of Chicago Press.

Oberg, E., F. D. Jones, and H. L. Horton (1984) *Machinery's Handbook,* 22d ed. New York: Industrial Press.

O'Neill, P. L. (1990) Torsion in the asteroid ray. *J. Morph.* 203: 141–50.

Orwell, G. (1937) *The Road to Wigan Pier.* London: Victor Gollancz, Ltd.

Ovid (P. Ovidius Naso) *Metamorphoses.* tr. F. J. Miller (1966). Cambridge, MA: Loeb Classical Library.

Owen, D. H. (1978) The psychophysics of prior experience. Pp. 467–524 in P. K. Machamer and R. G. Turnbull, eds., *Studies in Perception.* Columbus: Ohio State University Press.

Ozanam, M. (1862) Dissolution de la soie par l'ammoniure de cuivre. *C. R. des Séances de l'Academie des Sciences* 55: 833.

Pantin, C. F. A. (1964) Homeostasis and the environment. Pp. 1–6 in G. M. Hughes, ed., *Homeostasis and Feedback Mechanisms: Symp. Soc. Exp. Biol.* 18.

Papanek, V. (1971) *Design for the Real World.* New York: Random House.

Parkes, E. W. (1965) *Braced Frameworks: An Introduction to the Theory of Structures.* Oxford, UK: Oxford University Press.

Paturi, F. R. (1976) *Nature, Mother of Invention: The Engineering of Plant Life.* New York: Harper and Row.

Pearce, P. (1978) *Structure in Nature Is a Strategy for Design.* Cambridge, MA: MIT Press.

Petroski, H. (1985) *To Engineer Is Human: The Role of Failure in Successful Design.* New York: St. Martin's Press.

———. (1991) Still twisting. *Amer. Sci.* 79: 398–401.

———. (1992) *The Evolution of Useful Things.* New York: Random House.

———. (1994) *Design Paradigms: Case Histories of Error and Judgment in Engineering.* New York: Cambridge University Press.

———. (1995) *Engineers of Dreams: Great Bridge Builders and the Spanning of America.* New York: Alfred A. Knopf.

Pettigrew, J. B. (1908) *Design in Nature.* London: Longmans Ltd.

Phillips, H. F. (1885) Experiments with currents of air (anonymous report). *Engineering* 40: 160–61.

Pierce, J. R. (1961) *Symbols, Signals and Noise: The Nature and Process of Communication.* New York: Harper and Row.

Pool, R. (1989) Making new materials with nature's help. *Science* 250:1389.

Prakash, S. (1987) *Geometry in Ancient India.* Columbia, MO: South Asian Books.

Prescott, C. B. (1884) *Bell's Electric Speaking Telephone: Its Invention, Construction, Application, Modification, and History.* New York: D. Appleton and Co.

Pritchard, J. L. (1961) *Sir George Cayley, the Inventor of the Airplane.* London: Max Parrish.

Raibert, M. H., and I. E. Sutherland (1983) Machines that walk. *Sci. Amer.* 248 (1): 44–53.

Raup, D. M. (1966) Geometric analysis of shell coiling: general problems. *J. Paleontol.* 40: 1178–90.

———. (1992) *Extinction: Bad Genes or Bad Luck.* New York: W. W. Norton.

Reif, W.-E. (1985) Morphology and hydrodynamic effects of the scales of fast swimming sharks. *Fortschr. Zool.* 30: 483–85.

Reynolds, T. S., ed. (1991) *The Engineer in America.* Chicago: University of Chicago Press.

Riley, J. J., M. Gad-el-Hak, and R. W. Metcalfe (1988) Compliant Coatings. *Annu. Rev. Fluid Mech.* 20: 393–420.

Robinson, J. R., and E. C. Frederick (1989) Scaling of foot dimensions. *XII Congr. Int. Soc. Biomech. Abstr.*, p. 1074.

Rogers, E. M. (1983) *Diffusion of Innovations,* 3d ed. New York: Free Press (Macmillan).

Rolt, L. T. C. (1959) *Isambard Kingdom Brunel, a Biography.* New York: St. Martin's Press.

Rosenzweig, M. L., and R. D. McCord (1991) Incumbent replacement: evidence for long-term evolutionary progress. *Paleobiol.* 17: 202–13.

Roth, V. R. (1996) Cranial integration in the Sciuridae. *Amer. Zool.* 36: 14–23.

Rouse, H., and S. Ince (1957) *History of Hydraulics.* New York: Dover Publications (1963 reprint).

Runnegar, B. (1992) Evolution of the earliest animals. In J. W. Schopf, ed., *Major Events in the History of Life.* Boston: Jones and Bartlett Publishers.

Saidel, B. A. (1993) Round house or square? Architectural form and socio-economic organization in the PPNB. *Mediterranean Archaeology* 6: 65–108.

Salehpoor, K., M. Shahinpoor, and M. Mojarrad (1996) Electrically controllable artificial PAN muscles. Pp. 116–24 in A. Crowson, ed., Smart Materials Technologies and Biomimetics. *Proc. Int. Soc. for Optical Engineering,* vol. 2716.

Salisbury, J. K. (1950) *Kent's Mechanical Engineering Handbook,* 12th ed., vol. 2. New York: John Wiley.

Salvadori, M. (1980) *Why Buildings Stand Up: The Strength of Architecture.* New York: W. W. Norton.

Sarikaya, M., and I. A. Aksay (1995) *Biomimetics: Design and Processing of Materials.* Woodbury, NY: American Institute of Physics.

———, C. E. Furlong, and J. T. Staley (1994) Nanodesigning and properties of biological composites. *Advances in Bioengineering. ASME BED Publication* 28: 47–48.

Schlesinger, W. H., J. T. Gray, D. S. Gill, and B. E. Mahall (1982) C*eanothus megacarpus* chaparral: a synthesis of ecosystem processes during development and annual growth. *Bot. Rev.* 48: 71–117.

Schlosser, L. B. (1980) Papermaking and the industrial revolution: the search for new fiber. *Amer. Book Collector* 1 (6): 3–12.

Schmidt, P. R., ed. (1996) *Culture and Technology of African Iron Production.* Gainesville: University Press of Florida.

Schmidt-Nielsen, K. (1984) *Scaling.* Cambridge, UK: Cambridge University Press.

———. (1997) *Animal Physiology: Adaptation and Environment*, 5th ed. Cambridge, UK: Cambridge University Press.

Schneider, M. S. (1994) *A Beginner's Guide to Constructing the Universe.* New York: HarperCollins.

Schober, J. (1930) *Silk and the Silk Industry.* London: Constable and Co.

Schoenheimer, R. (1942) *The Dynamic State of Body Constituents.* Cambridge, MA: Harvard University Press.

Schopf, J. W. (1992) *Major Events in the History of Life.* Boston: Jones and Bartlett Publishers.

Shadwick, R. E. (1994) Mechanical organization of the mantle and circulatory system of cephalopods. *Mar. Behav. Physiol.* 25: 69–85.

———, C. M. Pollock, and S. A. Stricker (1990) Structure and biomechanical properties of crustacean blood vessels. *Physiol. Zool.* 63: 90–101.

Singer, C., E. J. Holmyard, and A. R. Hall, eds. (1954) *A History of Technology*, vol. 1. Oxford, UK: Clarendon Press.

Sleeswyk, A. W. (1981) Hand-cranking in Egyptian antiquity. Pp. 23–37 in A. R. Hall and N. Smith, eds., *History of Technology*, vol. 6. London: Mansell Publishing.

Smeaton, J. (1759) An experimental enquiry concerning the natural powers of water and wind to turn mills, and other machines, depending on a circular motion. *Phil. Trans. Roy. Soc. Lond.* 51: 100–174.

Smith, K. K., and W. M. Kier (1989) Trunks, tongues, and tentacles: moving with skeletons of muscle. *Amer. Sci.* 77: 28–35.

Smith, M. R., and L. Marx, eds. (1995) *Does Technology Drive History? The Dilemma of Technological Determinism.* Cambridge, MA: MIT Press.

Snow, C. P. (1959) *The Two Cultures.* Cambridge, UK: Cambridge University Press.

Song, S.-M. and Waldron, K. J. (1989) *Machines That Walk: The Adaptive Suspension Vehicle.* Cambridge, MA: MIT Press.

Sotavalta, O. (1953) Recordings of high wing-stroke and thoracic vibration frequency in some midges. *Biol. Bull.* 104: 439–44.

Stanley, S. M. (1987) *Extinction.* New York: Scientific American Library.

Steadman, P. (1979) *The Evolution of Design: Biological Analogy in Architecture and the Applied Arts.* Cambridge, UK: Cambridge University Press.

Stix, G. (1994) Robotuna. *Sci. Amer.* 270 (1): 42.

Sutton, O. G. (1949) *The Science of Flight.* Harmondsworth, UK: Penguin Books.

Swetz, F., and T. I. Kao (1977) *Was Pythagoras Chinese? An Examination of Right Triangle Theory in Ancient China.* University Park: Pennsylvania State University Press.

Tamm, S. L. (1982) Flagellated ectosymbiotic bacteria propel a eukaryotic cell. *J. Cell Biol.* 94: 697–709.

Taylor, C. R., and E. R. Weibel (1981) Design of the mammalian respiratory system. I. Problem and strategy. *Respir. Physiol.* 44: 1–10.

Tennekes, H. (1996) *The Simple Science of Flight.* Cambridge, MA: MIT Press.

Tenner, E. (1996) *Why Things Bite Back: Technology and the Revenge of Unintended Consequences.* New York: Alfred A. Knopf.

Thompson, D'Arcy W. (1942) *On Growth and Form,* 2d ed. Cambridge, UK: Cambridge University Press.

———. (1961) *On Growth and Form.* Abridged, J. T. Bonner, ed., Cambridge, UK: Cambridge University Press.

Thompson, T. E., and I. Bennett (1969) *Physalia* nematocysts: utilized by mollusks for defense. *Science* 166: 1532–33.

Thomson, K. S. (1991) *Living Fossil: The Story of the Coelacanth.* New York: W. W. Norton.

Tokaty, G. A. (1971) *A History and Philosophy of Fluid Mechanics.* New York: Dover Publications.

Toms, B. A. (1948) Some observations on the flow of linear polymer solutions through straight tubes at large Reynolds numbers. *Proc. Int. Rheological Congr., Scheveningen, Netherlands* 2: 135.

Triantafyllou, G. S., M. S. Triantafyllou, and M. A. Grosenbaugh (1993) Optimal thrust development in oscillating foils with application to fish propulsion. *J. Fluids and Structures* 7: 205–24.

Tributsch, H. (1982) *How Life Learned to Live.* Cambridge, MA: MIT Press.

Tylecote, R. F. (1992) *History of Metalworking.* London: Institute of Metals.

Urry, D. W. (1993) Molecular machines: how motion and other functions of living organisms can result from reversable chemical changes. *Angew. Chem. Int. Ed. Engl.* 32: 819–41.

Usher, A. P. (1954) *A History of Mechanical Inventions.* New York: Dover Publications (reprint, 1988).

Vaughan, A. (1991) *Isambard Kingdom Brunel: Engineering Knight Errant.* London: John Murray.

Vermeij, G. J. (1993) *A Natural History of Shells.* Princeton, NJ: Princeton University Press.

Vincent, J. F. V. (1990) *Structural Biomaterials*, rev. ed. Princeton, NJ: Princeton University Press.

Vincenti, W. G. (1988) How did it become "obvious" that an airplane should be inherently stable? *Amer. Heritage of Invention and Technology* 4(1): 50–56.

———. (1990) *What Engineers Know and How They Know It: Analytical Studies from Aeronautical History.* Baltimore: Johns Hopkins University Press.

Vitousek, P. M., C. M. D'Antonio, L. L. Loope, and R. Westbrooks (1996) Biological invasions as global environmental change. *Amer. Sci.* 84: 468–78.

Vogel, S. (1970) Convective cooling at low airspeeds and the shapes of broad leaves. *J. Exp. Bot.* 21: 91–101.

———. (1978) Organisms That Capture Currents. *Sci. Amer.* 239 (2): 128–39.

———. (1984a) Drag and flexibility in sessile organisms. *Amer. Zool.* 24: 37–44.

———. (1984b) The thermal conductivity of leaves. *Can. J. Bot.* 62: 741–44.

———. (1988) *Life's Devices: The Physical World of Animals and Plants.* Princeton, NJ: Princeton University Press.

———. (1989) Drag and reconfiguration of broad leaves in high winds. *J. Exp. Bot.* 40: 941–48.

———. (1992) *Vital Circuits: On Pumps, Pipes, and the Workings of Circulatory Systems.* New York: Oxford University Press.

———. (1993) When leaves save the tree. *Nat. Hist.* 102 (9): 58–63.

———. (1994a) *Life in Moving Fluids: The Physical Biology of Flow*, 2d ed. Princeton, NJ: Princeton University Press.

———. (1994b) Nature's pumps. *Amer. Sci.* 82: 464–71.

———. (1994c) Second-rate squirts. *Discover* 15 (8): 70–76.

———. (1994d) Dealing honestly with diffusion. *Amer. Biol. Teacher* 56: 405–7.

———. (1995) Better bent than broken. *Discover* 16 (5): 62–67.

———, and W. L. Bretz (1972) Interfacial organisms: passive ventilation in the velocity gradients near surfaces. *Science* 175: 210–11.

von Kármán, T. (1954) *Aerodynamics.* New York: McGraw-Hill.

Wainwright, S. A. (1995) What we can learn from soft biomaterials and structures. Pp. 1–12 in M. Sarikaya and I. A. Aksay, eds.: *Biomimetics: Design and Processing of Materials.* Woodbury, NY: American Institute of Physics.

———, W. D. Biggs, J. D. Currey, and J. M. Gosline (1976) *Mechanical Design in Organisms.* London: Edward Arnold (reprint, Princeton, NJ: Princeton University Press).

Walcott, C., J. L. Gould, and J. L. Kirschvink (1979) Pigeons have magnets. *Science* 205: 1028.

Walsby, A. E. (1980) A square bacterium. *Nature* 283: 69–71.

Wayman, M. L. (1989a) Native copper: humanity's introduction to metallurgy? Pp. 3–6 in M. L. Wayman, ed., *All That Glitters: Readings in Historical Metallurgy.* Montreal: Canadian Institute of Mining and Metallurgy.

———. (1989b) On the early use of iron in the Arctic. Pp. 99–100 in M. L. Wayman, ed.: *All That Glitters: Readings in Historical Metallurgy.* Montreal: Canadian Institute of Mining and Metallurgy.

Wells, C. S., ed. (1969) *Viscous Drag Reduction.* New York: Plenum Press.

Went, F. W. (1968) The size of man. *Amer. Sci.* 56: 400–13.

Westneat, M. W., and P. C. Wainwright (1989) Feeding mechanism of *Epibuus insidiator* (Labridae; Teleostei): evolution of a novel feeding mechanism. J. *Morphol.* 202: 129–50.

White, L., Jr. (1962) *Medieval Technology and Social Change.* New York: Oxford University Press.

Wiener, N. (1950) The *Human Use of Human Beings: Cybernetics and Society.* Boston: Houghton Mifflin.

Williams, T. (1882) *The Eddystone Lighthouses (New and Old).* London: Simpkin, Marshall and Co.

Wilson, E. O. (1980) *Sociobiology,* Abridged ed. Cambridge, MA: Harvard University Press.

———. (1984) *Biophilia.* Cambridge, MA: Harvard University Press.

———. (1992) *The Diversity of Life.* New York: W. W. Norton.

Wilson, G. (1855) *What Is Technology?* Edinburgh, UK: Sutherland and Knox.

Wilson, J. F., D. Li., Z. Chen, and R. T. George, Jr. (1993) Flexible robot manipulators and grippers: relatives of elephant trunks and squid tentacles. Pp. 475–94 in P. Dario, G. Sandini, P. Aebischer, eds.: *Robots and Biological Systems: Toward a New Bionics.* NATO ASI, Series F, vol. 102.

Wilson, P. N. (1975) J. G. A. Kitchen, 1869–1940, and his inventions. *Trans. Newcomen Soc.* 45: 15–43 (for 1972–73).

Winfield, D. L., D. H. Hering, and D. Cole (1991) *Engineering Derivatives from Biological Systems for Advanced Aerospace Engineering.* NASA CR-177594; Research Triangle Institute, NC.

Woodbury, R. S. (1960) The legend of Eli Whitney and interchangeable parts. *Technology and Culture* 1: 235–53.

Worden, E. C. (1911) *Nitrocellulose Industry.* New York: D. Van Nostrand Co.

Wright, O. (1953) *How We Invented the Airplane.* New York: David McKay Co.

Zimmermann, M. H. (1983) *Xylem Structure and the Ascent of Sap.* Berlin: Springer-Verlag.

INDEX

Page numbers in *italics* refer to illustrations